Studio Monitoring Design

I dedicate this book to Janet Payne, about whom I can say no more

Studio Monitoring Design

A Personal View

Philip Richard Newell

Focal Press
An imprint of Butterworth-Heinemann Ltd
Linacre House, Jordan Hill, Oxford OX2 8DP

⟨R A member of the Reed Elsevier plc group

OXFORD   LONDON   BOSTON
MUNICH   NEW DELHI   SINGAPORE   SYDNEY
TOKYO   TORONTO   WELLINGTON

First published 1995

British Library Cataloguing in Publication Data
Newell, Philip Richard
  Studio Monitoring Design: Personal View
  I. Title
  621.3893

ISBN 0 240 51407 6

Library of Congress Cataloguing in Publication Data
Newell, Philip Richard
  Studio monitoring design: a personal view/Philip Richard Newell.
  p.   cm.
  Includes bibliographical references and index.
  ISBN 0 240 51407 6
  1. Sound – Recording and reproducing. 2. Sound studios – Design and
  construction. I. Title.
  TK7881.4.N49
  621.389'3–dc20   95–35572
                                                             CIP

Printed and bound in Great Britain by Clays Ltd, St Ives plc

Contents

About the author

Philip Newell was born in Blackburn, England, in August 1949. He was educated at Billinge Grammar School, but was rejected when applying for university places on Tonmeister or musical acoustic courses due to having passed the 'wrong' subjects. Somewhat let down about having received no previous guidance as to the 'right' subjects, as the school considered his musical interests (and rock music in particular) to be 'only a passing phase', he left school a year early, despite being top in Chemistry, English Language and General Studies. He had been studying music privately for five years, so had seen no necessity to repeat the subject in school.

Newell then worked for three years from 1966 for Mecca, the leisure company, travelling round the ballrooms working on live sound and lights, along with some private recording of 'resident' bands. At this time, most visiting artistes used the in-house sound systems, so Newell found himself in a 'dream come true' situation, mixing the live sound for such artistes as The Who, Booker T and the MGs, Wilson Pickett, Sam and Dave, Junior Walker and the All Stars, Family, and many other top artistes of the time.

Whilst gradually spending more and more time recording, he was asked in 1969 to design and build a four-track 'demo' studio for a London bandleader who was starting a small record company. The project ballooned, however, and by mid-1970 had moved into premises where a 4000 sq. ft eight-track studio came into existence. In late 1970 he moved to Pye Records, which was then one of Britain's major recording companies, and had just installed the latest 16-track equipment. At Pye he spent his time equally in the studios and the mobile recording unit, which still allowed enough live work to keep him happy.

The mix of work at Pye was very broad, not only working with groups such as The Faces, The Who and Free along with engineers such as Brian Humphries and Glyn Johns, but also recording brass bands, church organs, Welsh male-voice choirs, Scottish bagpipes, cabaret shows at the 'Talk of the Town' and classical piano recitals at the Wigmore Hall.

In late 1971 he was invited by Tom (Tubular Bells) Newman to join a fledgling Virgin Records organisation as chief recording engineer, arriving at The Manor Studios three weeks before its opening. Newell missed his live work, however, so began assembling an eight-track facility in a truck in the summer of 1972. This proved to be an entirely unintentional thorn in Richard Branson's side, and in early December that year Branson asked

Newell to give him all that existed of the eight-track unit in exchange for providing the money to build Europe's most sophisticated facility, and offering Newell shares in the new company. Newell's answer was a swift 'OK', and the Manor Mobile went on the road in July 1973 as what was believed to be the world's first fully operational, permanent, 24-track recording vehicle.

Shortly after, Newman left Virgin, and Newell was invited to become the Managing Director, and later Technical and Special Projects Director, of the whole, rapidly expanding, Virgin Records recording division. During this time, he was not only responsible for the rebuilding of The Manor Studios, the building of a second 2 × 24-track mobile unit, The Townhouse Studios, and many other facilities, but also continued to work as a recording engineer, general coordinator, and producer of albums for artists such as 'Gong' and Mike Oldfield. During the eight years until 1981, his recording work was very extensive, including Captain Beetheart, Jack Bruce, The Duke Ellington Orchestra, The Warsaw Philharmonic Orchestra (in Warsaw), Gary Glitter, Don McClean, Tony Bennett, The Small Faces, Sylvie Vartan (in Paris), and everything from Dizzy Gillespie to the Buzzcocks.

During the mid-1970s he had learned to fly at Oxford Air Training School, and in 1977 Richard Branson paid for his conversion training on to seaplanes, in readiness for the development of a recording studio project on the then uninhabited Necker Island which Branson had just bought in the British Virgin Islands. It was to be a 'tax haven' studio, in the days when top British artistes were paying 93% tax on their earnings; but Margaret Thatcher, on becoming Prime Minister, cut back the maximum tax rates in 1980 and the need for the Necker studio receded.

Partly due to this disappointment, partly due to being badly beaten up by 'Sex Pistols' minders (after trying to stop two of the band from wrecking a studio), partly due to having bricks thrown at him by Jah Wobble for similar reasons, and partly also due to becoming more seriously involved in the seaplane operations, over the course of 1981/82 Newell sold his shares in Virgin and concentrated on the flying, and operating what was then the only licensed seaplane aerodrome in Northern Europe, of which he was the licensee.

During the period 1973 to 1983 he became the owner of a full, multi-engine instrument rating, and became an instructor and then examiner on certain types of seaplanes. Newell also flew for films and air displays, and it was during this time that he acquired the knowledge and experience of aerodynamics which led him in the mid-1980s into close collaboration with the Institute of Sound and Vibration research (ISVR) at Southampton University, initially through contacts in the aerodynamics department.

His semi-retirement from the music business for two years allowed him time to think in a more detached way about many related problems, and in late 1984 he was invited to rebuild one of the studios for Jacobs in the South of England; since when, he has worked almost continuously on recording studio and monitor design projects.

His 'live room' of Townhouse 2 in London (where Phil Collins recorded the drums for 'In The Air Tonight'), spawned an almost endless stream of requests for 'stone rooms', which continues to this day.

Philip Newell is still actively involved in recording work, though mostly of live performances, and has a continuing association with work at the ISVR. He is a member of the US and Canadian Seaplane Pilots Associations, a member of the Audio Engineering Society, The League Against Cruel Sports and British Mensa, and is a regular contributor of papers to the Institute of Acoustics. He is of course also well known for his many articles in international sound recording publications. Since March 1988, he has worked independently of any commercial organisations, and in that time has been an invited speaker at numerous universities, not only to Tonmeister students, but also students of music, and even engineering acoustics; somewhat ironically, this has been to some of the very courses which rejected him whilst still at school.

At the time of writing, he has been involved in the design of over 100 studios, and has worked in one capacity or another in 32 countries and territories.

Acknowledgements

I would like to acknowledge the support and inspiration of the following, during the course of writing this book.

Sally Nettleton – for being Sally, or perhaps despite being Sally.

Marta Izquierdo, Professor of Molecular Biology at the Autonomous University of Madrid, specialist in genetic engineering, and a close personal friend. Our friendship, however, never tempered her critical tongue, which drove much fuzzy thinking out of the early drafts.

Ruth Shaw, who is simply 100%.

Alexander J Weeks, whose idea it was to write this book, and who was the first typist of my manuscripts. Without the support of Alex in the early stages, the whole idea would never have borne fruit. He was still there at the end, dealing with publishers.

The Institute of Sound and Vibration Research at the University of Southampton, England, and especially to Professor Frank Fahy, whose intuition and foresight brought Keith Holland and I into a close collaboration of great mutual benefit.

The ISVR was the hub around which so much of this work revolved, and I can whole-heartedly recommend their enormous problem-solving capacity to any individual or organisation with a seemingly intractable acoustic problem.

Finally, I would like to thank Gillian Payne, who kept the machine rolling, between the inspiration of Alex Weeks and the dedication of Janet Payne.

To you all, I offer my deepest appreciation.

Philip Newell

Lisbon
February 1995

Introduction

A studio monitor system can be defined as the whole signal path from the output of a mixing console to the ear. This path consists of electrical, electro-mechanical, and acoustic components before finally arriving at the human auditory system. The paths to the ear obey the laws of physics, so, within the limitations of the technologies involved, they should be definable and predictable. Once *at* the ear, however, the signal meets with a system of perception which is unique to each individual listener. The effect of this individuality introduces a degree of subjectivity, which creates, in different persons, different hierarchies of priorities. These in turn have repercussions on the choices made in terms of the necessary compromises which must be made when any system is designed.

So, whilst there is a solid basis in the physics of electro-acoustics, there are complications created by human subjectivity, which exists even in the most experienced of listeners. Subjectivity also exists in the expectations of users of monitor systems, both in the way in which they use them, and the balance between accuracy, flattery and convenience which helps them to reach their own individual recording goals. In an interview in *EQ* magazine in June 1993, the highly experienced recording engineer George Massenburg said, 'I believe that there are no ultimate reference monitor systems, and no "golden ears" to tell you that there are. The standards may depend on the circumstances. For an individual, a monitor either works or it doesn't . . . Much may be lost when one relies on an outsider's judgements and recommendations.'

About a hundred years earlier, Baron Rayleigh, one of the true giants of the history of acoustical science, stated, 'The sensation of a sound is a thing *sui generis*, not comparable with any of our other sensations. Directly or indirectly, all questions concerned with this subject must come for decision to the ear, as the organ of hearing: and from it, there can be no appeal.'

The above quotations from two eminent persons show the degree to which human perception must be taken into account, no matter how 'hard' the physics of the subject. Studio monitor systems sit at the crossroads between the artistic, scientific, technical, and human aspects of the recording process. Sometimes the degree of mutual exclusivity of the individual requirements of the above components can be alarmingly high, but nevertheless, without the monitoring systems, there is effectively no viable recording process, so the problems *must* be addressed. In this book I shall try to fill in many of the gaps which exist between the multitude of text books available on the individual component parts of monitor systems. What is more, I shall try to do so in a language that is as accessible as possible to all concerned, whether their paths to the recording studios have been via the musical, technical or scientific routes. A general understanding of the principles involved should lead to a more practical *and* creative environment for all involved in the recording process.

The concept of a studio monitoring system as a single entity does not allow for a random 'pick-and-mix' of favourite components. If optimal performance is expected, then the system must be a very carefully balanced choice of the component parts which are capable of achieving the best *overall* performance in any given set of circumstances. If I may make an analogy with the world of Formula 1 motor racing, the championship does not necessarily go to the team with the most powerful engine, the best chassis, the best suspension system or the best tyres. Nor does it always go to the team with the best driver. The championship usually goes to the team with the best overall compromise in the balance of all the component parts, carefully set up, circuit by circuit, and driven by a driver not only possessing tremendous skill, but also a deep understanding of the whole subject.

Unfortunately, the highly professional and considered approach which goes into preparing for a Formula 1 season does not always prevail in studio monitor system selection. All too frequently, somebody's favourite amplifier is matched with generally well liked loudspeakers, and then sited in a room of arbitrary characteristics. Far too often this is the case, with monitor systems not being given their due priority in the specification of a recording chain. The decisions about monitoring are not simple ones, and even for specialist designers, much deep thought is usually required about each and every installation.

As will be seen, conflicting orders of priorities will elicit different interpretations from individual designers. Inevitably, there will be an element in this book of my own personal beliefs and experiences, but that should not prevent the reader from gaining an understanding of the process as a whole, and an appreciation of the ways in which other individuals may vary their approaches to achieve their own optimum compromise realisations.

There are many text books available which discuss at great length the physical operation and subjective human perception of loudspeakers. Such books are usually aimed at the domestic hi-fi, general sound reinforcement, and technical/academic camps. When one considers studio monitoring, however, a very different set of criteria becomes apparent. At the very least, the criteria are arranged in a considerably different order to those other categories listed above, where one either looks for the most pleasing sound, the most persuasive overall sound; or where one is concerned more with the purely technical aspects of the systems.

In the case of studio monitoring, the main aspects of system performance are geared towards enabling the users to detect things which may become problematical on other systems; to give the users the widest possible scope for artistic interpretation of what is being recorded; and also to be capable of producing the desired atmosphere in the creative environment. Somewhat surprisingly to many people, many well used studio monitor systems do not rate particularly highly on the fidelity ratings used by many hi-fi reviewers and magazines. To lay persons this may seem disturbing, until they can appreciate that the function of such systems is not overridingly to create wonderful sounds in the studio, but to enable the studio to be used to produce recordings which can create wonderful sounds when taken home by the record-buying public.

Although it will always be very worthwhile to aim for maximising the fidelity of a monitoring system, other requirements such as large size to enable reliable production of high SPLs at low frequencies, ruggedness in daily use, ease of servicing, consistency of operating parameters, and many other factors may to some degree negatively affect fidelity. A super-fidelity system which fails twice a week is of little use in studios. In reality, in the studios one is not listening to the fidelity *per se*, but to the aspects of a recording which may affect its future enjoyment on a domestic high fidelity system. Obviously, a low fidelity monitor system would be of little use, as it would not have the resolution to detect many of the more subtle aspects which may in turn affect the fidelity of the final product. On the other hand, highly important as it is, maximising fidelity to the nth degree is not *the* criterion of absolute importance in the monitoring environment.

It is in the light of these compromise points, together with the aspects of overall human perception which dictate much of the order of these compromises, that the forthcoming discussions will take place. There are essentially three phases to this discussion: the objectives and limitations, the practical realisation, and the variability of human perception and behaviour. It is a highly complex world of interrelated concepts which frequently seem totally unwilling to settle down into any absolute order; hence it is only by *understanding the variables* that satisfactory and reliable progress can be accomplished.

Possibly the most formidable obstacles to orderly progress are the human problems. Perception would appear to vary wildly, not only between individuals at any given time, but also within single individuals at different points in time. Moods change perception, yet perception in turn can change moods. The range of human hearing is truly enormous, but if this were not sufficient a problem in itself, it is further compounded by the fact that it is by no means entirely linear. Almost certainly, the non-linearities vary from one individual to another, and for this reason as much as any, monitor system design criteria should not be drawn too far down the road of personal, idiosyncratic preference. Having said that, system designers are themselves human beings, so personal leanings are unavoidable; but as we shall discuss later, as no one system can be all things to all people in all circumstances, a degree of variety in approach need not necessarily be considered to be a bad thing.

That studio monitor loudspeaker systems are a class unto themselves is borne out by the fact that so few loudspeakers straddle the professional/domestic divide. Monitoring loudspeakers are rarely found in other environments, not because of price, as many 'high end' hi-fi systems cost more than many comparable monitor systems, but because they are highly specialised, highly evolved tools. They are only a means to an end. The pathways *to* that end are not as obvious as one may initially expect, so it is those pathways which we shall explore in some detail.

Chapters 1 to 5 deal largely with the general aims, and both the technical and human limitations which compromise those aims. Chapters 6 to 14 discuss the individual component parts of the monitor chain, and how those parts may influence the overall performance. The practical realisation of the compromises are the subjects of Chapters 15 to 21, which serve to show the typical thought processes which lead from the philosophies and components, to three types of control rooms and three types of monitor systems. The examples are of professional systems in daily use, so there is actual 'proof of the pudding'. When all of the sections are taken together, a more global understanding of the whole process should begin to become apparent, which will hopefully be of help to anybody embarking on such a journey for themselves.

Where possible, I will use descriptions of actual systems and rooms, and by looking at specific cases, the relevance of the general theories will hopefully be more readily understood. For many people, it is much easier to grasp a concept when a reality is presented as an example, rather than by asking the uninitiated in particular to understand abstract ideas. The idea of putting together this book was spurred on by the many responses to my magazine articles, asking where more information of the sort contained here could be found. I hope that the presentation and language is sufficiently accessible to all interested readers.

Chapter 1

Monitoring – the tortuous way ahead

1.1 Underlying diversity

Perfect monitoring conditions are not going to exist within my lifetime. Although I am not anticipating my demise for three or four decades, I am very confident that the previous statement will hold true. Monitor systems are very diverse in their construction and performance, this diversity being a function of the different points of compromise chosen by the different individuals concerned with their design and usage.

Whilst I cannot claim to have heard every single model of loudspeaker, amplifier, crossover, or room in the entire world, I have certainly heard a very great number of different systems. What I *can* say, however, is that I have never yet heard *any* two different systems which sounded exactly similar on a wide range of musical programme. On many occasions I have been called to studios to try to improve the sonic compatibility between the large and small monitor systems. After a certain degree of adjustment, especially when using the music being recorded in the studio at the time of the adjustments, or the producer's favourite CD, a reasonably high degree of compatibility can usually be achieved. In almost all cases, this adjustment is made by some form of amplitude correction to one or more drive units in one or other of the systems. Unfortunately, despite the compatibility which may apparently have been achieved, a different type of musical programme may well reveal alarming differences in what had previously been adjudged to be highly compatible systems.

The key to the above observation is in the fact that the adjustments which had been made to improve the compatibility of the systems had been amplitude adjustments to the drivers of one or both systems; whereas amplitude is only one of the properties which define the characteristics of a system. A different type of music to that upon which the systems were originally adjusted may be more revealing in terms of the phase characteristics, or the non-linear distortions such as harmonic and intermodulation distortions. Furthermore, it is quite possible that the amplitude adjustments involved may have compensated for system differences in the most dominant region of the first piece of music, whereas those self-same adjustments may have disrupted a sensitive range in a subsequent piece of music. The interaction potential is infuriating. It is not possible to define which parameters are most important or to what degree their errors can be tolerated, as dependent upon drive signal, the priorities for a subjectively 'natural' reproduction can shift within surprisingly wide limits.

In general terms, and *only* in very general terms, phase is of less importance than pressure amplitude to steady state types of programme such as sustained reverberant organ chords or bowed violins, but as music is generally composed of very complex waveforms, all of the above distortions, both linear (amplitude and phase) and non-linear, have differing degrees of subjective effect upon the signal. Indeed, the phase responses which would be of minor effect on the sustained organ chord would be more noticeable on more transient signals, whereas the transient signals would tend to mask other problems such as resonances, which the organ chord could excite in a most objectionable way. Toole and Olive stated in a 1988 paper[1] that on an instrument such as an organ, the detection or otherwise of minor amplitude irregularities could be entirely dependent upon the amount of reverberation on that signal.

Over a hundred and fifty years ago, Ohm, and later Helmholtz, carried out experiments which appeared to show that the ear was 'phase deaf'. Their experiments had been carried out on sine waves, which are *very* steady in nature and not representative of normal musical programme, but nonetheless, quite unbelievably, their work is still quoted in some circles as implying the relative unimportance of phase responses in terms of audibility.

1.2 Thresholds of perception

Phase responses have an enormous bearing upon the waveforms of transient signals, changing considerably the way in which the shock of the transient excites the ear. Again, an argument had been put forward implying that if the waveform distortion changed the envelope only within the integration time of the ear, then the effect would not be perceived, but once again our perception systems seem to be far more subtle than often previously believed. The concept of integration time states that if the ear samples sounds in periods of time of length x, then it would not matter whether ten units of a sound occurred simultaneously, or whether each was separated by a period of one-tenth of x, or even any combination in between; the ear would still perceive ten sound units within that time window of length x, so the perception of any combination would be one and the same for all cases. Again, experiments have been performed under controlled conditions which would seem to bear this out, yet like the work of Ohm and Helmholtz on phase deafness, the experiments did not relate to the complexities of *musical* programme. In Fig. 6.2 (see Chapter 6) the two waveforms shown differ only in the phase relationship of the component frequencies. If the whole figure represented waveforms of less than the integration time of the ear, then classic theory predicts that they should sound the same, but they do not.

The picture which begins to unfold is, 'show me an experiment which proves that an effect is inaudible, and I will show you a set of circumstances which clearly show it to *be* audible.' Michael Gerzon published a series of papers, articles and discussion documents which dealt with attempts to define limits of the audibility of various distortions and combinations of distortions.[2] He strongly pointed out that some very low level effects which are by no means always subtle in perceptual terms cannot be measured, purely because we do not yet have the equipment or techniques. Gerzon's article on 'Why Do Equalisers Sound Different' quotes some of Harwood's work from the 1950s at the BBC. This work was on loudspeakers, but as loudspeakers and equalisers are both filters – they both limit or modify the response – the points are relevant in both cases. The implications of this work were that if severe colouration could be detected from resonances 40 dB below the main signal, then that − 40 dB signal (1% of the total signal) represented an accuracy in terms of amplitude of ± 0.1 dB, and in phase of ± 0.6 degrees. Gerzon goes on to suggest that a factor of 10 improvement on these specifications would still produce audible colouration on *some* programme material, that is, accuracy to ± 0.01 dB in amplitude and ± 0.06 degrees in phase. In support of this he cites work by Dr Roger Lagadec, the head of digital development at Studer International, Switzerland, who in the early 1980s detected audible colouration caused by amplitude response ripples in a digital filter at levels only just above ± 0.001 dB.

Work is currently being undertaken, notably in Canada, which is leading towards the conclusion that for a true sense of 'being there', phase, and to a slightly lesser degree amplitude response, must be maintained down to a frequency of 0.001 Hz. Clearly, we can *hear* neither 0.001 Hz nor 100 kHz, yet our *tactile* senses can be shown to respond in these regions. Conduction from the cheek bones at 100 kHz has been detected as significant, and the 0.001 Hz, which represents one cycle roughly every 16 minutes, is in the 'weather frequency' band where one 'feels it in the bones' that a change is on the way. Much more work still needs to be done on the integration by the brain of the audible and tactile sensations, but it is of little doubt that they are very important in terms of 'natural' reproduction.[3] For many years I have advocated electronic crossover and amplifier frequency responses from DC to 100 kHz and nothing has recently come to light to cause me to change my mind.

At very low frequencies, phase does seem to play an important part, possibly to a greater degree (if you can compare apples to bananas) than amplitude. Gerzon once again claims that in the naturalness of perception, relative phase accuracy down to 15 Hz is probably more important than amplitude accuracy down to 5 Hz; but here, once again, integration of the senses plays an important role. As will be explained in more detail in Chapter 3, there is evidence of some rooms, especially with hidden treatments (i.e. not visible) just not looking as though they should sound like they do. The brain, expecting a different sound from previous experience of the visual cues, suffers an uneasy confusion, and can be left somewhat uncertain about the 'rightness' of the sound. Similarly, severely asymmetrical rooms can cause some people difficulties with stereo centring.

1.3 The great valve (vacuum tube) debate

Ironically, whilst we seem to be moving towards the criteria of 0.001 Hz to 100 kHz frequency bandwidth, ± 0.05 degrees phase response from 10 Hz to 40 kHz and a ± 0.01 dB pressure amplitude response over the same range, until such 'perfection' is achievable, there is a certain element of 'the wider you open the window, the more dirt blows in'. I was brought up in studios using valve (vacuum tube) monitor amplifiers, but with the advent of the early transistorised Crown DC 300s in the late 1960s, we soon changed from our Radfords, Quads, Leaks and Pamphonics, never to return. For very many years I viewed the valve amplifiers rather in the same light as vintage cars – nicely made and nice to have, but not really suited to the requirements of today's daily use. Although I would still not advocate valve amplifiers for monitoring purposes, as I doubt that such technology will lead us much further down the road to improved systems, the affection which some people have for valve amplifiers could partly arise from them 'limiting the amount of dirt which can blow in'. In other words, sometimes a genuine photograph of a person, especially with a cluttered background, may well be deemed less easily recognisable than a crude caricature of the person in a cartoon. Valve amplifiers may well allow through more of the essence and less of the dirt. Whilst I accept that this is not the whole reason for the widespread love of valves, it is almost certainly a part of it.

There are definitely circumstances under which objectively 'worse' can lead to subjectively 'better' when more *essential* detail is left exposed, even at the expense of conventional 'accuracy'. In some situations, 'improved' specifications can lead to a deterioration in perceived or preferential concepts of accuracy, until a much lower point is reached in the generation of spurious distortion products, where the benefits of further reductions will once again become apparent. For example, a simple, portable transistor radio may well have alarming non-linear distortion figures for its amplifier, yet it may sound tolerable for everyday use; its loudspeaker's poor response bandwidth rendering benign many of its more annoying characteristics. Improving the overall technical performance solely by opening up the bandwidth of the loudspeaker may render the whole unit most unpleasant to listen to; a further improvement in technical performance by reducing the amplifier non-linearities may be required before the whole unit could once again be deemed to be as subjectively acceptable as in its original state. There is also a hideous convolution of the different distortion products which conspire to mask each other in a very programme-dependent way, such that it cannot be said in any general terms whether ± 1 dB in the pressure amplitude response is any more or less important than 1 degree in phase accuracy, or than an extra 0.1% of harmonic distortion. Even within the harmonic distortion, no absolute, programme-independent ratio exists in terms of the relative importance of the individual harmonics.

1.4 Low level awareness

The sensitivity of the ear to low level effects has again been dealt with at some length by Michael Gerzon,[4] not only in terms of perceived colouration, but also in the context of very low level or even subliminal mixing of instruments in a musical balance. Such low level signals can either have a positive effect on the 'feel' of the music, or would be conspicuous by their absence. On the same subject, Canadian scientist Stanley Lipshitz reported that on one test to determine the audibility or otherwise of the insertion of a capacitor into a test circuit, one listener was consistently scoring a very high degree of accuracy in detecting its inclusion. Much hard searching revealed a buzz, produced by the switch and wiring detecting airborne interference when the capacitor was switched in; this buzz was *80 dB* below the programme. Screening of the circuitry removed the buzz, and rendered the insertion of the capacitor undetectable. The above highlights not only the extreme rigour with which subjective listening tests must be carried out, but also the probability that many subjective differences deemed to be audible in cables, amplifiers and other links in the monitoring chain may not be inherent, but may be due to outside interference. An example of such circumstances would be where an amplifier was deemed to be more 'gritty' than a seemingly inferior amplifier with reduced bandwidth. It may well be that the wideband amplifier was subject to more radio frequency interference in the test set-up; removing the test to a more screened location may well have reversed the subjective preference.

The brain notwithstanding, what the ear alone can respond to is worth note. It has a range of hearing of around 120 dB, which represents a power differential from the quietest perceivable sounds to the threshold of pain (120 dB) of 10^{12} or one million, million times. At the threshold of hearing, 0 dB, the distance of movement of the eardrum is around one-hundredth of the diameter of a hydrogen molecule, or about one ten-thousandth of one-millionth of an inch. After exposure to a bright light, the eyes take some considerable time to recover the ability to see in low light levels: after exposure to a loud sound, the ear/brain also takes time to regain its sensitivity to low level sounds. In the latter case, however, after exposure to a burst of sound at 120 dB, the ear can recover its 0 dB sensitivity at the threshold of hearing in around half a second[5] – in other words within the reverberation time of an average domestic living room. It has also been shown that some people who cannot hear a 20 kHz tone, can under some circumstances detect the effect of amplitude changes of as little as 0.1 dB at 20 kHz.

In the first paragraph of this chapter, I mentioned the 'gross' imperfections of current monitor systems. In the light of the sensitivity of our sensory systems, the performance of any currently available loudspeaker is *certainly* grossly imperfect; the wild advertising claims of many manufacturers do nothing to further help the industry in general to understand just where we need to be going, nor how far we still have to go. Being so far away from perfection, it is little wonder that arguments rage over which systems are the 'best'. In current reality it is merely a case of choosing a system, the imperfections of which are most subjectively benign with regard to the programme material being monitored, and which also offend as little as possible the sensibilities of the individuals involved in the recording and reproducing process. As the variables are so great, it is inevitable that a wide range of monitor systems will exist in order to provide a 'best fit' compromise to differing circumstances.

1.5 Prospects for further development

We are looking for sonic accuracy usually via loudspeakers which are little more than pieces of metal and cardboard, somewhat arbitrarily glued to chassis and coils, and mounted in less than perfect magnetic fields. It is little wonder in the light of the previous paragraphs that we have not achieved perfection.

Given today's technology, we cannot even make any *vague* claims towards true accuracy. In the near future, digital signal processing may lead the way forward to more controlled performance, but even here, we do not yet know just how far we will need to go before we can claim sufficient *subjective* improvements, nor at just what level of imperfection reduction this subjective improvement will begin to occur. Certainly the low frequency phase accuracy problem seems to be tailor-made for attack by digital filtering methods.

At the moment, all recorded music is intended for playback over loudspeaker systems of one sort or another, including headphones. Our currently great distance from our goal of audio perfection would imply that we must bear in mind the object of the exercise, which in turn means that we should choose our monitor systems as we would choose any other tool to do a job: we should select the tool which either ourselves or our clients deem to be the most efficient in use, whilst enabling us to achieve our objectives to the greatest possible degree. This implies compromise, different compromises require different choices, and different choices require a variety of products to choose from.

So where does that leave us? To stop seeking perfection, no matter how elusive it may be, would be defeatist and against human nature; but as for practical advice to those who must use monitor systems as part of their daily work, my advice would be the same as that which I would give a tennis player in selecting a racquet: 'Use the one which works the best for you under any given set of circumstances.' Having said that, a tennis player may use a different racquet on grass courts than on clay courts; similarly an engineer or producer may choose different monitor systems for classical or rock music. There is no reason why anybody should feel unhappy about this, as technologically that is as far as we have currently progressed in sound reproduction via conventional hi-fi or monitoring systems – it is a reality of life in the 1990s.

In terms of the older, now outmoded concepts of using monitor equalisers to attempt to achieve compatibility – if you cannot equalise a Shure microphone to sound like a Neumann, or a Neumann to sound like a Schoeps, or a Schoeps to sound like an Electro-Voice or an AKG; or for that matter a Neumann U87 with a KM84, also from Neumann; then why expect to be able to equalise two *monitor systems* to any degree of compatibility? It was the enormous ravine which we still must bridge in our search for perfection which drew me to write the opening sentence of this chapter, 'Perfect monitoring conditions are not going to exist within my lifetime.'

The whole concept of monitoring is influenced by so many variables that we must, of necessity, be dealing to a considerable degree with an art form. Within that art, there are without doubt numerous sciences, the greater understanding of which will help in the further advancement of the art. Even with active control, I am prepared to stake my reputation on my belief that we will never see, or rather hear, anything even approaching true sonic fidelity with microphones or loudspeakers of any type which we are aware of in the 20th century.

Audiophiles contribute much, but their search is that of a Holy Grail, certainly on any *presently* known path. If we do not seek, then surely we will not find. On the other hand, however, some claims to accuracy which emanate from certain circles verge on the absurd. We are dealing with a limited technology, but nonetheless a technology which is remarkable and capable of creating much pleasure. If music is being created to be played back over loudspeakers, then recording and mixing to achieve the best *from* those imperfect systems is still a justifiable aim. Nevertheless, the search for improvement must still continue.

Unfortunately, we cannot mix music recordings to the parameters of each and every commercially available loudspeaker system, and certainly not to each and every ear/brain combination. I would suggest that we either mix to a standard, or produce loudspeakers to a standard. The other option of standardising ears and brains is not available. However, standards are not agreed on very easily. During audible similarity experimentation at Southampton University with Dr Keith Holland, we were shaken to the core by the unfolding realisation that people, many of whom were experienced in the recording industry, could not even agree on whether two sounds were *similar*, let alone 'right'. In the light of this, wherever I have referred in this book to accuracy, it must be construed as 'the greatest accuracy achievable with current technology', but the term 'accuracy' must itself carry some degree of personal variability.

If ears and brains are so sensitive to small electro-mechanical differences in such a way that those same differences are differently perceived by different individuals, then only by the absolute re-creation of the original sound field could any general agreement be expected. I cannot envisage the realisation of this goal even in intellectual terms, let alone electro-mechanics. In many ways, the key chapters in this book are 3, 4, 5 and 13. The other chapters deal with the achievable goals from the rational application of the technology which we have available, but these four chapters show just what an uphill struggle we are facing as we attempt to progress towards better monitoring conditions.

In 26 years of working in studios, I have never refused to use a studio because of its monitoring. Obviously, I have had preferences and have been inclined, whenever possible, to choose to use studios which I liked for very many reasons, monitoring included. To this very day, however, I am being called in by studio owners or operators to 'sort out' their monitoring because producer F or engineer G say that they will only book the next six months if the monitors are changed. Given that *all* monitors are wrong, when such producers or engineers claim that the monitors which they demand are 'right' whereas the studio's in-house system is 'wrong', the problem usually lies not in technology, but in ignorance. Monitors – *all* monitors – are musical content dependent, recording technique dependent, storage media dependent, personality dependent, ear dependent, brain dependent, room and equipment dependent . . . ad infinitum.

1.6 Practical examples

A large London studio once called me in to 'sort out' a very expensive monitor system of American origin. A currently very 'hot' band were in the studio and threatening to pull out. I duly looked at the problems, consulted with the designer and manufacturer, and proceeded to 'sort out' the system.

About a week later, the maintenance engineer from the studio told me that for the first time in his career, he had seen a positive comment in the fault book saying, 'The monitors are wicked!' He was very happy; so was the studio manager, and so were the band in question – they booked another month. Some months later, a producer of some stature who I had known for over 15 years went to the studio to discuss the possibility of 4 months' work. He listened to the monitor system and said that he thought it sounded hard, and would only book the studio if they hired in a certain other type of monitor loudspeaker. It so transpired that the monitors which he wanted existed in abundance in my garage, as I had taken many of them in part exchange for my own systems. Nobody else seemed to want them, but for the producer in question, they worked!

Was the first monitor system in the studio hard, or was it the recorded music that was hard? Studio designer Tom Hidley once sent me a special compilation CD of recordings which *he* used to set up systems. Almost all of the recordings were of acoustic or electro-acoustic instruments, recorded with only minimal amounts of equalisation. He contends that only by this means can monitor systems be assessed, as *any* recording from an electronic instrument can only be subjectively assessed as sounding however it sounds through whatever loudspeakers which it may be played – there is no 'real' reference to check the sound against. I tend to agree with him that in some areas, 'harder' computer generated sounds and digital recording have created a tendency towards 'softer' monitoring. It certainly seems to have led to less of a consensus on whatever may be deemed to be 'right'.

People have cited the use of certain types of horn loaded monitors, having poor mid-range directivity spread, as being the root of much of this, by virtue of producing a reflected sound field which can be light on mid-frequencies, leading to a mid–light overall monitoring balance. Other manufacturers and designers, notably of American origin, have cited the softer sound of many direct radiator systems as leading to harsher mixes, as any 'harder' horn would ostensibly lead to a softer mix. No doubt there is an element of truth in both of these viewpoints, but what I believe to be the true culprit is the excessive use of signal processors and effects, often in conjunction with bad monitors, either with or without horns.

Good monitor systems have good impulse responses which keep intact the leading edge information of the musical signals. In order to achieve this, both amplitude *and* phase responses must be very good indeed. Widespread signal processing can lead to some very chaotic phase information, especially when time delayed, cross-panned, re-processed, and whatever else. The greater the definition of the system, the more probable it will be that any unwanted anomalies become apparent. Indeed, this is one reason why many good monitor systems do not automatically become expensive domestic hi-fi: they can be unpleasant to listen to on poor quality programme – they are too ruthless, especially at the high sound pressure levels for which they may be optimally balanced.

Conversely, the use of small monitors in particular, such as used in many 'home studio' set-ups, and which may be pleasant enough for home consumption, do not have the accuracy of resolution to make immediately apparent the aggression which can be introduced into a recording by extensive signal processing. There are not many compact, high-resolution monitor systems available from specialist manufacturers, which would show *immediately*, at the recording stage or during the subsequent mixing, when the processing was becoming the root cause of much harshness. Unfortunately however, to many people seem to begrudge having to pay £2000 ($3000) or more for such systems, so armed with their cheap monitor systems, they remain blissfully unaware of the problems. When playing the production CDs through their own systems, which in many cases would be the self-same system on which they were recorded and mixed, they would merely be participating in a self-fulfilling prophecy, that the mix was 'right'. People who work on bad monitors are merely burying their heads in the sand, taking a lazy way out, and leaving the record buying public to suffer because of the lack of professionalism being shown in the recording studios.

The general performance of loudspeakers is remarkable considering just what crude devices they are, but a general acceptance of their inherent limitations by all concerned would certainly appear to be the more professional approach. However reasonable that last suggestion may seem, in the highly charged and fiery environs of the world of music recording, prevailing reason may well be a *very* tall order. Nonetheless, forewarned is forearmed, so on we go!

References

1 F. E. Toole and S. Olive, 'The Modification of Timbre by Resonances: Perception and Measurements', *Journal of the Audio Engineering Society*, Vol 36, No 3 pp 122–142 (March 1988)
2 Michael Gerzon, 'Why Do Equalisers Sound Different?' *Studio Sound*, Vol 32, No 7 pp 58–65 (July 1990)
3 Khanna, Tonndorf and Queller, *Journal of Acoustical Society of America*, Vol 70 (1981)
4 Michael Gerzon, 'A Question of Balance', *Records Quarterly Magazine*, Vol 2, No 3 pp 48–53 (1987)
5 Neal F. Viemeister, 'An Overview of Psychoacoustics and Auditory Perception', AES Preprint, 'Sound of Audio' 8th International Conference, Washington DC (1990)

Some persons mentioned in this chapter

Professor Floyd Toole and Dr Sean Olive were both researchers at the National Research Council of Canada, in Ottawa. They have collaborated on many 'landmark' papers on electro-acoustics, acoustics and human perception. At the time of writing, Professor Toole is also the International President of the Audio Engineering Society (AES), and Dr Olive is manager of subjective evaluations at Harman International Industries, California.

Michael Gerzon is a Fellow of the AES, and is a mathematician, trained at Oxford University. He is considered by many to be one of the finest theoreticians in the field of sound recording and psycho-acoustics. He is a patent holder on the SoundField microphone concept, and the Ambisonic system. Indeed, he was the person who first pulled together the concepts behind Ambisonics into a viable, working system. In the mid-1980s Michael Gerzon was the first proposer of the concept of dithered noise shaping for digital audio systems.

Dr Keith Holland is a Research Fellow at the Institute of Sound and Vibration Research, and as well as being involved in aspects of sound/ music reproduction, has done much pure acoustic research for Rolls Royce and British Aerospace amongst many others. He is also on the staff of the University of Surrey, where he lectures in electro-acoustics to Tonmeister students.

Chapter 2

The state of the art

2.1 Reciprocity

In an article in the July 1990 edition of *Home and Studio Recording* its author Ashley James stated: 'The majority of broadcasters and classical recording engineers have tended to use hi-fi type loudspeakers for some years now, leaving pop and film studios as the defenders of the older technologies.' Well, there are very valid reasons for those states of affairs to exist, but they only very tenuously amount to the defence of older technologies. Leaving the film studios out of the equation for the moment, the main difference between classical and rock performance and recording is that the *reproduction* in the studio control room of classical/acoustic music *seeks to emulate the original acoustical performance*, whereas the *live performance*, especially of electronically generated rock/popular music, *seeks to emulate the original studio recording*. In classical music, classically recorded, the original performance exists as a real entity in time and space. Much popular music never existed as a single performance at any one time. It is created *on* loudspeakers for performance *by* loudspeakers, and, as such, may be totally 'unnatural' in timbral content.

2.2 Origination of sources in time and space

Even in the realms of acoustic instruments, a bass drum rarely sounds like the recorded version of a bass drum. Recorded bass drums are now usually caricatures of acoustic bass drums; microphone types and positions are chosen to develop this stylisation of what is now *expected* for a recorded bass drum sound. Only with a listener's head *inside* or immediately in front of the bass drum would that listener have any hope of perceiving a natural version of the recorded instrument. Indeed, listening from that position, the balance of the remaining drum kit and other instrumentation would be severely distorted from the recorded balance. *Modern music has often no acoustic counterpart*: it exists *only* in recorded form.

In many cases, each instrument in modern music exists in its own reverberant space, a space entirely artificially created by electro-mechanical or electronic means. No attempt is even made to construct a realistic, integrated, homogeneous environment for the ensemble. Conversely, classical acoustic recording uses *one space* in which the entire ensemble is recorded, and it is the illusion of being *in that space* which the recording and subsequent reproduction seek to capture. From the purist's point of view, only stereo pairs or SoundField microphones are acceptable, as even the addition of one spot microphone for highlighting a particular instrument introduces a spacial distortion, since no listener could be at the phantom source of the stereo pair *and* at the location of the spot microphone at any one instant.

Adding to this spot microphone spacial distortion is the fact that the positional location of the sound on playback relies largely upon the differences in *time* and *phase* of the signals collected by the left and right microphones, which are subsequently transmitted to the left and right loudspeakers. The spot microphone would be amplitude panned into the overall sonic picture via the panorama potentiometers (pan pots) of the mixing console. Immediately, we are here faced with two entirely different mechanisms of inter-aural, left/right localisation: time and phase differences for the stereo pair, and amplitude differences for the spot mic. *This never happens in nature!* Accurate reproduction of this situation would, in a technical sense, by definition be unnatural. The only reason for the spot microphone being used in the first place was that for one reason or another, the 'natural' recording of the stereo pair failed to convey the true experience of being there, so we are immediately in the realms of creating the most natural illusion of reality, rather than the absolute fidelity to that reality. In many instances we are dealing with audio-visual sensory integration. Were we actually to be present at the recording, our eyes would be drawn to the soloing violin in such a way that its spacial image would be most effectively fixed, and we may well subjectively hear it to be louder than its true proportional balance to the other instruments. Without such visual cues, it may be the spot microphone which is relied upon to reinforce the status of the solo instrument.

2.3 Re-creation of the illusions

As a consequence of much of this, we are rarely, if ever, dealing with absolutes from a loudspeaker, but more with illusionary perspectives. Given that *all* loudspeakers are very far from theoretical perfection, our choice of a loudspeaker's performance parameters is largely dictated by the most important aspects of the characteristics of any particular music or instrumentation. In the case of a stereo pair recording of a classical ensemble, we are required to pay great attention to accurate time and phase alignment, smoothly falling low frequency response with no abrupt phase anomalies, a generally even directivity/frequency characteristic, and a relative freedom from non-linearities. The price which one must pay in order to achieve these desired characteristics is usually in terms of overall realisable sound pressure levels, resulting from the small physical size usually required to achieve coherent phase/time characteristics, and wide, even directivity.

Larger systems find it difficult to meet many of these criteria, but are necessary in order to develop realistically high sound pressure levels (SPLs), especially at low frequencies, which are required for much modern recording. Here, two important factors come into play. Firstly, a violin at 120 dB is amplitude distorted, as no violin in real life could achieve 120 dB at say, 3 metres. At these higher levels, our brains will perceive greater proportional amounts of high and low frequencies in close approximation to the classic human ear sensitivity/frequency curves of Fletcher and Munson (Fig. 3.1; see Chapter 3). Similarly, a well hit bass drum at 70 dB at 3 metres will sound thin, as the lower amplitude level would suppress the highs and lows compared to the real life sound of standing 3 metres from the actual drum during the performance. Whilst sight must not be lost of the subjective impact of a recording under domestic circumstances, it is necessary to be able to monitor the performance in the control room at a real life level (if there *is* a real life reference available), in order to ensure that the signals passing to the recording systems are *capable* of being representative of the producers' objectives. Reference to 'near-field', or more correctly 'close-field', systems and the obligatory 'taking home of a cassette' will deem the processes to be compatible or otherwise with typical domestic usage. For this reason of real world SPL compatibility, popular/rock music monitoring systems do require 120 dB plus capabilities at three or four metres.

2.4 Evolution of separate approaches to monitoring environments

The second reason for the high SPLs, and the maintenance of those high SPLs down to very low frequencies, is one of our aforementioned tenuous links with the technology of the past. Nonetheless, it is now far too dyed in the wool for there to be any chance of washing it out, even if such a chance *were* to be deemed desirable. The *development* of electric and electronic music in particular has largely occurred around the monitoring systems as they have existed. They have taken into account the strengths and weaknesses of existing systems. Many of the larger systems had prodigious low frequency capabilities which were of little relevance to most classical/acoustic recordings but were born of the need to solo bass drums and bass guitars to check for rattles and squeaks. That low frequency capability has been the very source of inspiration for a new found art form in the electronic, and particularly digital domain. Much of this music exists only as synthetically generated waveforms. Never did these sounds exist in one space at one time until perceived for the first time in a mix. Positional, spacial discrimination is almost always via left/right amplitude differentials as dictated by the position of the pan pots of the mixing console. The stability of position of these amplitude-dependent signals is easily disrupted in conditions of high ambience, so control rooms for rock/pop music have tended to become less and less reverberant to help to support these phantom images. Possibly the extreme of this development is represented by the rooms conceived in the mid-1980s by American designer, Tom Hidley. From the loudspeakers' point of view, such rooms are virtually anechoic, whilst the subtle positioning of reflective surfaces which the loudspeakers cannot 'see', give life to general conversation behind the mixing console.

An effectively dead control room has, in turn, had two significant effects on the development of monitoring systems for use in these rooms. Firstly, the room gives no 'help' to the subjective loudness of the monitors, therefore the monitors themselves are required to produce higher axial SPLs. Secondly, the axial response becomes the most dominant factor of the loudspeaker's performance, which in turn has two subsequent repercussions: the concentration of power in a more narrow frontal arc reduces the 'wasted' power which would be driven into the side walls and ceilings by a wider directivity system, and, consequently, systems can be designed largely for axial performance, as directivity irregularities are of little concern – side wall reflexions effectively do not exist.

In a concert hall, much of the spaciousness of its acoustic is generated by lateral reflexions from the side walls. For these reflexions to be effective, they *must* arrive from the sides of our listening position, hence they cannot be effectively reproduced by two loudspeakers in a frontal location in a relatively anechoic room. Allowing for the fact that most music will ultimately be reproduced in a room with a certain degree of ambience, the microphone positioning for classical acoustic recording is usually somewhat nearer to the instruments than would be desirable as the best listening position in the concert hall itself. The listening room will usually add some early lateral reflexions, hence going some way to restoring a sense of spacial reality. Such side-oriented early reflexions are not desirable from the point of view of pin-point positional discrimination, but when sitting in an actual concert hall with a generally accepted 'good' acoustic, closing one's eyes will create some difficulty in determining precise locations of the different instruments. This, however, is usually of little consequence, as the enjoyment of such music usually depends to a far greater degree on the 'ambience' or sense of spaciousness, than on any requirement for pin-point spacial localisation. One tends to listen to the ensemble.

In real life, one does not achieve 120 dB at 30 Hz in the middle of a classical concert hall when listening to an orchestra, therefore such requirements from a monitor system do not arise. Smooth directivity is a fundamental requirement in order that any side wall reflexions are consistent with the tonal balance of the direct sounds from the instruments, as would be the case in a 'good' live hall. The limitations on maximum SPL requirements, and particularly the low frequency transient demands, have allowed extremely good loudspeakers for classical work to be realised in manageably sized boxes, which are relatively tolerant of listening room variations. In appropriate rooms, these can produce excellent and most realistic sensations of the reproduction of concert performances. It can be seen from the preceding discussion that in many areas, the requirements for rock and classical work soon begin to drift into mutual exclusivity. Such 'never the twain shall meet' aspects reveal themselves even further in the aspects of compatibility with domestic listening environments, and also with the use of horns in the regions above 600 or 1000 Hz.

It has also been suggested by design consultant and author Martin Colloms, amongst others, that the compromise points in loudspeaker design would differ depending upon whether the source was digital or analogue. Analogue systems do not respond to such low frequencies as digital systems, and also possess less low frequency phase accuracy. Consequently, smooth low frequency extension would seem to be a greater necessity to meet the demands of digital recordings than would be the case for analogue monitoring. On the other hand, loudspeakers for use with digital recordings could possibly be allowed to possess slightly more colouration than that inherent in analogue recording systems. Even the slightest additional colouration on

Table 2.1 Areas of design parameter priorities on two-loudspeaker stereo perception for rock/pop and classical/acoustic recording generalisations

Rock	Classical
High output SPL capability, say 120 dB at 3 metres	Lower maximum output SPL, say 105 dB at 3 metres
More tolerant of time/phase distortion if no acoustic source exists	Minimum time/phase distortion for good localisation and natural timbre
Extended low frequency performance at high SPLs, holding relatively 'flat' as far down as possible	Can tolerate generally lower low frequency SPLs and tailing off responses, as long as smooth roll-offs exist in terms of both amplitude and phase
Require relatively dead rooms with minimal lateral reflections in order to support strong, phantom, amplitude-panned images. Axial response more important than off-axis response	Require rooms with lateral reflexions, hence need to have smooth, even, directivity plots in order to achieve due sense of spaciousness
Whilst harmonic distortions should remain as low as can be achieved, this should not be done at the expense of transient headroom. The stylised, highly transient sounds of electronic music will often be more audibly tolerant of harmonic distortion than of amplitude limiting	Require low harmonic distortion as the more steady sounds and subtle ambient information of classical music, along with the lower general levels of transients, make harmonic distortion perception much more of a problem than high level transient headroom limitations

a loudspeaker intended for monitoring analogue material may prove unacceptable. Consequently, if we were faced with the possibility of a modification which could much improve the very low frequency response of a loudspeaker at the expense of only a slight increase in colouration, the relative amount of use of either digital or analogue could dictate whether, overall, the modification would be deemed to be desirable or not.

In the reproduction of classical recordings, spacial *localisation* is frequently less important than the *sense* of spaciousness, for the simple reason that pin-point localisation is not a significant factor in the live performance. Conversely, in much modern recording, the stereophonic positional interplay of the different instruments is a *very* important feature of the art forms. It is exciting and stimulating, and can greatly enhance the subjective sensation of a performance. Unfortunately, as we have already discussed, the positioning of such electronically generated or spot mic'd sounds is entirely achieved by the left/right amplitude differentials via the pan pots of the mixing console; such phantom images are highly unstable in reverberant surroundings. Indeed, it only needs reproduction in a room of relatively modest reverberance in order for a large proportion of the sound stage to collapse into the nearest loudspeaker as soon as the listener moves only inches off axis; possibly only sounds which are panned to an extreme side remaining in position.

As no studio can hope to simulate the average listening room, simply because no such average would be relevant with the spread of room characteristics being so great, the goal of control rooms for rock music has been tending towards maximising spacial localisation, in order to give the artists and production staff the best vehicle for *their* interpretations of *their* creations. Whilst no domestic environment will be expected to exactly match the control room sensations, at least the production staff will be aware of the *potential* of the recording, and, hopefully, those who really seek to do so will be able to achieve a very great degree of the spacial sensations in their home environments. Those who do not wish to take full advantage will no doubt still enjoy much of the effect on the systems which they possess. It is worth remembering that most people who buy music recordings are not *that* interested in 'ultimate' sound. Furthermore, by far the majority of people will never experience anything even approximating to the sounds achievable in a control room. Granted, this does little to help the search for 'perfection', but it is a commercial reality. Most people are quite content with a reasonable recording of music they like at a moderate price. Again, though, we should not use this as an excuse not to strive for the best results achievable.

2.5 Headphones

Headphones now represent a very great proportion of the listening systems for such music, and whilst the spacial sensations are largely within the head and entirely different from loudspeaker sensations, nonetheless relative spacial sensations from the acoustically dead control room will translate well under most circumstances. Headphones are in fact an entity virtually to themselves. Other than with some binaural dummy head recordings, all recordings produce totally unnatural effects of sounds localised within the head. This is largely due to the fact that as sounds arriving from all positions in 4 pi space (a solid sphere around the head) can occur in real life, if the ear receives signals which do not relate to 4 pi space, as is the case from headphones on non-binaural material, there is only one place left for the brain to perceive the source of the sounds to be – inside the head. Even with binaural recordings we are still very limited, as truly effective reproduction would require recording via a dummy head and torso, cast from the individual listener, as none of us are used to listening through somebody else's pinnae (outer ears) located on somebody else's head and body. Once again, mutual exclusivity rears its head: one cannot monitor accurately for headphones on loudspeakers and vice versa. Binaural recordings will only work on headphones; conventional stereo frontal sound stage recordings will only work on loudspeakers. There are some very clever boxes in development which seek to resolve some of this paradox, but they are most definitely not yet universally accepted systems (see Chapter 18).

2.6 Horn loudspeakers

Another area of conflict between the classical and rock/pop monitoring concepts is the use of horns in the mid and high frequency regions. In general, horns have been shunned by the classical recordists, yet one, the Tannoy Dual Concentric, has been widely used, particularly in the UK, for over forty years in both the classical and popular camps. It is probably true to say that most monitor systems using horn loaded compression drivers have borrowed much technology and design from the worlds of cinema and

Automatic impedance measurement with Solartron 1200 analyser

Date: 21-7-88
Sample type: 1A
MIC. positions: 30 & 55 mm

$C_0 = 346.6$ m/s
—— Resistance
—— Reactance

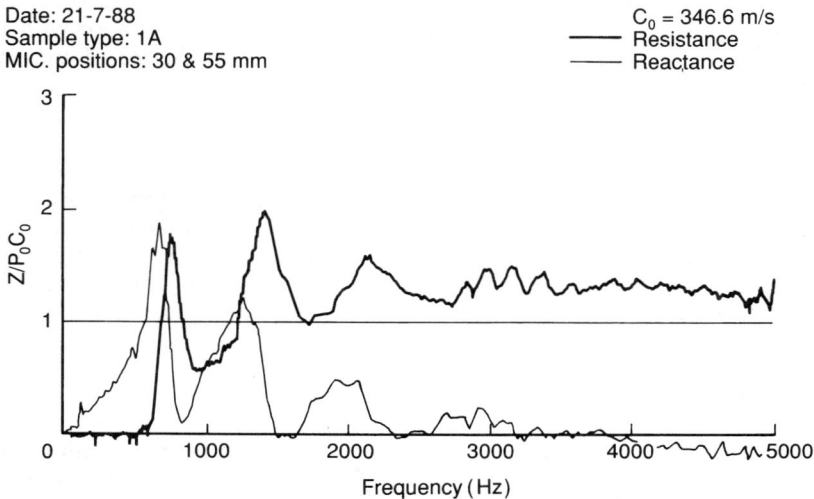

Frequency (Hz)

Figure 2.1 Throat impedance of a typical, well used mid-range horn of American origin

public address/sound reinforcement. In these industries, the prime objectives have been high SPL capability and tight directivity control, rather than subtle timbral neutrality. Horns *have* been designed specifically for studio use, though they are relatively few and far between; but more of this in Chapters 10 to 12.

Figure 2.1 shows the typical throat impedance curve of a horn designed principally for sound reinforcement use. The throat impedance is a reasonably good guide to the 'frequency response', or more correctly the pressure amplitude response, of a horn when connected to a 'perfect' driver. The irregularities are a function of reflexions as the pressure wavefront strikes the air, where the mouth of the horn terminates, and an irregularity in the acoustical loading impedance is manifest due to the abrupt change in cross-sectional area from the horn to the room. Such an effect is also evident upon the sudden change in cross-section from a direct radiator (cone, plate, or dome) as it too meets a rapid cross-sectional area change from the radiating surface to the area of the front baffle. In actual fact, a direct radiator is an extreme case of a conical horn – there is no technically clear dividing line between the two. In the direct radiating cases, however, there is no effective distance from the diaphragm to the air, so no time delay exists: there is no distance for the reflective wave to travel. In these instances, the poor impedance match gives the radiating surface little to push against as the air moves out of the way very easily. Consequently little work is done, which explains the poor efficiency of the direct radiators, typically less than 5 per cent.

Thought of in terms of a boxer, he will do more work hitting a heavy bag than swinging punches into thin air, which explains the efficiency of a horn with its heavier load on the radiating diaphragm. The punch-bag can of course swing back and forth, superimposing its presence on the boxer's punching in either a positive or negative manner (the reflected throat

Automatic impedance measurement

Date: 30-5-89
Sample type: AX2
Speed of sound: 346.5
Microphone positions: 30 & 55 mm

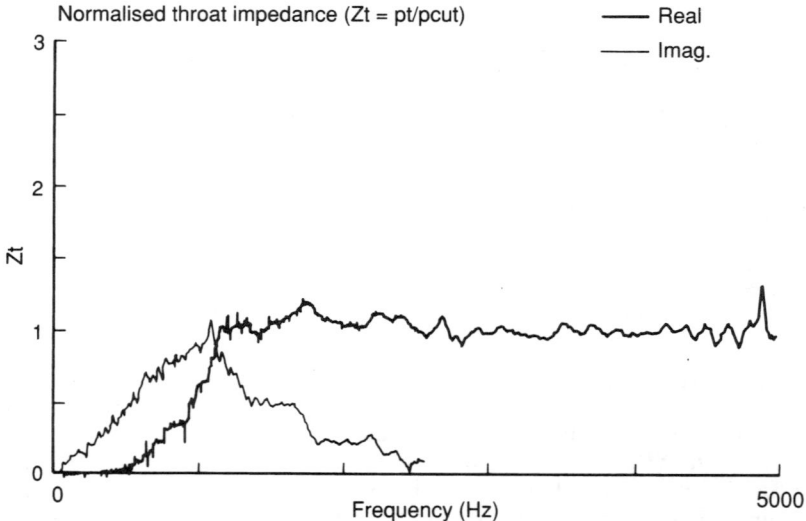

Figure 2.2 Throat impedance of AX2 axisymmetric mid-range horn showing smooth response and vice-free cut-off region

impedance irregularities), but on the other hand, both boxer and direct radiator are more liable to rupture themselves by punching at full power into thin air.

Figure 2.2 depicts the vice-free throat impedance of a horn developed as part of the aforementioned research programme. Figure 2.3 shows the smooth directivity control. Ironically, although very different in physical construction, the shape is not too far removed from the axial symmetry of the Tannoy 15″ horn, which was one of the few horns to be readily accepted in classical recording circles. It is worth noting here that driver and horn are critically matched. Any old driver just will not do. Figure 2.4 shows the response of the same horn, utilising a different driver which was well received when mated to a different type of horn.

Constrained layer diaphragms with highly damped resonances, along with lower compression ratio phasing plugs, can make great inroads into other horn caveats, but we have now proved conclusively that the horns need not be responsible for significant response non-linearities. For anybody unfamiliar with the innards of compression drivers, the diaphragm usually faces a perforated block which serves to channel the pressure variations caused by the movement of the diaphragm, and deliver them to the throat of the following horn in a phase coherent manner, free from the small path-length distances which might exist if all points on the diaphragm were not exactly equally distant from the horn throat. The compression ratio is a

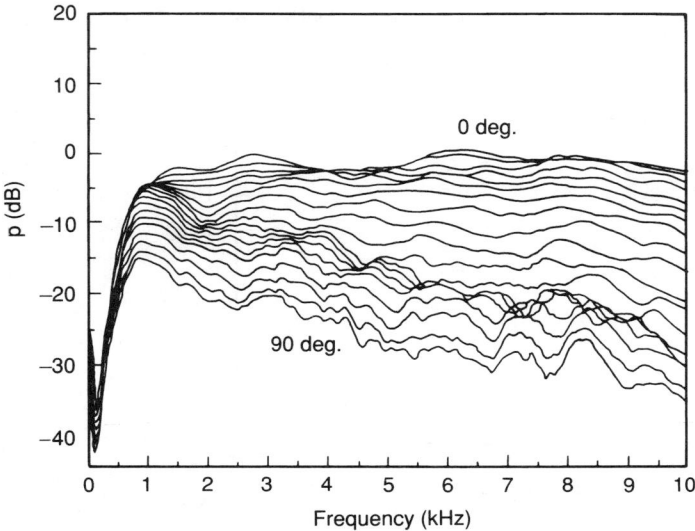

Figure 2.3 Excellent axial and off-axis frequency response performance of AX2 axisymmetric horn. Felt diffraction rings around the perimeter can go some way to linearising the 7 to 9 kHz region of the 70° to 90° performance if required

function of the degree to which this 'phasing plug' and driver throat restrict the free passage of air, and cause the pressure variations in the phasing plug to exceed that in the diaphragm cavity.

In general, compression driver/horn combinations have three major advantages and one significant drawback when compared to direct radiators. The advantages are higher achievable sound pressure levels, and tighter directivity control which can be an advantage in concentrating sound only where required. Plus, when well designed, they can be the sources of excellent spherical/spheroid expanding wave sections, which approximate much more closely to a point source than the pistonic radiation from a cone or dome. The effects on lobing in the polar radiation patterns are shown in Fig. 2.5.

Currently, the major area in which compression horns lag direct radiators is in the area of non-linear distortions. Indeed, the lowest harmonic distortions are produced by electrostatic loudspeakers, which can sound exceptionally sweet on classical music but which show little advantage on rock music. Personally, my honest preference in high level rock orientated systems is for horns above 1 kHz in 'monitor dead' rooms, whilst for classical/acoustic reproduction, I prefer direct radiators all the way from top to bottom in slightly more reverberant rooms. There is an element of self-fulfilling prophesy here, as in many cases each type of music will have been mixed on such loudspeakers in such environments, to sound good on those systems. I am not, nor ever have been, an advocate of horns *per se*: it is merely that at high levels on rock music, they do have advantages which can outweigh the drawbacks. Papers presented to the AES 8th convention on the 'Sound of Audio', by George Augspurger[1] and Richard Cabot[2] discuss in some detail the concepts of harmonic distortion perceptions. Keith Holland and I

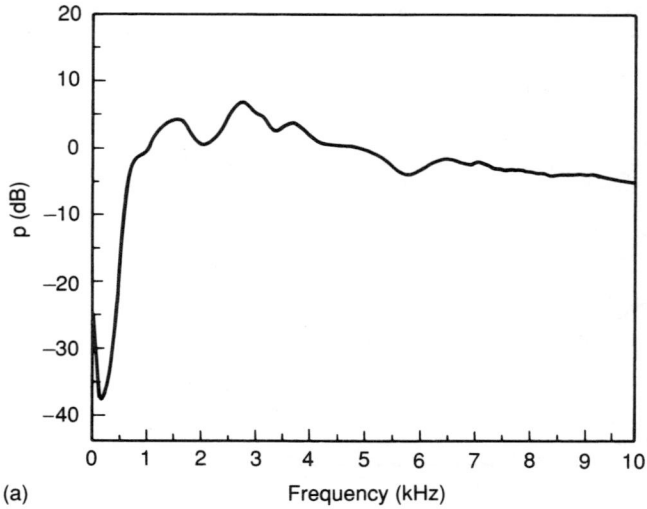

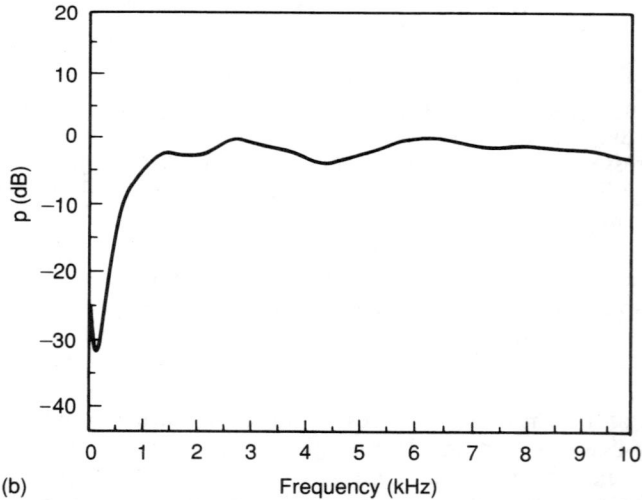

Figure 2.4 (a) Axial response of AX2 axisymmetric horn driven by a driver which is noted for its smooth response on certain other horns. (b) Axial response of same horn utilising TAD TD 2001 compression driver

conducted four months of listening tests at the ISVR in 1989, the conclusions of which tended to indicate that harmonic distortion was more noticeable on 'steady state' sounds such as flutes and bowed violins than on transient sounds such as explosive snare drums. There were exceptions, however, such as the enormous tolerance of harmonic distortion before it becomes noticeable on the 'steady state' sound of a sustained fuzz guitar. In some modern music, harmonic distortion is inherent, and yet, if it *is* inherent in the original signal, it cannot *be* distortion!

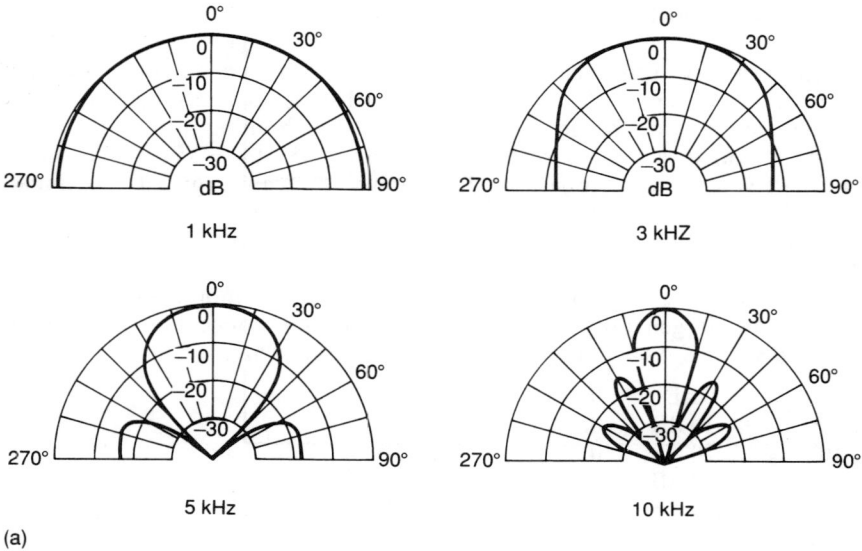

(a)

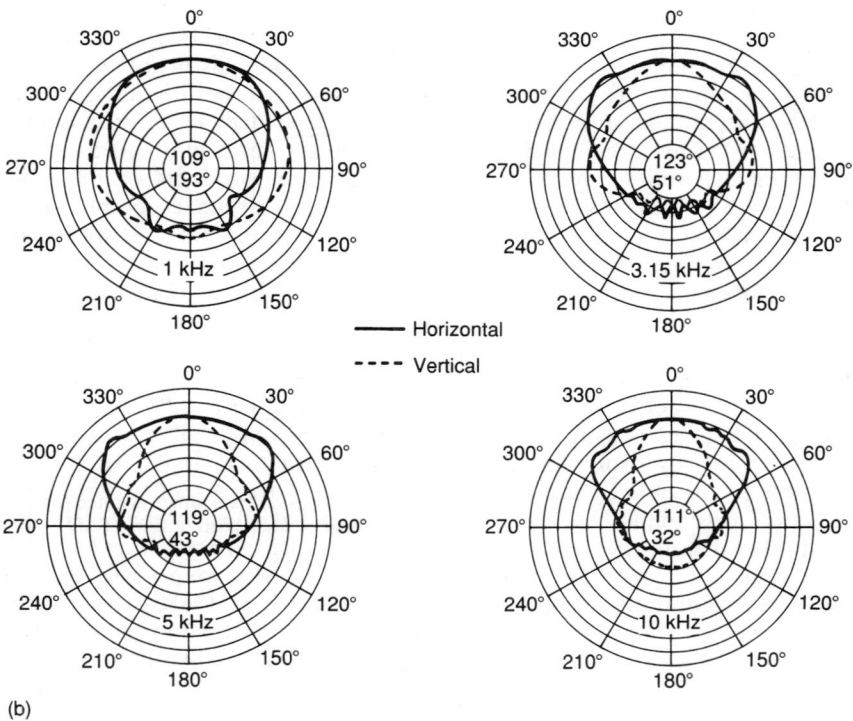

(b)

Figure 2.5 (a) Typical, lobed radiation pattern of a piston radiator ($4\frac{1}{2}''$) (after Beranek). (b) Typical radiation pattern of spherical expanding wave as produced by a well designed horn (Electrovoice)

2.7 Perceptibility of harmonic distortion

In the listening tests mentioned above, one surprising outcome was that some loudspeakers could be deemed by the listeners to be subjectively very similar on a wide range of signals, whilst exhibiting up to 20 dB of difference in measured harmonic distortion (-55 to -35 dB); yet many other drive units of comparable distortion were deemed grossly dissimilar. The relevance of harmonic distortion is very dependent upon signal content, and also upon the secondary and tertiary effects of that distortion, or even the root cause of which that distortion is itself a secondary or tertiary effect. All available evidence suggests that the harmonic distortions in themselves, if generally below 1%, are not that noticeable on complex musical signals. As a guide, the per cent to dB ratios are:

100%	-0 dB	0.3%	-50 dB
30%	-10 dB	0.1%	-60 dB
10%	-20 dB	0.03%	-70 dB
3%	-30 dB	0.01%	-80 dB
1%	-40 dB	0.003%	-90 dB
		0.001%	-100 dB

Clearly, if at all possible, harmonic distortion should be minimised in all systems. It is only when the distortion reduction forces compromise in other areas of performance that any questioning of distortion reduction should be relevant. The main problem is that harmonic distortion audibility depends upon so many factors that absolute standards for audibility threshold levels are almost impossible to set down. Dependent upon programme material, listening level, spacial effects in both the recording and listening environments, and many other factors, distortion products of different harmonics at different levels can be either benign or unacceptable. Two systems with closely equivalent steady state distortion characteristics may have wildly different perceived performances under dynamic conditions.

I remember discussing the relative merits of bass drivers with a technical person from a reputable manufacturer of high quality loudspeakers. I encountered a semi-fanatical pursuit, dedicated to the minimising of harmonic distortion as the overriding priority in driver design. It was difficult to get the point across that, certainly to me, some other excellent drivers of inherently higher distortion sounded more natural. After all, it was shown some time ago by Moir that at 60 Hz, 7.5% 3rd harmonic distortion was inaudible from a loudspeaker, and that at 80 Hz, over 40% 2nd harmonic distortion was inaudible. Given these facts, what does it matter if the measurable distortion of a driver at 80 Hz is reduced from 2% to 1.8%? It might look good on paper, but in terms of perceived sound quality it is totally irrelevant, unless it has some further, related, side effects.

2.8 On and off axis distortion significance

Another difference which is greatly relevant is off-axis distortion as a function of axial distortion. When Keith Holland was operating the blind testing equipment in the aforementioned tests, I remember well his looks of stunned consternation when subjects repeatedly deemed certain loudspeakers to be similar. Keith was roughly 90° off axis to the test drivers whilst operating the test equipment, and it was grossly apparent that just about no commonality of performance existed at those angles. Once again, this can be dealt with in most circumstances by room design, but a wildly differing off-axis spectral response almost certainly will result in a different total harmonic distortion percentage in the direct and reflected sounds for loudspeakers in more reverberant rooms. The differing distortions will be perceived as originating in different locations in time and space from the axial signal. Under these circumstances, the masking effect of the direct signal can be reduced by an order of magnitude, and distortions can become apparent which are quite inaudible in rooms with a more dead acoustic. This highlights yet again the almost insurmountable task of designing a loudspeaker without knowledge of the room in which it will be used, or designing a room without knowledge of the loudspeakers to be used in that room. For serious monitoring, they *must* be matched; moreover, they should perhaps be optimised for the type of musical programme and recording philosophy for which they will be most frequently used.

2.9 Optimisation

I am not implying that one type of music cannot be enjoyed on anything other than a system optimised for that music. Such is clearly not the case. Music of all types gives tremendous enjoyment to an enormous number of people on a very wide range of reproduction systems. Popular music in particular frequently contains so many different sensory illusions that there would seem in most cases to be at least something for everybody liking that music, available from whatever system upon which it may be reproduced. Conversely, in order to be able to assess as many of those desirable effects as possible, a control room monitor system for such music is best optimised for the ability to show up the maximum number of both the desirable and undesirable characteristics of that type of music; and not to compromise those requirements for subtleties which will not be utilised.

2.10 Film industry monitoring

I referred earlier to film studios and their link with past and current technology. In the past, cinema systems were horn loaded not specifically for higher achievable SPLs in the auditoria themselves, but more due to the fact that greater efficiency was *required* when amplifiers of 30 to 50 watts were just about the limits of valve (vacuum tube) technology. Cinema remixing theatres were built to replicate as closely as possible the actual theatres in which the films would be shown. Such an environment allowed engineers to optimise the sound track for actual performance conditions. Even today, a normal distance for a film sound engineer (recording mixer) to sit would be around 25 ft (7–8 m) from the monitor loudspeakers, as the sense of distance is an important factor in the perception of the overall sensation. At the levels required for realistic effects in the cinemas, if the people in the back rows are to hear what is happening, the SPLs produced by the loudspeakers behind the screen can be relatively high. Diaphragm failures during public performance can be *embarrassing*, and given the cost of film recording operations, diaphragm failures during film mixing can work out *very expensive.*

Without me wishing in any way to cast a shadow over their *quality*, consider the fact that one well known monitor system manufacturer recently ran a series of advertisements, the whole thrust of which was that his soft-dome mid-range diaphragms were now 'user-replaceable'. There would have been no point whatsoever in running the advertisements if diaphragm failures had not become a significantly common occurrence in conventional studio control rooms to warrant comment. There is therefore good reason why, given current technology, the cinema world still very largely continues to use horn loaded mid-range and high frequency systems.

Film soundtracks are now frequently released for home consumption on video or CD. In an exactly similar manner to the way in which music studios use 'close-field' domestic references for checking compatibility, there is nothing to prevent the film recording engineers from doing exactly the same thing. First and foremost the main monitor system, be it for classical, rock/pop, or film recording and mixing, is a tool for exploring to the maximum degree the most important aspects of the essence of each type of recording. Differing priorities lead to different systems.

Despite the remarkable physiological similarities and commonality of ancestry, gorillas and humans are now separate species which can no longer mate. In a similar way, monitor systems for classical, rock, and cinema have gone their separate ways to lead their own lives to best advantage. I refer to the works of the late Richard Heyser, the doyen of Time Delay Spectrometry, who stated to the effect that in order to fully enjoy the intended *illusion* of a recording, it is necessary to willingly suspend one's belief in reality. *All* recording and reproduction via *two* loudspeakers is illusory: it will take powerful signal processing systems and multi-loud-speaker technology before we will ever be likely to see the three main groupings of classical, rock, and cinema in absolute agreement.

2.11 Human conflicts

The purist engineer in me does not relish this incompatible state of affairs. On the other hand, having been a record producer and engineer for many years, I know the excitement which has been created in both the acoustic and electric domains when using what *I* would deem to be appropriate systems. If by the use of those systems I have been able to better convey the illusions and emotions in each domain to the record buying public, then, for now, 'vive la difference!'

Having said that, I entirely agree with the aims of both camps. I have to: ultimately they are the same. It is merely that one group of loudspeaker designers is attacking the problems from the 'get the distortions in order then increase the SPL capability', whilst the other group have opted for the 'now that we've got the SPL and the frequency range, let's reduce the distortions'. Given the limits of current technology, each path has its own merits dependent upon circumstances; as yet, however, neither can be all things to all people.

References

1 G. L. Augspurger, 'Loudspeakers in Control Rooms and Living Rooms'. Preprint, AES 'Sound of Audio' 8th International Conference, Washington DC (1990)
2 Richard C. Cabot, 'Audible Effects vs. Objective Measurements in the Electrical Signal Path', Preprint, AES 'Sound of Audio' 8th International Conference, Washington DC (1990)

Chapter 3

Different priorities, differing approaches

3.1 Background

It will probably take a significant major break-through in either technology or operating practices to bring about any far-reaching change in the status quo of studio monitoring. There are far too many vested interests, which could not stand too much upheaval at this late stage in the development of the recording industry. When marketing departments have to keep share-holders and bank managers happy, then all too frequently idealism is flushed away by the philosophy that 'we make what we make and we *are* going to sell it!'

Monitoring design is still a combination of art and science: the art allows for the adjustment of those parameters for which the science has, as yet, no solution; whilst the science provides the solid base from which those adjustments can be made.

In a paper presented to the 8th Audio Engineering Society (AES) conference on The Sound of Audio in Washington DC (May 1990)[1] Floyd Toole made the following statements in the opening section of his presentation:

> Stereophonic reproduction attempts to reconstruct, in the minds of listeners, replicas of the timbral and spacial effects of acoustical events that have occurred at earlier times and other places . . . assuming that the necessary information has been properly encoded in the recording, the replication can be successful only to the extent that the loudspeakers are capable of reproducing the appropriate sounds, and that listening rooms are capable of conveying those sounds to the ears of listeners . . . In practice, much of the musical pleasure survives the gauntlet, but it is clear that there are arbitrary variations in the record/reproduce process that should not be there.
>
> Stereophony is an encode/decode process. When a listener sits in front of a pair of loudspeakers in a room, the sounds arriving at the ears are the only information that the perceptual system has to work with. If the sounds are different in different circumstances, the percep-tions will be different. The only real solution is to control these variations and, ultimately, to standardise the important factors. At present, there are no industry standards for either loudspeakers or

rooms, the design objectives are inconsistent and methods of control are not uniform . . .

Some of the difficulty stems from the restriction to two channels. This limits the directional and spacial effects that can be presented to a group of listeners . . .

Conventional stereo cannot re-create all of the directional impressions that may have been part of an original live performance, unless the original 'performance' was a studio creation monitored through a pair of loudspeakers in a control room . . .

Another difficulty with stereo reproduction is that the eyes see an acoustic space that is not consistent with the auditory illusion presented by the recording. The awareness of an incompatibility between what is seen and what is heard is difficult to ignore.

At the emotional level, there is the lack of a sense of occasion. One is at home, instead of at an elegant concert hall (or jazz club, or rock concert) dressed up, surrounded by similarly motivated strangers, experiencing a performance by real artists on a stage – whom one can see.

The above statements by Professor Toole outline with extreme lucidity many of the limitations of the studio monitoring and subsequent domestic reproduction processes.

3.2 Variables

Lack of standardisation through the recording chain stems from the fact that the market has, does, and probably will continue to support an industry which has no common source and no common destination. There are three factors which continue to sustain this state of affairs, namely:

1 Anti-trust laws
2 Tail chasing
3 Personal preferences

3.2.1 Anti-trust laws

Although these are laws of the USA, that country, having been at the forefront of the world's cinema industry development, has ensured that its internal difficulties have had far-reaching effects. Back in the 1930s and 1940s, cinemas were generally owned or licensed by the major film studios, only showing the films from those studios with which they were affiliated. Cinemas and film mixing theatres were designed to correlate well, and films were optimised for presentation in appropriate theatres. Some time around the 1950s, anti-trust (monopoly) laws forced the demise of this state of affairs to allow free enterprise, such that any cinema was free to show any film. Standardisation was lost, and optimisation for the different systems was no longer practical. In the world today, free enterprise and market forces seem to be the spreading philosophy. Under such circumstances, imposition by *anyone* of standards for music recording and reproduction would be unthinkable in any mandatory way. Free enterprise tends to create diversity rather than commonality, and it is almost certain that whoever was losing out from any standardisation would surely take legal action for supposed restriction of ability to trade. Obstacle number one is now firmly in place.

3.2.2 Tail chasing

The capacity for the evolution of self-fulfilling prophesies has been great. A typical example of tail chasing would be such that for whatever reason, a certain monitor system becomes popular in the early stages of the recording industry. A range of microphones with complementary deficiencies become widely used, as the subjective effects of the combination are deemed to be acceptably natural. The use of those microphones becomes so widespread that some subsequent monitoring systems are deemed in turn to be 'unnatural' by virtue of the playback of recordings made by the earlier combination. Monitor systems which *are* deemed to be acceptable often tend to have similar complementary defects. Vested interests at all times seek to exploit their ground, and so try to perpetuate their share of the market. New microphones, as with the new loudspeakers, are often under pressure to toe the original erroneous line.

I am not saying that such errors are gross, as if so, very simple measurement would have cried foul at a much earlier stage. However, there is little doubt that these processes *have* taken place and that there are many people who *would* say that they have led to gross repercussions on quality. Currently, there are unacceptably wide spectral differences in the recorded sounds that reach the music buying public. Some of the deviation is between company and company, where each have had large selling albums for whatever reason. All too often the 'magic' selling point has been attributed to the sound, so subsequent releases by that company are tailored to match. Some differences exist in mixes from one studio to another, which given their disparity is understandable. Other differences exist from country to country, often reflecting trends which relate to whatever monitor system is most popular in any one country at any given time. An enormous section of the industry will follow whatever sonic trend is currently selling well, absolutely regardless of any reference to accuracy or neutrality. They have the right to do this of course – they are breaking no laws.

The above tendency is understandable if not desirable, especially in the light of the fact that a very large proportion of the industry is geared to producing and selling *musical* experiences that will travel well, as opposed to purely sonic experiences which may require specialised, high quality reproduction equipment. After all, it is the *music* business as opposed to the sound business. Whatever the desirability of improving the standards, we must accept that not everybody in the industry is as passionately concerned with those standards as some of the other people. The business, kudos, and ego aspects of jumping on the current bandwagon as quickly as possible renders suppression of tail chasing to be a very difficult task indeed.

3.2.3 Personal preferences

Who precisely hears what is as enigmatic now as it has ever been. In the classical world alone, whether producers opt for Blumlein pairs, M + S microphones (middle and side), coincident pairs, spaced pairs, SoundField microphones, stereo pairs with spot microphones, close microphone multi-channel recording with ambient pairs, or whatever other microphone system anybody would choose to use, it is almost inevitably to some degree down to a personal preference. There is no doubt that some choices are forced by suitability due to restrictions in the recording environment, but nonetheless, subjective tastes have their very strong say in the matter.

Quite simply, some people glean more enjoyment from the dynamics and presence of close microphone techniques, some wallow in the spaciousness of stereo pairs, whilst others opt for just about all points in between. I do not believe that anything can, or necessarily even should, be done about this. People listen to music for their own personal pleasure. The problem for the recording industry is that no one combination of loudspeakers, amplifiers, crossovers or rooms can be optimised for the most lifelike sensory illusions for all of the choices of recording technique. When one adds to that equation the idioms and particular technical demands of the electronic/rock music industry, then have pity on the designers of reproduction systems which are required to serve all tastes.

3.3 Current dilemmas

We do not as yet have the technology to produce such all-encompassing systems. We can produce 'Jacks-of-All-Trades', we can produce 'Jacks of Most Trades and Masters of Some'; unfortunately, 'Masters of All Trades' as yet elude our grasp.

The classical concept of studio monitoring design has been to achieve 'the closest approach to the original sound', as the Acoustical Manufacturing Company so aptly stated in their advertising literature shortly after World War II. Unfortunately, only the absolute, definitive re-creation of the original sound field would be free of compromise, and even then, it may not ultimately achieve its objective. As this goal is neither achievable now, nor likely to be achieved in the foreseeable future, compromises must be made. It is these compromises which lead to endless subjective argument in the hi-fi press as to which loudspeakers or amplifiers are 'the best'; the lack of correlation between measured and perceived responses only serves to fan the flames of the debate. Ironically, in order to achieve a realistic perspective on the problem, it would seem prudent to begin with something that is tantamount to a conclusion.

The 'absolute' goal of the re-creation of the original sound field need not necessarily lead to the ultimate monitoring system for the following three reasons. Firstly, the very introduction of the listener into that sound field would disturb it in such a way as to render it no longer original. In other words, if the listener's acoustically absorbent body were present in the original sound field, then the reverberation in that field would have been different from the field recorded. Secondly, with much, modern, electronically generated music, an original sound field never existed. Thirdly, other sensory cues such as vision can have a very great bearing on overall perception such that what may be *accurate*, may not always be perceived as *real*. It is *reality* as opposed to pure accuracy which must be the goal of efforts to achieve practical monitoring designs – reality in this sense being the overall transmission of an illusion which the producer intended to create.

A magazine editor asked me jokingly if I was implying that the introduction of an odour of camel sweat would be desirable when listening to dance music from nomadic Arab tribes. Somewhat to his surprise, I answered 'probably', but that this would depend upon whether the listener had previously experienced such smells in the context of the music. If so, then it is possible that a less definitively accurate system could 'sound' more natural and realistic with the addition of the odour, than a more definitively accurate system without.

3.3.1 Sensory integration

Whilst we deem ourselves to be 'listening' to a monitor system, in reality tactile senses are usually playing a relatively large part in our perception of events. The inability of headphones to vibrate one's chest cavity is a serious drawback in their ability to 'accurately' portray rock music or the proximity to the bass drum of a marching band. In a concert hall with what may be generally accepted to be a 'good' acoustic, it would be very difficult, if blindfolded, to locate with any degree of accuracy the position of a violin on stage from a listening position half-way down the hall. Visual cues can 'lock-in' our auditory localisation in such a way that ears alone cannot achieve. Herein lies one of the major caveats in monitor design: whose reality are we dealing with?

A well known studio designer was somewhat bemused to find that a control room which he had built with a $\pm 1\frac{1}{2}$ dB third octave pressure amplitude accuracy was not being so well received as an earlier room, utilising the same amplifier/loudspeaker configuration but with only a ± 3 dB third octave pressure response. He could not understand why this should be the case, especially as the new room definitely sounded better to him. Caipura and Deutsch[2] published findings on the use of monitor equalisation which have a great bearing on the above designer's dilemma. They suggested that experience can allow us to expect certain characteristics from certain shapes, sizes and surface treatments in any given room. Upon entering a room, the brain can put a correction curve on to our perception in order to accommodate for the anticipated effect of the room. Electronically superimposing the inverse of the room response upon the loudspeaker system with monitor equalisers will 'double correct' for the room; once electronically and once by the brain itself. The resultant would be the perception of an equalised monitor system which nevertheless offered a relatively flat response to a conventional analysing system.

The $1\frac{1}{2}$ and 3 dB rooms discussed earlier were of a design in which virtually all of the visual surfaces, other than the floor and the front walls, were fabric-covered frames concealing a great depth of acoustic control measures. The rooms as seen by the eye were most definitely not the rooms as perceived by the ear. It is highly probable that in the two rooms referred to, the visual cues for the ± 3 dB room more closely matched the auditory expectation from that space than did the $\pm 1\frac{1}{2}$ dB room, which contained considerably more complex acoustic treatment behind the scenes in order to achieve its very creditable pressure amplitude response. Listening blindfolded under similar circumstances, it is feasible that the $1\frac{1}{2}$ dB room could be restored to its rightful position as being perceived to be more sonically accurate than the 3 dB room, but this opens the door to some very interesting questions. If a person's sight began to fail over a course of months as opposed to years, then assuming that the listener's hearing remained constant, would the preferred working environment shift from one studio to another as the visual relevance of the design began to be a lesser proportion of the overall sensory stimulus?

3.4 The arbitrary nature of compromise

Imagine the case of a chamber orchestra in an auditorium, which would generate a complex sound field within its performing environment. Any attempt at capturing the 'natural' sensation of being there would almost inevitably involve some form of binaural recording, where the positional cues for the listener, either live or recorded, are essentially relying on interaural time and phase differentials. Spot mic'ing of any instruments to be highlighted, or at the extreme, close microphone multi-tracking of the entire ensemble, would immediately destroy any attempt to re-create or capture an original sound field. One would need to be a very strange animal indeed, with a couple of dozen ears on long antennae, in order to hear simultaneously what is going on six feet from every instrument or section in the ensemble, which is after all what the close microphones are picking up. To help to make this unreality somewhat more plausible, overall pairs or ambience microphones may be used in an attempt to weld the whole thing together. Beautiful it may well sound, but accurate it most certainly is not. Reproduced music is an illusion – no more!

3.4.1 The ideal

The ideal monitoring system would have a frequency response from DC to infinity with neither linear distortion (phase and amplitude) nor non-linear distortions (harmonic, intermodulation, rattles), and would be sited in an acoustically neutral environment. The sound source would almost certainly *not* comprise of two loudspeakers as is conventional today. As this ultimate goal is way beyond the current recording practices of the 1990s, we must inevitably compromise our results. The very wide disparity in sound sources as discussed above, together with the consumer reproduction systems ranging from headphones, motor vehicle systems, conventional domestic hi-fi and audiophile systems, not to mention the myriad of other reproduction systems, render many of the design compromise points mutually exclusive. If quality was all-important, it would seem more prudent to mix for 'hi-fi' and make compensations within the equipment of car stereos and the like, rather than compromise at the mixing stages. Unfortunately, this would involve extra cost in a very cost-conscious consumer equipment market, and also involve the fitting of a switch for older, non-compensated recordings with the attendant risk of consumer confusion. Commercial concerns are the problems here. No one system can be truly representative of the wide range of domestic systems and environments, and no system designed for the phase and time subtleties of stereo pair pick-up from an orchestra would be capable of delivering the earth-shaking low end demanded by reggae artistes, synthesiser bands, or film soundtracks. We *are* faced with multiple systems for reference, whatever the *ideal* goal.

3.5 Hidden objectives

In another paper presented to the 8th AES Sound of Audio conference in Washington DC,[3] David Moulton pointed out very strongly an underlying shift in a large portion of studio monitoring usage: the generally unacknowledged fact that certainly for a large percentage of the time, the monitoring system of a modern recording studio has become an extension of the sound production process: an extension of the computer-based instrumentation. The loudspeaker itself has become, without formal recognition, the predominant musical instrument of our time. Music is being created *on* loudspeakers for playback *by* loudspeakers. There is no reality, there is no reference, other than that perceived in the mix in the studio of origination. Given the disparity in monitor systems, that must be arbitrary.

Even the usual stereo pair of loudspeakers is following a pragmatic convention. It is relatively easy to site a pair of loudspeakers in roughly symmetrical locations in most rooms. Much of the commercial failure of quadraphony was due to the fact that most rooms have a door in at least one corner, or a window in the centre of a wall, each precluding a permanent symmetrical location for a four-loudspeaker set-up. Having said this, however, there is no magical reason why two-loudspeaker stereo *should* be capable of the ultimate audiophiles' Utopia. It cannot! Amongst others, Michael Gerzon[4] has been strongly pushing the concept of three-loudspeaker stereo as a greatly superior format, even when derived from a two-channel source. Much loudspeaker evolution has been dominated by the constraints of the possibilities from the two sides of the groove of a vinyl disc. A third channel is now easily realisable from a three-channel digital storage medium, so, unshackled, we may soon begin to see more rapid moves towards multi-loudspeaker systems for the 'high end' of the market. Indeed, Gerzon's ingenious proposals would allow for total inter-format compatibility, either from three-channel stereo masters, or from the conventional two channels.

3.6 Capturing the illusion

Music is essentially a medium for transmitting emotions from one being to another. The emotion or sentiment can be between persons, between a person and a place, a thing, or whatever. Music has little meaning if it carries no emotional content, so the ability of a monitor system to reproduce or aid in the creation of the music's emotional content is a very high priority in its design. A technically poor loudspeaker will rarely be able to fulfil the demands of full emotional transmission, yet, conversely, a technically excellent loudspeaker will not *necessarily* faithfully carry the intended illusion and feeling. Two monitor systems, remarkably similar when measured in terms of conventional performance parameters, may be very markedly different in their ability to deliver the nuances of an artistic performance. It is in this area where we run into the real conflicts of opinions. In previous pages, we discussed the possibility of a person with gradually failing sight changing his or her preference between two control rooms as the visual input into the total sensory equation was gradually reduced. In a different way, two measurably similar loudspeaker systems may be deemed better or worse, one to the other, entirely down to the fact that one listener may have a preference for the subtle timbral characteristics of one of the two, especially when reproducing the music and instruments to which that person usually listens. Should that listener proceed to listen to music of a type which he or she would typically avoid as a matter of taste, it is not inconceivable that the decision would be reversed in terms of subjective timbral neutrality.

Just to recap, let us suppose that a listener's preference in music was for recordings of string quartets using stereo microphone pairs. Time and phase accuracy would be fundamental performance requirements for the reproduction of the string quartet in order to achieve timbral neutrality and spacial discrimination from the stereo pair recording. Reproduction of such a performance may have to be achieved at the expense of the generation of high sound pressure levels and some bass extension. Anyhow, there is no need in terms of definitive accuracy to be capable of reproducing music at any level above that which can be achieved by the real thing. A point worth repeating is that a violin at 120 dB could not possibly be accurate because a violin can never generate 120 dB in real life. Reproduction at higher levels will, as a function of our hearing responses, cause us to perceive the instrument to be top and bottom heavy at larger than life SPLs. The classic Fletcher–Munson curves illustrate this point well (Fig. 3.1).

In order to approximate more closely to a co-incident source, thus enabling time and phase accuracy over a reasonable listening area, the loudspeaker unit may have to be reasonably small, which in turn will place limitations on the maximum SPL, particularly at low frequencies. So this system may well, in an appropriate room, sound remarkably neutral whilst reproducing the string quartet, but what of the rock band with heavily predominant bass guitar and bass drum? Almost certainly, the recording would be via a close microphone technique with positional imaging being panned in the amplitude domain. Phase and time performance characteristics would still be important for timbral neutrality, but any subtle degradation in mid-range tonality may well be a very small price to pay for the powerful bass extension at the naturally high SPLs of some rock music. In order to as accurately as possible re-create the sensation of sitting ten feet in front of a drum kit in an ambient room, a very high SPL capability is a must for a suitable loudspeaker's performance characteristics.

From the two instances cited above, it should be clear to see how one loudspeaker could be deemed to be more natural and accurate on one type of music, whilst another loudspeaker, maybe much disliked on that selfsame music, could in turn be deemed to be more accurate and natural on different programme material. Given the imperfections in the systems, the compromise points for reproducing different music can be forced to very

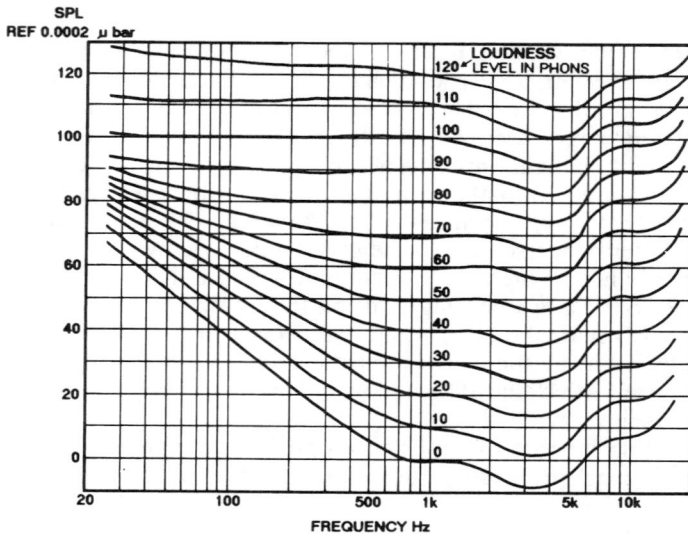

Figure 3.1 Classic Fletcher and Munson contours of equal loudness for pure tones, clearly showing higher hf and lf levels being required for equal loudness as the SPL falls. In other words, at 110 dB SPL, 100 Hz, 1 kHz and 10 kHz would all be perceived as roughly equal in loudness. At 60 dB, however, 10 kHz and 100 Hz would require a 10 dB boost, in order to be perceived as equally loud to the 1 kHz tone

different locations, due to some of the mutually exclusive constraints of the requirements to create the best illusions. High SPLs demand large physical size, whilst time/phase accuracy demands small size, to enable compact location of the drivers for a coincident source approximation. The larger systems can of course be optimised for time/phase coherence, but only for one point in space; elsewhere, alignment errors will normally be greater than those for a well aligned, physically smaller system.

3.7 Inability to be representative of all end-user systems

Purists may grimace at the 'different monitors for different music' concept, but we must be aware of all of the underlying reasons. The enormously wide range of end user systems will ensure that whatever subtleties lie within a mix, on disparate systems, some subtleties will be highlighted and some will be degraded. The main object of the monitoring environment is to allow the producers, engineers, and artistes to be aware of the existence of as many subtleties as possible in order to make them available to the creative process, in the sure and certain knowledge that they will not all be manifest on any one domestic system. It is now effectively *de rigueur* to take home the final mixes for assessment on several systems before the recording is available for general release. Here again, a system which has been optimised for the subtleties of classical, stereo pair recording of acoustic music will probably fail to fully reveal the impact of electronically generated, amplitude panned, hard-hitting dance music. The twisting around of priorities is probably most effectively highlighted by remembering the fact that the reproduction of classical/acoustic music via loudspeakers is intended to *emulate the live performance*, whilst the live performance of electronically generated, studio produced music seeks to *emulate the original recording*; which usually never existed as a real-time entity. Both to some degree or other fail to fully realise their objectives.

3.8 Summing up

There are certain limitations which are common to some extent in all loudspeakers. Much modern studio created music will be perceived in total *only* over loudspeakers: once again, the total overall performance never existed, and perhaps *could* never exist in real life! There has been an insidious acceptance of loudspeaker limitations which has led to such limitations even being exploited to advantage in the recording/reproducing processes. Music created *on* loudspeakers for performance *by* loudspeakers.

We therefore have differences of opinion on musical type, musical performances, relative balances, dynamic close microphones or spacious pairs; and even almost unmeasurably small subtleties of texture, which may be of great significance to one listener whilst totally passing by another. We all hear different things: we all respond to different combinations. We are all put together differently. Standardisation of the major parameters is a very worthwhile goal, but absolute agreement upon the order of priorities will probably not exist until either human beings are cloned, or loudspeakers and rooms can be produced with responses several octaves beyond the currently conventional audio limits, and with power response errors in the region of one part in 10^{12}.

References

1 Floyd E. Toole, 'Loudspeakers and Rooms for Stereophonic Sound Reproduction', Preprint of AES 8th International Conference on 'The Sound of Audio', Washington DC (1990)
2 Dennis R. Caipura, 'Audio Fidelity: The Grand Illusion', *Recording Engineer and Producer*, Vol 20, No 11 pp 42–46 (Nov 1989). Article relating to work of Dr Diana Deutsch, University of San Diego, USA
3 David Moulton, 'The Creation of Musical Sounds for Playback Through Loudspeakers', Preprint of AES 8th International Conference on 'The Sound of Audio', Washington DC (1990)
4 Michael Gerzon, 'Three Channels, The Future of Stereo?', *Studio Sound*, Vol 32, No 6 pp 112–125 (June 1990)

Chapter 4

Loudspeakers, rooms and phase responses

4.1 One system

From the point of view of accurate studio monitoring, loudspeakers and the rooms in which they will be installed cannot be designed in isolation. They are part of one and the same system. This is true from the points of view of phase and directivity responses, drive unit positions, sensitivity and many other parameters.

4.1.1 Test conditions and loading

It is frequently overlooked that the listening room has an enormous effect on the loading of a loudspeaker. This loading in turn, affects electro-mechanical impedances, back-e.m.f.s (electromotive forces), resonances, and many other things.

There is currently a growing tendency for modern music to be recorded and mixed in control rooms with seemingly ever-decreasing reverberation times. Nevertheless, even the rooms with very low RTs do possess some, at least at very low frequencies. It has been customary to measure the responses of loudspeaker systems in anechoic chambers. This was not only for comparative consistency, as to all intents and purposes one anechoic chamber responds (or rather does not respond) like any other, but also because in conventional rooms there were always severe difficulties in measuring loudspeaker responses without the superimposition of room reflexions. The alternative to the anechoic chamber used to be the free-field, where loudspeakers were mounted outside, clear of buildings, and on a pole 30 feet (10 m) in the air, preferably above grass, snow, or some other relatively non-reflective ground covering. This technique worked well except for the inconveniences of rain, snow, wind and general extraneous noises.

Fortunately, modern, computerised techniques, with 'time windows' for gating the measurements, are now allowing the majority of necessary measurements to be made in the actual rooms of use. In many ways, the modern system is perhaps also the most relevant of the methods, as it allows much separation of the individual characteristics of the loudspeakers and the rooms, whilst still measuring the loudspeaker under actual room-loading conditions. This is important because a room in which a loudspeaker is placed is also a loudspeaker cabinet. If a large loudspeaker system, say a low frequency cabinet, is placed in a control room, one naturally tends to consider the cabinet as providing the loading for the drive unit(s). On the other hand, if one were to remove a drive unit, climb inside the cabinet, then have the drive unit replaced, effectively the cabinet would become the listening room, and the control room the cabinet: certainly from the listener's point of view. The two spaces are coupled, and changes in the real room, which viewed from inside the real cabinet would amount to changes in the new, large cabinet, would in turn affect the loading on the loudspeakers. For this reason, measured performance in free air is not always going to translate accurately into performance in any given room. Thought of this way, it helps to reinforce the concept of the room and the loudspeaker system, which in turn is coupled to amplifiers and crossovers, as all being component parts of one integrated system.

4.1.2 Natural vs reproduced sound

If a large gong is struck with a beater, a very large range of frequencies will be produced. Some build up further as resonances, but most, to some degree, are there from the moment of impact. The low frequencies radiate from a large area, with wave motions rippling through the entire surface area of the instrument. High frequencies travel at the appropriate speed of sound through the metal of the instrument, and are radiated into the air in a very complex manner. In fact, the radiation pattern from the instrument is very complex indeed, and its complexity is an important part of the gong maker's craft. Should a person then move in front of the instrument, say through an arc of 20° or 25°, there would be little significant change in the sound, despite the fact that the phase relationships of the different frequencies would have gone through chaotic changes. This sort of effect has no doubt had some bearing on the fact that for so many years, many aspects of phase relationships and wave envelopes were considered to be inaudible, and that it was the overall amplitude/*frequency* response which was of overriding importance for the accuracy of reproduced sound.

Nonetheless, it is probably the changing phase relationship which would be responsible for most of whatever change in timbre existed whilst moving in front of the instrument. Whilst the amplitude/frequency response or the power spectrum of the instrument would remain largely the same, the waveform shape would change with position. This waveform change with position can never be accurately recorded then reproduced via a loudspeaker.

Once we record that gong, we begin to have troubles. The physical requirements of drivers for low and high frequencies are almost diametrically opposed: large propagation area and high mass for the bass drivers; point source and lightness for efficient response of the high frequency units. For this reason, the practical difficulties in obtaining an adequately linear response from 20 Hz to 20 kHz from one, single loudspeaker drive unit is, to this day, all but impossible. Inevitably we resort to multi-element systems, so we now have a situation where the point of origin of any part of the sound is entirely dependent upon its frequency. 10 kHz will come from a very small source, the high frequency driver, and nowhere else. This bears no relationship to the origin of the sound from the gong. A person passing across the face of the loudspeaker, will no longer receive a recognisably chaotic wavefront. High frequency arrival time will vary dependent upon the relative location of the drivers; frequencies will no longer be jumbled in their origin as in the case of the instrument itself. Even in a co-axial loudspeaker system, the centre of propagation of the sound will be somewhat further forward or further backwards for the high frequencies, relative to the low frequencies. This is because the HF diaphragm is not co-located with the centre of propagation of the low frequency driver. Indeed, it could not be, as the precise centre of propagation of almost any electromagnetic driver moves forwards and backwards with frequency. True, by using crossovers which are aligned in time, they could be co-located at the crossover frequency, but only at that frequency. At other frequencies, there could still be a time shift. Even if digital delays were incorporated, the frequency-dependent propagation shifts would render less than perfect results.

As we see from travelling in front of the gong, the absolute shape of the waveform is not a mandatory requirement for our recognition of the sound as 'natural', but we have already flown in the face of nature by distributing our sound sources in the loudspeaker with respect to frequency. The fact that great disturbances to the relative phase of differing frequency components do not render the gong any less gong-like, does not mean that we can ignore the effects on arrival times in the reproducing loudspeakers: there are big differences in the method of propagation of the natural and reproduced sounds. From the loudspeakers, the time advance or delay of all high frequencies, due to relative driver positions when moving in front of the loudspeaker, is totally unlike the chaotic phase relationships when moving in front of the gong.

4.1.3 Time integrity

The time of arrival at the ear of the different frequencies of a transient wave-front has a great bearing on the perceived timbre of the sound. With the low frequency components of a wave-front reaching us before the high frequency components, despite the system having a 'flat' overall frequency response, the sound may be perceived by the brain as a bass heavy signal. The inverse is also true, irrespective of the fact that the steady state pressure amplitude response is 'flat'. On signals with low transient content, the effect will not be so noticeable. Here, then, we could have different loudspeakers with flat frequency responses, sounding similar on steady state signals, but with significantly differing characters on transient signals. The forward or backward positioning of the HF units can give a top or bottom heavy transient characteristic, dependent purely on driver locations.

Certain temporally aligned designs go to great lengths to ensure in-phase, on-axis signals at the crossover points, but quite often chaos reigns at either side. This is one effect of phase shifting/time delaying, either by electronic means or by physical positioning, where the prime criterion is the effect of phase on the steady state amplitude response of the entire system. Phase cancellation at the crossover points of a 24 dB/octave, or 18 dB/octave system, is restricted to a relatively sharp dip at the crossover frequency.

Indeed, you *can* measure it, but the inability of the ear to detect high *Q*, moderate amplitude troughs is considerable. Rule one: 'Do not compromise your overall phase response for a dead flat amplitude/frequency readout!' I strongly feel that an adherence to a policy of minimum *overall* time displacement is of greater importance than absolute phase integrity at the crossover points. Given that the impulse response is dependent upon amplitude *and* phase, I believe that the adherence to maintaining the impulse waveform is the prime consideration.

Here, some horns can produce a stumbling block. In many designs, one encounters a distance of eighteen inches from the driver mounting flange to the mouth of the horn, in addition to a six-inch throat tube in the driver itself. In round figures, we have a sound velocity of around one thousand feet per second in air at sea level – around one millisecond for every foot! Given that the diaphragm in the above horn/driver combination is two feet from the mouth, that is a 2 ms time delay if the horn mouth is mounted on the same baffle as the other drivers.

What can be done to correct this is certainly not to project the horn two feet out from the baffle. If we did, we would have lost our integrated plane of the baffle for minimising reflexions, whilst also giving everybody something to bang their heads on as they walked past. As yet, we have no means of moving time forwards, so we cannot 'time advance' the signal to the horn diaphragm. We could insert a phase shift in the crossover, giving the horn a 2 ms equivalent phase lead at the crossover point, but this, largely, only straightens out the steady state frequency/amplitude graph. Two octaves up, we could be two feet out of sync again. We could put a 2 ms time delay, analogue or digital, in the crossover outputs to the amplifiers for the other drivers, but then we would be introducing more electronics, more filters, and, potentially, more problems. I believe that this leaves us with a big problem on our hands if we do use a horn/driver combination with two feet between the diaphragm and the mouth.

However, with well chosen driver positioning, coupled with the judicious choice of crossover parameters and loudspeakers positioning, we can achieve *minimum overall phase* loudspeaker systems sounding even more natural than designs where undue attention has been paid solely to crossover point phase integrity. It is the *time integrity* of the leading edge which is usually more relevant to sonic neutrality than precise amplitude response.

4.1.4 Crossovers, lobing and directivity

All too frequently, too much attention is paid to one aspect only. A constant *directivity* is often the main goal of phase aligning. Time discrepancies in the crossover region cause a tilting in the axial response of vertically aligned drivers. With 18 dB and 24 dB/octave and higher order filters, the region of mutual propagation of any two drivers is much reduced, so in these crossovers, the lobing and tilting effects are far less apparent. The lobing is a function of phase shifts in the crossovers at the crossover points only, and should be dealt with there. The overall time/frequency response of the *entire system* should be considered a genuine priority, in order that a system may be as neutral on transients as it is on steady state programme material.

These seemingly confusing and sometimes contradictory sets of circumstances are usually born of attempts to build loudspeakers for use in average rooms. The compromises are therefore chosen by the individual designers based on their own sets of important criteria. Room reverberation times and the presence of early reflexions will have an enormous bearing on the priorities for any loudspeaker design. If the room is reverberant and is troubled by early reflexion problems, then overall power response and an even directivity/frequency performance will be very important – possibly much more important than absolute axial response integrity. If the room has a very short reverberation time, and is free of early reflexions, then the axial impulse response would be the prime target to aim for.

4.1.5 Loudspeaker/room interface

Given the complexities of the widely different directivity characteristics of the different instruments, together with the fixed directivity/frequency pattern of any single monitoring system, we can currently only hope to approximate to the characteristics of the original sound field. To some degree, however, all loudspeakers suffer from the same drawbacks. They are, therefore, at least reasonably representative of each other; but what can have an enormous bearing upon the performance of a loudspeaker is the interface with the room. It is totally unjust to criticise *any* reasonable loudspeaker system after evaluation in just one room! Figures 4.1, 4.2, 4.3 and 4.4 clearly show the effects of rooms and room positions on the responses of various loudspeakers. It is worth carefully considering the above figures, and noting just how similar are the responses of the two *different* loudspeakers in Fig. 4.4, which were located in the same position in the same room, when compared to Figs 4.1 to 4.3 which represent the differences shown from *similar* loudspeakers, but in different positions and rooms. This again reinforces the case for considering the room and the monitor system, as being parts of one and the same entity.

Due to the directivity characteristics of the loudspeakers, the effective space into which they radiate varies as a function of frequency. This in turn affects the on-axis power response, and the subjective frequency balance in the room. The distance to the adjacent room boundaries also has a *time* function, in that the closer the room boundaries are to the loudspeakers, the faster will be the arrival of the early reflexions from those boundaries to the listening position: see Fig. 4.5. The faster the arrival of the early reflexions, the less distinct the imaging becomes, so the overall definition suffers. It is for this reason that larger rooms frequently sound less coloured than their smaller counterparts, as the first reflexions arrive later due to the greater distances travelled in larger rooms.

4.1.6 Loading

At low frequencies, where the wavelengths are long, the loudspeakers will radiate in an almost omnidirectional pattern. If the loudspeaker is a freestanding box, and has been optimised for 'free-field' conditions, the low frequencies will increase when the loudspeaker is placed in, or against, a wall. Frequencies below 200 Hz can no longer radiate in all directions, and so will be concentrated back into the room. Side walls, ceilings and floors all have a similar effect, depending on their proximity to the loudspeakers. A 'free-field' aligned system can experience anything up to 18 dB of boost,

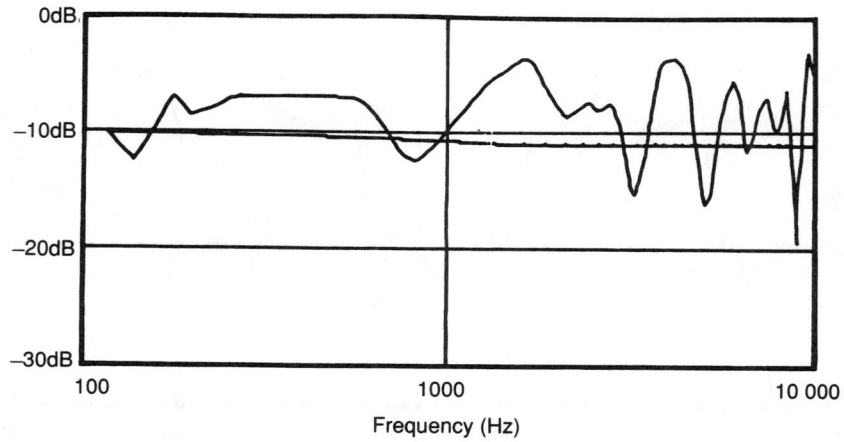

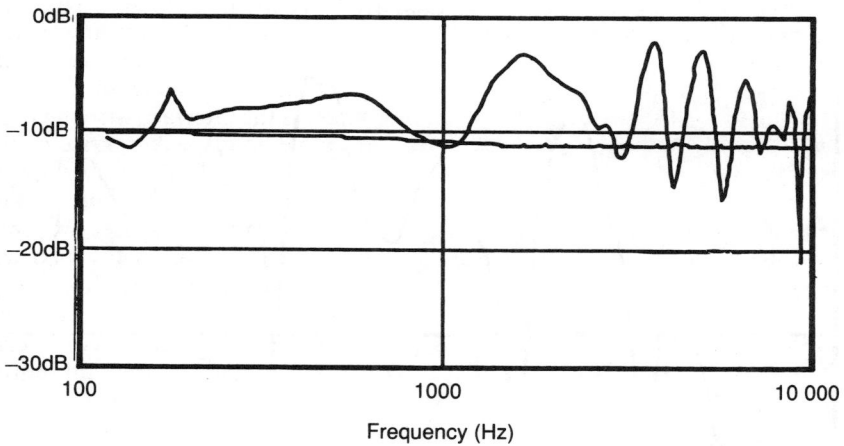

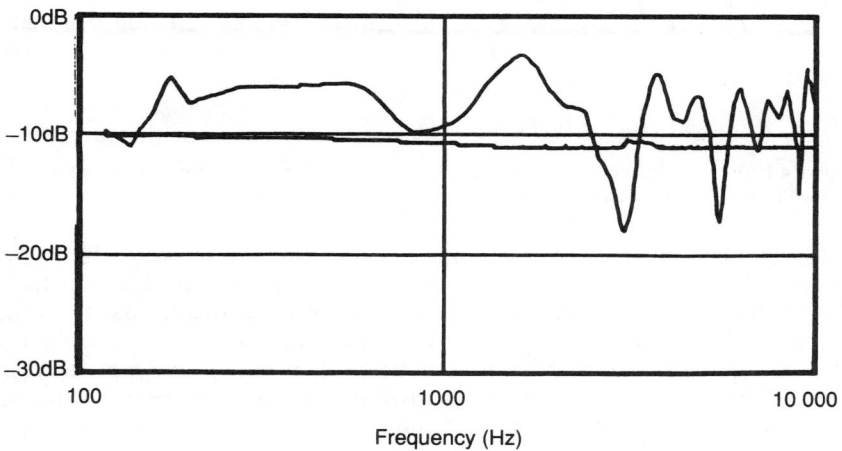

(a)

Figure 4.1(a) One loudspeaker measured in three rooms, on top of three different mixing consoles. Note the different high frequency responses due to varying reflexion patterns

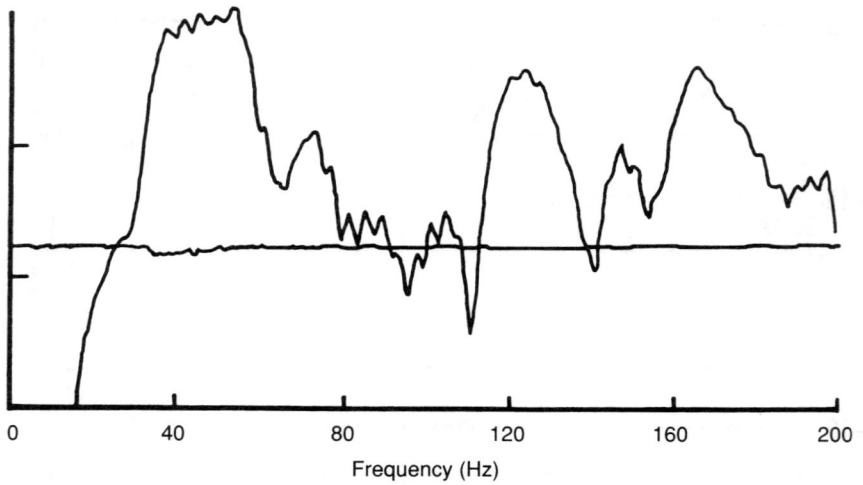

Frequency (Hz)

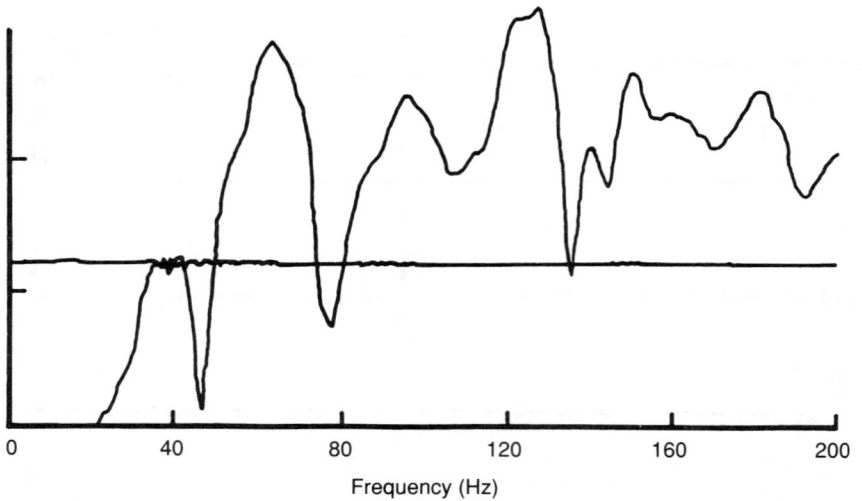

Frequency (Hz)

(b)

Figure 4.1(b) Low frequency response plots of two identical loudspeakers in two different control rooms. Measurements are taken at engineer's position

below 200 Hz, entirely dependent upon its location in a room. As far as the loudspeaker is concerned, it would be the selfsame loudspeaker, but from the listener's point of view, there would have been a drastic change in the perceived frequency balance and general sound character. As we have already seen in Figs 4.1 to 4.4, two dissimilar loudspeakers may sound more alike in one position in one room, than one loudspeaker in the two different locations. This may go some way towards explaining why many differing loudspeakers may appear substantially the same when measured, but have little in common on the reproduction of music in differing domestic circumstances.

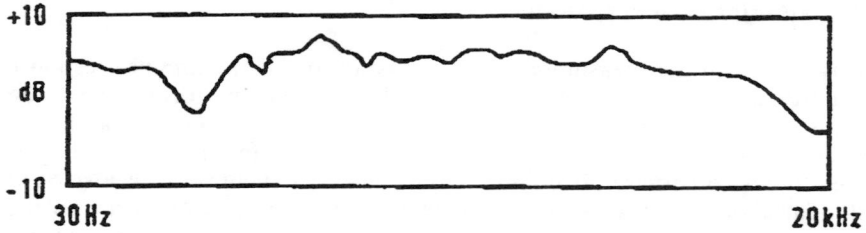

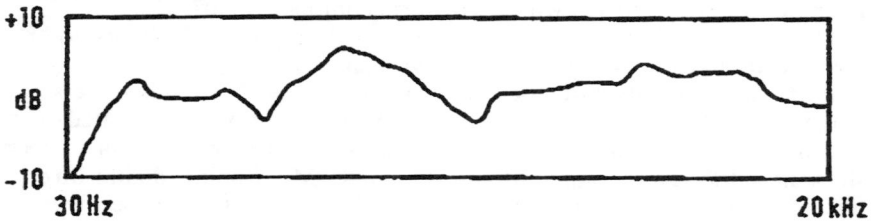

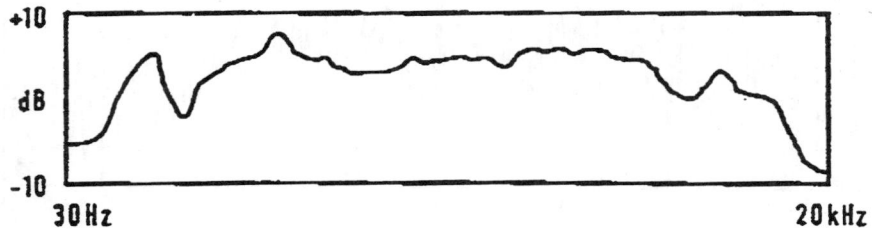

Figure 4.2 Response of one loudspeaker at three different locations in the same room. Differences are entirely due to room position

In a room, the spectrum analyser only tells us of the direct *and* reverberant total of the frequency distribution of the sound power in the room; and what is more, only at that microphone position. At realistic working distances, the room will tend to dominate! Moreover, up to one hundred different microphone positions would need to be computed and averaged to ever have any hope whatsoever of gaining a representative display below 100 Hz. Standing waves and resonant modes of the room may swamp the actual loudspeaker output, so such a measurement system tells very little about the loudspeakers themselves.

4.2 Standards of reference

It is only with the measurement or reconstitution of the impulse response that reasonably accurate assessments of loudspeaker performance can be made. Despite the fact that we are not in control of loudspeaker placements in people's homes, we *must* aim for some sort of standard of accuracy in the studios, otherwise all points of reference will eventually be lost. Any standards of reference must be a function of what is actually heard in the control room. From Figs 4.1 to 4.4 it can clearly be seen that no 'reference' standard can be given, neither to any particular set of loudspeakers, nor to any individual room. The overall performance of any monitoring system must be a function of the loudspeakers, crossovers, amplifiers *and* the rooms. It is absolutely futile to make statements to the effect that one amplifier is 'the best' or that a certain crossover is 'correct', without full reference to the systems of which they form component parts.

There seems to be some innate, human urge to pigeon-hole things; to find security in having one particular loudspeaker or amplifier to cling to. Unfortunately, reality just does not work that way. No loudspeakers are going to give good stereo imaging and punchy bass whilst situated in a ceramic tiled bathroom. I realise that this is an extreme case, but it does show how the reverberant field can dominate. Multi-point, digital inverse filtered, active control systems are already being developed which will tie together the room and all other components of the monitor system, and are discussed further in Chapter 18. These should take the hit and miss out of the interface, but such computerised, digital technology is unlikely to

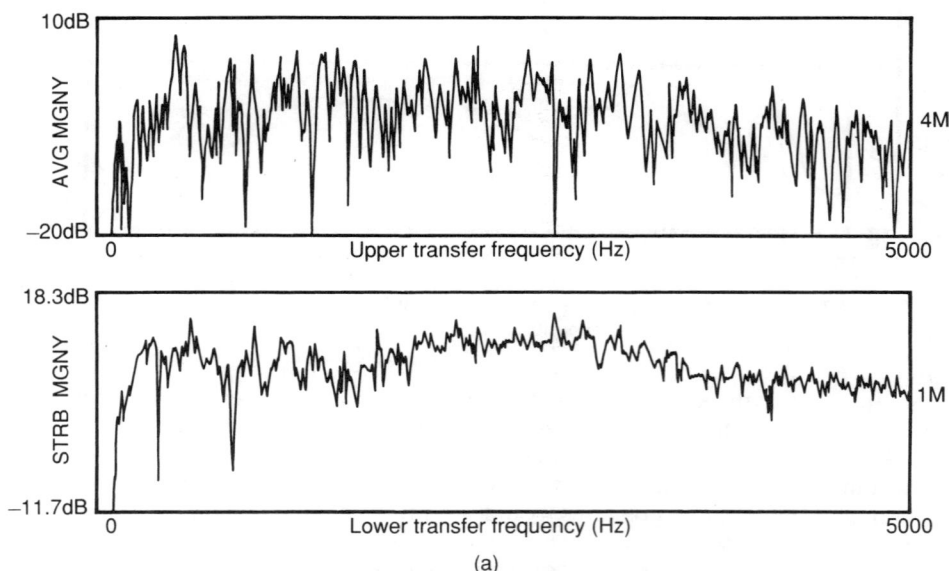

(a)

Figure 4.3(a) One loudspeaker at one location in one room. Lower trace measured at distance of 5 ft; upper trace measured at a distance of 15 ft.

Note the predominance of the room at low frequencies and the comb filtering in the upper trace caused by room reflexions

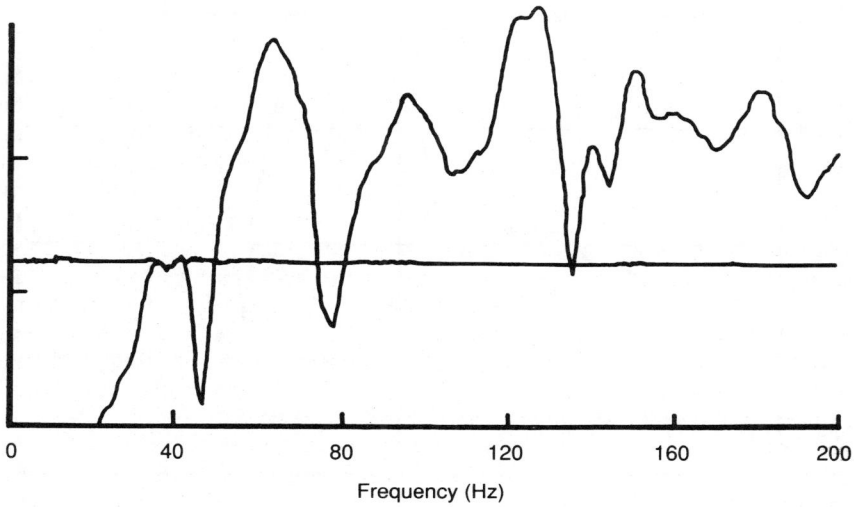

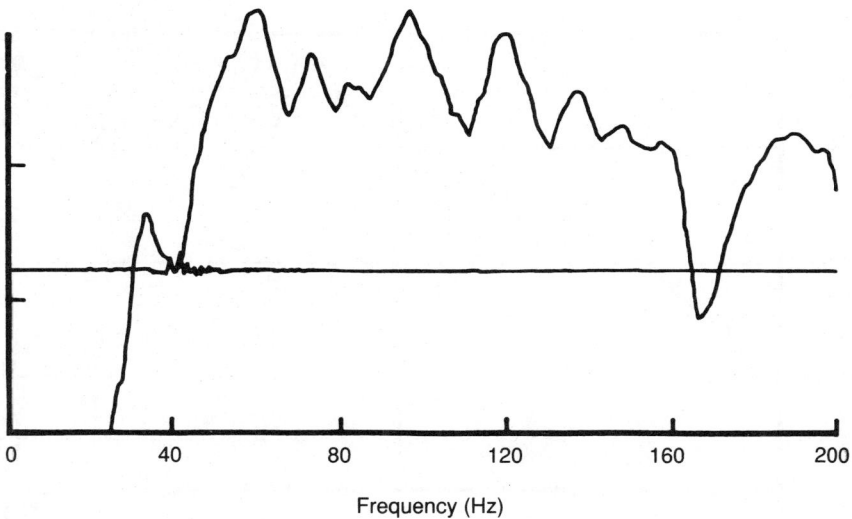

Figure 4.3(b) Low frequency response plots of one loudspeaker in one control room but in two different positions

become universal in the near future. In the absence of such digital reconstitution, the balance of impulse response to reverberant response can only be achieved by a very careful balance of the characteristics of the monitor system components and the room. In any case, current digital processing systems with only a 20 kHz bandwidth will soon be seen as unacceptable as the demands for greater digital bandwidths in the recording media become more intense.

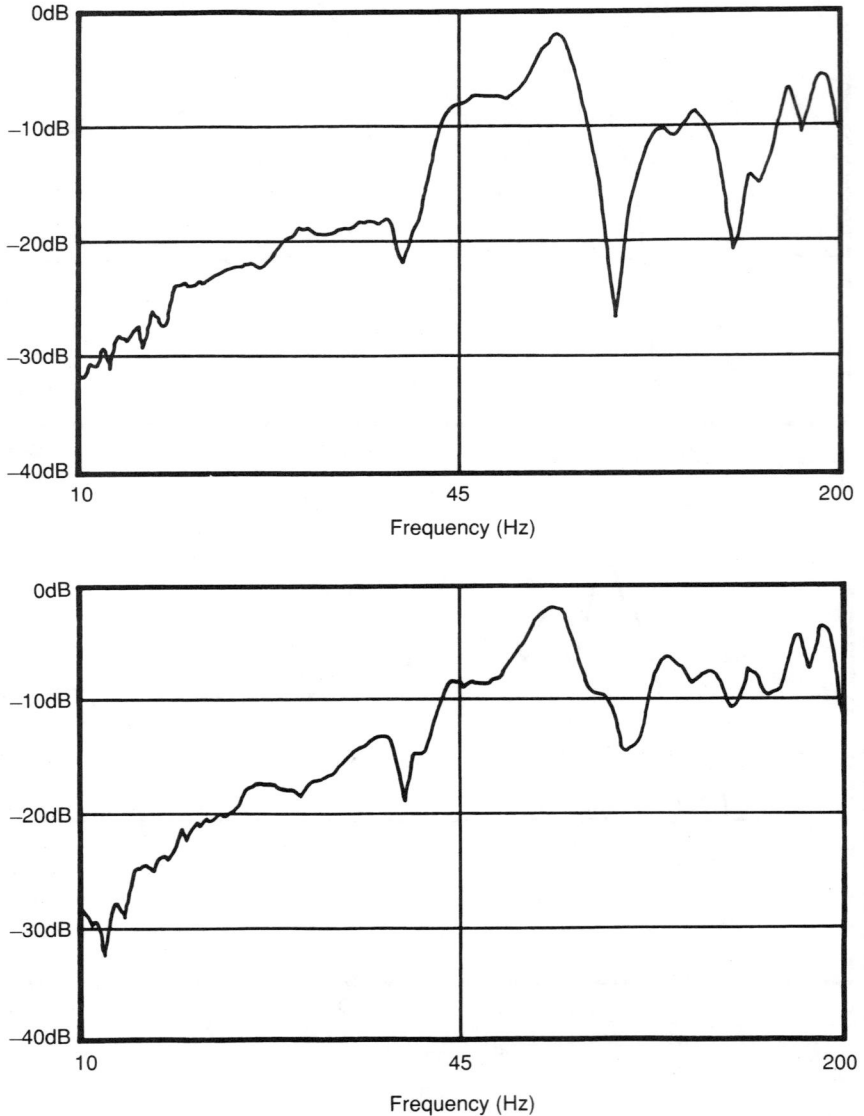

Figure 4.4 Low frequency response of two very different loudspeakers in the same location in the same room. Note the greater similarity than in Figs 4.1, 4.2 and 4.3

4.2.1 Changing criteria

The whole face of the recording industry is constantly changing in an unpredictable manner. Sooner or later it will become apparent that, in many circumstances, a more orderly approach will be required. When this time comes, it will be the studios which produce the most consistent, reliable results, and are staffed by the most objective and knowledgeable

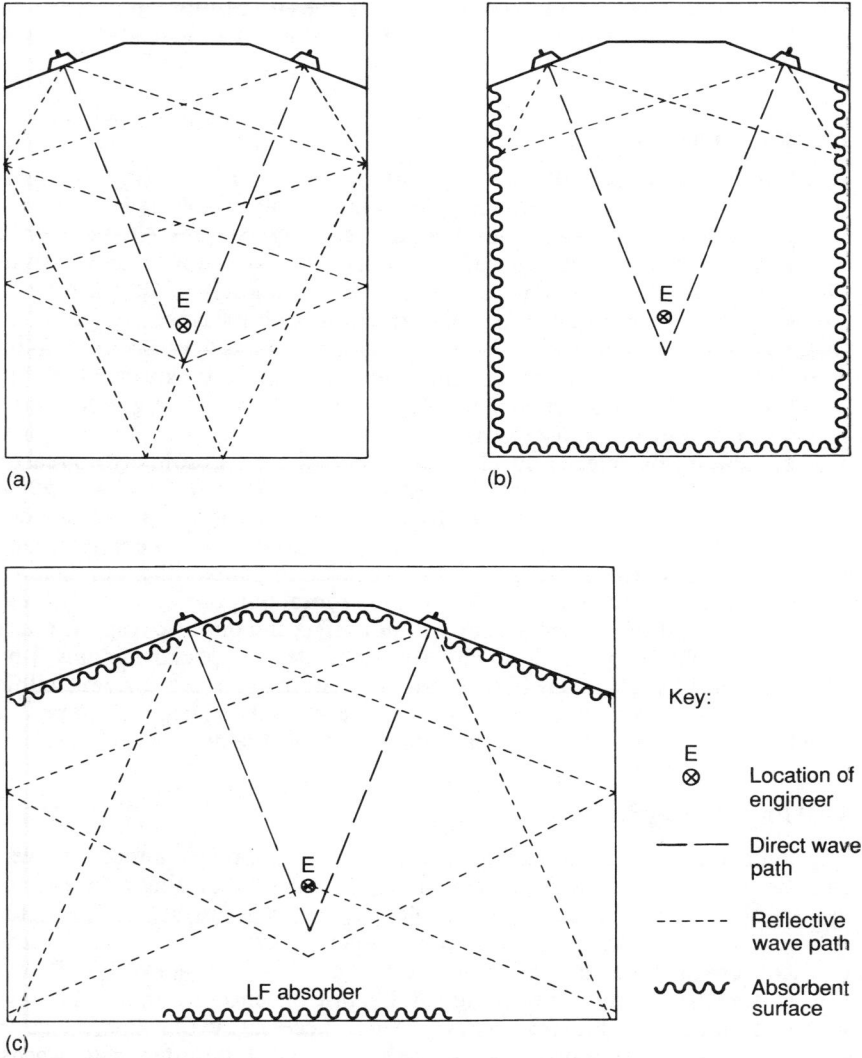

(a)

(b)

(c)

Key:

E
⊗
Location of
engineer

— —
Direct wave
path

- - - - -
Reflective
wave path

∿∿∿
Absorbent
surface

Figure 4.5 (a) Relatively 'live' room, with loudspeakers mounted at the narrow 'end'.
Reflexions can reach the engineer after travelling relatively short, additional distances
compared to the direct waves. (b) Relatively 'dead' room, absorbing reflexions at room
boundaries. This may produce an unnatural internal acoustic and below a few hundred Hz,
the room's absorbent properties will revert to those of the 'live' room (a). This can render an
uneven overall response. (c) Moving side walls further from the loudspeakers, and placing an
LF absorber behind the listening position, will prevent early reflexions. If the first reflexions
can be delayed 15–20 ms after the direct wave arrives, the brain will accept them as separate
signals, and not as blurring of the initial signal

personnel, who will be best suited to survive. A wider understanding and
more open discussion of the *truly* relevant aspects of monitoring must soon
be instigated. With the technology which we now have, the use of conventional one-third octave spectrum analysers and graphic equalisers are

outmoded. Furthermore, in the wrong hands, the end results can be about as accurate as a control room set up by the use of a hammer and monkey wrench!

4.2.2 Room reflexions

Fast Fourier[1] analysis of the impulse response gives reliable, easily interpretable representation of both the amplitude and phase characteristics of the monitor system. The Maximum Length Sequence System Analysis units (MLSSA) are also capable of excellent results in this area. Pink noise analysis will provide reverberation analysis of the rooms themselves. Whilst the two things must be considered in parallel, they must be measured in isolation. Once again, so much of the confusion in monitoring has been brought about by people making measurements who only *think* that they know what they are doing; usually measuring the wrong things or interpreting incorrectly. Temperature and humidity change have more effect on monitor response than any day-to-day electro-mechanical 'drift'. There is absolutely no reason why endless realignments should be required under normal circumstances. It is always a strong indication of the phase/impulse accuracy of a system when the desire to 'adjust' that system is a rare occurrence. Personnel and equipment brought into a control room will in all probability significantly affect the reverberation characteristics of the room, but unless placed *between* the monitors and the listeners, they will not affect the impulse response. It is the impulse/phase response which carries the stereo, spacial imaging, and which can only be effectively blurred either by room reflexions arriving within the first 15 ms or so of the direct sound, or alternatively by being swamped by high levels of reverberation within around 10 dB of the direct signal level.

4.2.3 Impulse integrity

The importance of the integrity of the impulse response cannot be over-emphasised, but it is difficult to achieve in practice. A single, full-range loudspeaker would be a very desirable piece of equipment; however, ten octaves from one driver is too tall an order in practice. Crossovers and physical separation of the drive units are something of a fact of life. Before I hear somebody say 'Time Aligned Dual Concentrics', they still have similar problems in practice. They are not necessarily the answer, as in many instances it has been a case of one step forward, two steps backwards. Crossovers suitable for monitoring purposes, unless they are extremely complex devices, are almost impossible to build with an entirely minimum-phase response; but with very careful dedication to the specific loudspeaker system for which they are designed, they can produce acceptably phase-accurate *systems*. They must be designed in the full light of understanding of the physical layout and phase/frequency characteristics of the loud-speaker drive units. Nothing in a monitor chain, from the desk output to the listeners' ears, can be fully developed in isolation. The overall system must really be considered one single entity, and that includes the space within the room!

There are four major parameters affecting tonality in any monitor system:

1 Pressure amplitude (frequency) response
2 Phase response
3 Non-linear distortions
4 Reflexions and reverberation

N.B. Resonances can be considered in the phase/amplitude domain.

Whilst the third parameter is primarily a function of the loudspeaker/ amplifier system, the first, second and fourth are also functions of the rooms/ loudspeaker/amplifier/crossover combinations. If the room has an uneven reverberation time, the peaks in its response will appear to hang over and colour or 'muddy' the sound at and around those frequencies. This cannot be controlled by graphic equalisers as any reduction in the energy from the loudspeakers, in an attempt to drive the room less at its reverberant peaks, will distort the impulse response, and hence the direct transient signal which carries so much of the definition to the ears. Room reverberation problems must be dealt with in the room. Any room reflexions will add to and subtract from the direct signal, dependent upon their amplitude and phase relative to frequency. Here we have two possible options: control the loudspeaker directivity, or control the room. One advantage of controlling the room is that loudspeaker drivers can be chosen for tonality, power handling, and other parameters, without as much consideration being given to directivity as would need to be the case in a room with more reverberation and early reflexion problems. Nonetheless, abrupt directivity changes within the loudspeakers should be avoided as other colouration problems may then arise.

4.2.4 Proprietary loudspeakers

Reflexions will also cause phase confusion, which in many instances will mask many subtle perceptions of spaciousness and position, not to mention the overall tonality. If a designer has control of a room, it becomes a much more manageable task to optimise the amplifier/crossover/loudspeaker system for 'accuracy'. Manufacturers who design and market loudspeaker systems for general use without any control of the listening environment are facing an uphill struggle. Having said that, I am not suggesting that such manufacturers should pack up and go home, but rather that users should pay more attention to the intended mounting regimes which those manufacturers may specify. Loudspeakers just *cannot* sound great in *any* room, so without close cooperation with the manufacturers, the users cannot realistically hope to achieve any degree of accuracy, unless they are very fortuitous in their environmental circumstances.

To sum up, below is a copy of part of a letter which I sent, some time ago, to another studio designer whom I had known for many years. All that I can add is that it takes for granted that the rooms and loudspeakers would be designed to be used together, and that the title was tongue in cheek – I was not claiming absolute originality!

4.3 Newell's law

Monitor problems must be fixed in the monitors, room problems must be fixed in the room!

The job of a monitor system is to drive the room with the desired signal. That desired signal is the sum of its instantaneous parts. I maintain that the 'transparency' and 'neutrality' of the perceived sound is almost entirely due to the integrity of the leading edges, which means an absolute minimum of amplitude *or* phase distortions. Room problems are mainly resonance problems – reverb time problems. The monitor system therefore needs to be accurate to the impulse; the room needs to be accurate in steady state conditions. Should a room have a peak or dip, the option to drive the room to a lesser or greater degree by means of monitor equalisation is not available. True, a peak in the room will be less objectionable if it is driven to a lesser degree by a correspondingly equalised monitor system, but nonetheless, to a slightly lesser extent, the resonant overhang or underhang will still exist. As the correct reconstitution of an impulse requires absolute fidelity to both amplitude and phase, then by definition it can no longer be accurately reproduced if any form of monitor equalisation is used to correct for a non-minimum phase, room problem.

Monitor equalisation *can* have an application when, for certain reasons, the monitor system itself displays small peaks or dips. If these deviations are of a minimum phase nature, then an equaliser exhibiting minimum phase[2] properties may under certain circumstances restore both the amplitude and phase accuracy of the entire system. Large deviations, however, are rarely of a minimum phase nature, and therefore whilst amplitude accuracy *can* be restored, phase accuracy most certainly cannot. Monitor equalisation can only be used for the compensation of monitor deficiencies of a minimum-phase nature: they must *never* be used to correct for a room reverberation time problem. We have, in the years past, been fools. If the impulse integrity is accurate upon arrival at the listening position, then once again, by definition, the amplitude *and* phase responses must be correct. Get this right and the rest will fall into place – neutrality, transparency, stereo imaging, fatigue-free listening . . .

I hope that with the above approach we will, in the near future, be working with a new degree of repeatable definition in terms of control room monitoring. Maybe, then, there will be a return of overall confidence in consistency. By no means, however, is it all in the hands of the studio designers, nor was it ever the sole fault of loudspeaker manufacturers. Demands upon control room monitor systems become increasingly complex. They must be capable of horribly accurate, nit-picking truthfulness at prodigious sound pressure levels. They need to be representative of reasonably 'average domestic conditions' which, even if that could be quantified, would be subject to fashionable change. They must also, ideally, be capable of inspiring the musicians to greater creativity; which sometimes means creating 'on stage' environments. This really calls for three or more monitor systems, yet in inexperienced hands, the more systems available, the more confusion is possible. The requirements are not practically realisable in any one system, hence the dual, close and far field systems evident in most present day control rooms.

Art, by definition, is beyond science. The recording business is expanding rapidly with more widely based points of view, and, in many cases, proportionately less in-depth knowledge. The days of tape operators waiting five years before getting any chance of sitting behind a mixing desk are long gone. This wider breadth of knowledge with less depth of understanding makes some pretty heavy demands on the available science. No room, and no monitor system, can be expected to be all things to all people, but we *can* work towards more fixed references. Accepting that the end result – music in people's homes – is an artistic interpretation, we can still aim for more consistency in the practical realisation of that goal. It would, however, appear that at the moment (1990) chaos rules!!

Notes

1 Fourier was a French mathematician of the 18th century who devised a mathematical procedure which translates data from the 'time' to the 'frequency' domain, and vice versa. The analytical ability of the process was very powerful, but as it required calculations to be made frequency by frequency, it was very tedious and time-consuming to put into practice. The advent of computers, however, made possible procedural modifications to allow a 'fast Fourier transform' (FFT) to be realised as a very practicable method of time/frequency analysis, and is now in very widespread use.

2 Pi spaces and minimum phase. When an omnidirectional sound source radiates into a free-field, it is said to be radiating into 4 pi space. Effectively, its radiation pattern will consist of spherical expanding waves, and sound pressure will be lost at the rate of 6 dB for each doubling of the distance from the source, once clear of the acoustic near-field. If the sphere of radiation were to be cut in half by a wall, and the loudspeaker were to be placed in the centre of the wall, radiating into a hemisphere, then it would be said to be radiating into 2 pi space. Should a floor then be placed under the loudspeaker, it would radiate into only a quarter of a sphere, which is known as pi space. If a wall was then built alongside the loudspeaker, the radiation space available would be only one-eighth of a sphere, or $\frac{1}{2}$ pi space.

   Each time the space available for radiation is halved, the 'pi' number of its space is halved, but at the same time, because the available power of radiation is concentrated into only half the previously available space, the radiated power in that space is doubled (increased by 3 dB). Consequently, if for a given input power a loudspeaker radiated one acoustic watt into a free-field, if it were then moved to a corner of a room where it was bounded by two walls and a floor, it would be radiating into $\frac{1}{2}$ pi space, or only one-eighth of the free-field space. The whole radiated acoustic watt, which previously dispersed spherically, would be restricted to one-eighth of its 4 pi radiation space, so instead of one-eighth of the sphere receiving one-eighth of the radiated watt, it would receive the whole watt: in other words, eight times the power. This translates (3 dB for each doubling) into a 9 dB increase in the radiated sound into the $\frac{1}{2}$ pi space, as compared to the 4 pi, free-field space.

   Radiating from the corner of a room, a loudspeaker can be considered to be operating into a waveguide, which effectively horn loads it. The extra resistance to the movement of the diaphragm caused by the restriction of the available space for radiation causes more work to be done by the driver, and hence its radiation efficiency will be increased, yielding a further boost in radiated output.

   These 'room' effects are uniform with frequency, and have equal bearing on both the measured impulse and steady state responses. They are said to be 'minimum-phase', and corrections of the amplitude response for these effects, when loudspeakers are positioned differently in a room, will also tend to correct any changes in the phase-response. Conversely, room effects caused by reflexions, modes and resonances, will be frequency dependent due to the time delay caused by the varying path lengths of the direct and reflected signals. At some frequencies they will add, at other frequencies they will subtract. As such, any corrections made to a loudspeaker and room's combined amplitude response, which is a function of the above effects, will not correct the phase responses. Such effects are said to be of a 'non-minimum phase' nature, and hence equalisation cannot be used to simultaneously correct both impulse and steady-state responses. This is covered in more detail in Chapters 7 and 9.

Chapter 5

Aural appraisal considerations

There are four general categories of use into which loudspeakers fall: sound reinforcement, domestic high fidelity, studio monitoring, and special purposes. In the first group, sound reinforcement, we can include all installations intended for the purpose of entertaining a mass audience; such as in cinemas, theatres, at music events or in audio visual presentations. The second group covers all units for the attentive listening to music within the domestic environment, possibly extending to better quality background music systems. The studio monitoring group includes all quality control applications, including radio and television studios, mastering rooms, transfer suites and copying rooms, as well as conventional, control room monitoring. The fourth category, special purposes, encompasses loudspeakers for televisions, radios, intercoms, paging systems, voice evacuation systems and all purposes where intelligibility or cost, as opposed to overall sound quality, are the fundamental requirements.

The four categories exist in their own rights, and have been developed within their own sets of criteria. There is no 'league table' of quality or prestige, no one category is in any way superior to the other. It is true to say that discoveries and developments in one field may spill over into, or influence, another, but the four categories still exist in isolation, and for fundamentally different purposes.

5.1 Purposes

Public address/sound reinforcement

Loudspeakers developed for sound reinforcement purposes in general are created to excite the audience – punchy bass, exciting highs, and a controlled penetration of the mid-range. The loudspeakers are designed to deliver the greatest intelligibility above the ambient noise of maybe several thousand people, and to create the most entertaining sound for the entire audience. Similarly with cinemas and other auditoria, flattery of the sound source can be a valid and worthy aim. After all, it is show business, and entertainment and presentation are all-important.

Special purposes

Special-purpose loudspeakers usually have one sole aim in mind for each application. A system for making announcements over the noise of a busy factory, of necessity, is designed to be most effective over the range of greatest intelligibility. Outputs over an area of the response which contributed nothing to the message would only serve to add to the unwanted noise. Though such a loudspeaker may sound very nasal on full range music, it is not necessarily an inferior loudspeaker. It may indeed be a very fine loudspeaker for the purpose for which it was intended.

Domestic hi-fi

Hi-fi loudspeakers for domestic use are in a very, very contentious area. The prime function of such a unit is to give the most pleasure to the person listening to them, and indeed also from that person's choice of music programme. The choice is extremely personal, just like the choice of the music itself, or for that matter, food, or colour, or choice of a car! In general, however, the very term 'hi-fidelity' does describe a major design aim.

Studio monitoring

Studio monitoring loudspeakers have often been approached on a 'big hi-fi' basis, but I believe that their predominant aims are accuracy over a wide range of frequency and volume levels, and the ability to resolve fine details and flaws in the sounds being monitored. The term 'monitoring' in itself implies assessment for control or awareness.

5.2 Domestic requirements

Why does any person like any particular thing? Why does one person prefer cabbage to cauliflower whilst another person prefers cauliflower to cabbage? Why are there so many cars to choose from in any given price range, with partisan groups supporting each and every model? The answer lies in the fact that we most certainly did not all come out of the same mould. Our DNA structures are as individual as fingerprints. This means that our sensory organs and brains are all individual to ourselves. The information arriving at our sensory receptors – eyes, ears, nose, tongues, fingertips and so forth – may well be the same, but how our brains interpret this information is probably different in every case. It is the brain's interpretation of any stimulus which determines our own emotional reactions.

Ears

Auditory perception is unique to each of us. The lynch-pin of our chain of hearing is the Basilar Membrane. This is a cluster of thirty to forty thousand fibres, vibrating in response to the sounds impinging on the ear drum. There is also a certain amount of random activity, which can be set off by the interaction of certain mechanical components of the ear. This mechanical stimulation of spurious product generation can cause the brain to believe that certain frequencies are present, which were not part of the original signal arriving at the ear drum. Hair cells detect the motion of travelling waves which pass down the basilar membrane in response to 'sounds', and nerve endings transmit these signals to the brain.

It would also appear that some of the component parts have 'gated' responses, not causing any sensation in the brain until a pre-determined threshold of vibration is reached. Any sound levels below this threshold will cause no sensation whatsoever to be produced. There are numerous people in whom the random responses of the inner ear can cause unpleasant clattering or other sounds, sometimes triggered by specific frequency bands, and sometimes above certain sound pressure levels. This can cause some people to have an inbuilt bias against 'loud music' or certain types of musical sounds. These are more extreme cases, but in general it can be seen that changes from person to person in the make-up of their inner ears can create wide variations in their opinions of what sounds 'good'.

Preferences

Size, cost, decorative appearance and many other peripheral factors influence hi-fi loudspeaker choice. One great obstacle in the way of selling Quad Electrostatic loudspeakers has been the typical response from an enthusiast's wife – 'Bring those heating radiator things in here and I'm moving out!' Faced with this choice, most men, quite rightly in my opinion, choose to keep the wife and buy smaller loudspeakers. By the way, the above is not intended as a sexist statement, it is just that in my whole life till the present time, I quite genuinely have never heard a man object to his wife bringing in larger loudspeakers. Seriously!!

Individual differences go some way to explain why certain people have strong preferences to certain instruments. This could explain why one person finds one instrument, combination of instruments, or style of musical score, exhilarating, whilst the same stimuli leave another person cold. Resonances in a certain frequency band may be exciting to one person, or distasteful to another person. Psychological factors can also play their part. Why should a series of notes, when played in one sequence, bring a person to tears, yet the same notes in a different order produce in the same person a great uplifting of the spirit? There are just so many emotional, psychological and physiological differences from one person to another that personal taste must be both accepted and allowed for. One cannot generalise or be dogmatic about matters of individual taste! Following this path, would a person with a strong liking for, say, saxophones, select for him/herself a loudspeaker which appeared to emphasise the saxophone's characteristics? Recording engineers do this all the time, selecting microphones for individual instruments which highlight the essence of that instrument. After all, a microphone is only a loudspeaker in reverse. If professional engineers can use this principle to achieve what *they* want, it would be rather arrogant to suggest that people buying hi-fi should not be allowed to use the same principle in selecting their home loudspeakers.

Differing individual compromises

If all hi-fi loudspeakers met a common standard, life indeed would be dull. The acoustics of people's homes do not conform to any standard, so should a person decide to buy a new three piece suite, new curtains and generally rearrange the lounge, it is quite conceivable that that person's much loved loudspeakers will no longer sound the same. Room positioning, size, and furnishing make huge differences to the listening environment. There is no reason whatsoever why the person mentioned above should not go out and buy a new pair of suitably different loudspeakers. If that change restores the accustomed and well loved sound to the lounge, then all is well. Houses in general are not built with acoustics as a primary consideration, nor under most domestic circumstances are the layout and choice of furniture secondary to acoustics. As long as there is variation in the human household, there is justification for differences in the sound of any loudspeakers designed for the home. To paraphrase a well known saying: 'Beauty is in the ear of the beholder', and there are so many variables which affect that perception of beauty. This makes something of a mockery of the 'audiophile' hi-fi system with $\frac{1}{4}$ dB accuracy to equalisation curves and no tone controls. To expect that this will give more accurate sound in the average lounge is ridiculous.

A vast number of records are completed in studios with which the engineers and producers are not familiar – they make mistakes. The above statement has a two-pronged thrust! Firstly, that recordings are not made to a fixed reference, hence the absolute necessity of providing tone controls on the home hi-fi – that is if you want to hear it as it really *should* have been, given listening room discrepancies at both ends of the record/reproduce process. The second point is that studio monitoring criteria do indeed require some considerable further consideration. I suppose a few words in the defence of the producers and engineers would be appropriate here. Their lives are not all plain sailing, as music is still an art form, thank goodness, and as such, the recording staff may be arbitrarily keeping twenty or thirty variables, artistic and technical, in creative balance at any given time. Sound quality is not of *absolute*, paramount importance. In certain circumstances, it may have to be compromised for the benefit of other, musical or artistic, considerations. The true goal is the best achievable, overall compromise. Perfection and the arts rarely mix. Art is an interpretation and representation of an imperfect society.

5.3 Studio monitoring objectives

The general requirements for a studio monitor are: very wide frequency response (even more so with the advent of computer musical instruments and digital recording), low distortion, low colouration, large dynamic range, the ability to resolve fine detail, consistency over a long period of use, and in the event of component failure, simplicity of maintenance in order to restore it to original performance.

On the subject of frequency response, we would be requiring 20 Hz to 20 kHz, even if only to keep unwanted phase shifts as far out of range as possible. Distortion and colouration due to resonances, non-linearities and reflexions should affect the sound quality to an absolute minimum degree; even when producing 120 dB in the control room! If inherent ear distortions prevent persons from hearing many resonance or distortion products at such levels, they will be heard in the decay tails of any reverberation in the control room. If they are present in the direct sound, then they will be present in the overhang. The dynamic range should allow peaks of 125 dB in the control room. Throughout the entire usable range, the essential character of the monitor should remain unchanged. Poor designs can give a noticeable change in timbre at high levels, as voice coils heat up and suspension non-linearities become apparent. Voice coil heating produces both mechanical and electro-magnetic changes, as resistance increases with temperature, and physical distortions take place. Once again, errors at peak levels, will cause errors in the reverberant decay tails!

5.4 Assessment problems

Two different loudspeakers of similar performance may well, under analysis, show their differences in measurable, quantifiable ways. It is very doubtful, however, whether a different manufacturer, given the tabulated data from the measurements, could contrive to produce a loudspeaker from the data alone which would actually *sound* like either of the originals. At this point in time, there seems to be only one instrument *truly* capable of assessing the actual performance of any loudspeaker – the ear. Having already established that all ears are different, we must look for an average of a cross-section of ears – but whose ears, and what are they listening for? The only effective reference point which we have to compare with any loudspeaker is a series of 'live' voices, instruments, or everyday sounds. Recorded signal sources are of less value, as the recorded quality depends largely on the monitoring system used for that recording, and/or the limitations of the microphones used. To minimise room effects, it would also seem preferable to have the sound source in the open air and the monitors under test in a relatively dead environment. Microphones used for the test would tend to be the instrumentation types, not normally used in recording precisely because of their 'blandness': engineers usually find them uninspiring in music recording, as their lack of colouration is often of no help when looking to highlight the characteristics of a particular instrument.

If recordings *are* used for listening tests, microphone positioning should be such that as much as possible of the sound of the instrument is picked up in equilibrium. Microphones on, say, an oboe, should be placed at a distance, and not in such a position as would favour the mouth of the instrument, the reed, or keying noise. A position too close to the reed would produce a sound balance which could only be heard with the ear in close proximity to the instrument. This would not sound natural listening some distance from a loudspeaker. Indeed, an oboist may well *not* be the person most suitable to make a judgement on this. Despite listening to 'live' oboes for much of his or her career, the oboist listens mainly from a position at the reed end of the instrument. Furthermore, the pre-conditioning of the expectation of the sound may be further affected by bias towards certain aspects *of* that sound. This can be connected with whatever aspect of the instrument first drew the musician to that particular instrument. A loudspeaker highlighting those aspects may well rate highly with that oboist, making judgement on taste, and not on accuracy.

This potential problem, pertaining to musicians as judges, was graphically described many years ago by G. A. Briggs in his book *Sound Reproduction*. Briggs described a BBC programme, 'Records I Like', which, on one occasion, featured the choice of the legendary orchestral conductor Sir Thomas Beecham. After selecting a pre-electric recording, Sir Thomas remarked how little the recording of the human voice had progressed over the previous forty years. Reading between the lines, on hearing this, Briggs all but fell off his chair, as to him, the recording sounded like a person bellowing down a horn, which was of course exactly what it was. The only reasonable conclusion to be drawn from this is that Sir Thomas, despite his lifetime in the presence of live music, tended to hear not the sound quality, but the performance – the emotion, the phrasing, the overall musicianship, and the delivery. The implication here is not that musicians are unsuitable jurors on loudspeaker selection, but that great care should be taken to make extremely clear just what the exercise seeks to achieve.

Studio staff, engineers and producers are not an automatic choice. Again care and attention to detail is called for. Many times in the past, a new 'revolutionary' loudspeaker drive unit has been fitted into a monitor system with initial, universal acclaim from all concerned. Unfortunately this has all too frequently been followed, some months later, by the request for the return of the old drivers with the comment 'the mixes are disappointing'. The new driver often produced comments such as 'listen to the harmonics on that instrument'. This frequently turned out to be a characteristic of the driver which subtly, and almost unmeasurably, emphasised certain details of the sound. Closer attention to the 'live' source, in many instances, revealed that the new 'harmonics' were not to be heard in such predominance on the instrument itself. If it is not on the instrument in that proportion, it has no place on the monitors. It must be mentioned however, that should the engineer or producer *choose* to highlight these details by recording technique, it is well within their artistic licence to do so. Under no circumstances, however, must they be unsolicitedly emphasised by the monitors alone. In contrast, the new driver may well be excellent in the sound reinforcement field, where an exciting sound can often be a very desirable attribute.

The reason for the 'disappointment' in the resultant mixes stems from the fact that monitor loudspeakers which sound 'exciting' in themselves can flatter an otherwise uninspired mix. When the tapes are played back elsewhere, what is heard is something more akin to the disappointing truth. A somewhat moot point arises here, in the instance of monitors being used during the recording as opposed to the mixing process. Control rooms are now being used for much of the recording; that is, musicians playing electric instruments in the control room. Under such circumstances, a little extra 'excitement' in the sound may inspire the musicians to greater heights of performance than would be achieved using 'neutral' monitors. I suppose this could conceivably portend a requirement for separate recording and mixing monitors, but at this point in time, that case would surely seem impractical as a general rule. It must be admitted that to some degree this situation already exists, with Yamaha NS10s and the like being used more and more frequently on mixdown. This, however, is probably a consequence of a lack of trust in unfamiliar, larger systems, rather than as a function of the main monitors having been designed specifically for recording. The unfamiliarity is an inevitable consequence of the current mobility of recording staff, from room to room, and system to system, with too few agreed standard conditions.

It is imperative, in any controlled tests, that the volume level perceived from the monitor system should be identical to that emanating from the 'live' instrument at the appropriate listening distance. There are two main reasons for this. Firstly, should a flute be used as the subject, it would be very obvious that it was a loudspeaker being listened to, were it to be heard at 120 dB at ten feet. The instrument itself could not produce such sound levels, and the distinction between 'live' and 'monitored' would then be plain for all to hear. Secondly, the effect of the Fletcher–Munson curves (see Fig. 3.1). Different sound pressure levels produce differences in perceived overall frequency response. The perception of the frequency extremes increase by a greater degree than do the middle frequencies as the overall level increases. Hence a flute, louder than a natural level, would be perceived to have extra highs and lows compared to the natural flute.

5.4.1 Microphone technique

Realistically, only something such as 'SoundField' microphones should be used for such an assessment. Basic microphones of other sorts are simple, pressure-detecting transducers. The microphone diaphragm is capable of moving in one plane only – in and out. A guitar string, when plucked, vibrates in the plane along the face of the instrument. It also vibrates in and out, towards and away from the face of the instrument, and longitudinal waves travel up and down, along the length of the strings. The complexity of the phase relationships of all the component vibrations is enormous. The human ears, in their own extreme complexity, detect far more information than a simple, single plane, microphone diaphragm. The conventional microphone can detect only a pressure change at the diaphragm – a gross oversimplification of the complete pattern. At least the 'SoundField' microphones and certain other array combinations go some way further in retaining the integrity of these highly complex phase patterns, so necessary in our perception and recognition of realistic sound reproduction.

Yes, loudspeakers also move in one single plane only, don't they! There is still a long way to go to the audiophile's Utopia. These restrictions in a loudspeaker's capabilities are, however, no justification for allowing a compounding of the problem from the microphone end of the system. A consequence of the profound importance of phase in realism is that it is probably only fair to compare or assess loudspeakers in stereo pairs, in order that appropriate use of realistic, phase-sensitive microphone systems may be used to maximum potential. A single loudspeaker would in general not be a fair assessment, as spacial sensations are of great importance to overall realism. The stereo imaging potential can also, obviously, only be realised and assessed in pairs, though it is true to say that a small number of performance aspects, perhaps very subtle colouration, *can* best be assessed on single loudspeakers.

5.4.2 The panel of jurors

The jurors must be selected from people who, whilst being familiar with live sound, know what they are listening for. To the best of their ability, pre-conception and bias must be set aside, being replaced by a ruthless concentration on facts. A monitor loudspeaker which either adds to or detracts from the original sound is not on course to achieve the stated goals.

Bland neutrality and accuracy are prime objectives. A neutrality which dictates that an exciting sound going in, produces an exciting sound coming out; a dull sound going in, produces a similarly dull result. If the original sound is lifeless, and is clearly heard to be lifeless, then the engineer and producer are in no doubt that if that effect is not wanted, something must be done to change the sound at source. Flattery at this stage will only lead to subsequent disappointment.

Jurors should not be overworked, and frequent cross-reference to the original sound must be made. Probably no more than two sets of loudspeak-ers can be compared with the original 'live' sound at any one time, as the positioning of many stereo pairs of large monitors could prove extremely difficult. A very wide range of source material is essential, as certain subtle defects may only become apparent on a limited range of programme material.

The statistical analysis of the jurors' perceptions is in itself quite a complex subject. The jurors will at best only provide a consensus – no absolutes, but this need not be as bad as it seems. The ultimate purpose of loudspeakers is to be listened to, so, should they pass a critical, orderly listening test, the results are just as valid as any printout. Back on the subject of printouts, repeatability is still a problem, as in most instances microphone positioning is hypercritical. It is extremely difficult to repeat exactly the reading from one day, when setting up the equipment as carefully as possible on another day. Although the *readings* may vary with slight microphone position changes, the loudspeakers would almost cer-tainly show no perceivable difference in sound quality.

5.4.3 Location

Whilst the previously mentioned proposal for assessment in 'dead' rooms probably holds true for precise, subjective, neutrality tests, the ultimate acceptance tests must be carried out in a variety of typical control room environments. Suitability to the actual rooms of use is a critical last step, certainly until we have more accepted room standards to work to.

5.5 Drive unit differences

In the Autumn of 1989, at the Institute of Sound and Vibration Research, four months were spent testing for certain loudspeaker characteristics. (For the full description of the listening tests, see Chapter 12.)

Nine different sounds were recorded digitally, some being computer generated whilst others were recorded either anechoically or outside, using measuring microphones of very low colouration. The levels were set on the comparator, by means of both measurement and listening, whilst reproducing a noise signal. When listening to a sound such as a flute, the levels were no longer equal. Minor response irregularities in the different drive units caused some to sound significantly louder than others, whilst a different sound could easily reverse the relative loudness. This immediately posed the question of the balance of instruments within a mix. For example, a flute and a triangle could be deemed to be of equal level when mixed, only to appear *triangle* heavy by 2 dB on domestic loudspeakers. A flute and bell mix could be similarly adjudged to be 2 dB *flute* heavy when reproduced at home. Extrapolating from this, a mix of triangle and bell adjudged to be *equal* in balance in the studio could be *4 dB triangle heavy* when heard at home, which would clearly be artistically unacceptable. Changing the mixing loudspeakers *or* the home loudspeakers could change or even reverse the situation.

The degree of sensitivity to these tests was quite outstanding under such controlled circumstances. White noise even sounded different on identical drive units from the same production batch, and was generally grossly different from one type of loudspeaker to another. From this, it is to be expected that percussive sounds such as snare drums will change subjective pitch when switched from one loudspeaker to another. Even where this is a function of amplitude response alone, equalisation is still no answer. Figure 5.1 shows the response of the low frequency end of a monitor system, before and after adjustment by a one-third octave equaliser. Whilst the *overall* levels have been adjusted, the characteristic curves of the loudspeaker's response – the bumps on the bumps – are clearly still superimposed on the plot.

The degree of difference between loudspeakers once the response is further affected by room characteristics is quite startling, strongly suggesting that with the phase and amplitude accuracy now made possible by digital recording techniques, control rooms which are more dead in nature may well be desirable where reference to just what is going to tape is concerned. That is, there are enough problems of fine detail resolution and low colouration in the loudspeakers themselves, without the further complication of room colouration.

5.6 Human differences

In the aforementioned tests at the ISVR, the loudspeaker inconsistencies were only a part of the problems – the 'big half' involved the listeners themselves. The tests were not asking which loudspeakers were the best, or which ones people preferred. The listeners were asked which of reference loudspeakers A, B, C or D, did they think was the most similar to a given sample. The listeners could tick the appropriate boxes marked A to D, or a box marked 'none'. If it was considered that the sample sounded equally similar to two or more from A to D, then two or more boxes could be ticked. The degree of variation was large. Some listeners ticked many multiple entries, very rarely resorting to the 'none' column, whilst others used the 'none' column almost exclusively and hardly ever used multiple entries. Some people were very clear-cut in their judgement, whilst others were very arbitrary. Just how similar is 'similar' appears to be a very

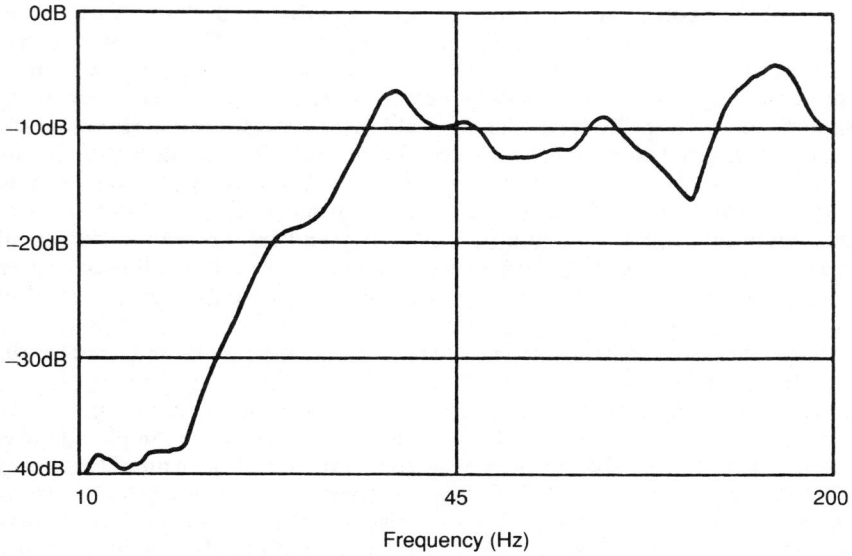

(a)

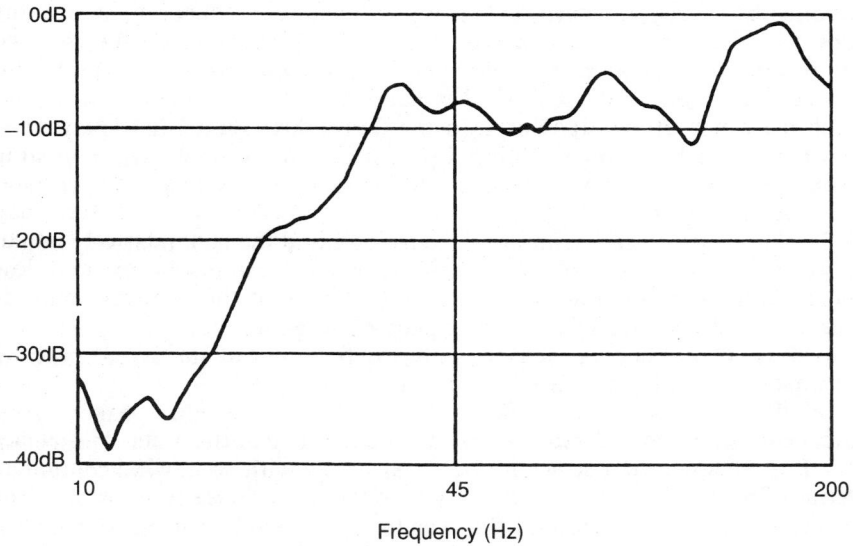

(b)

Figure 5.1 (a) Plot of response of a studio monitor bass driver after use of ½-octave equalisation. Note that despite the more uniform overall response than Fig 5.1(b), the characteristic 'signature' of the driver itself is still clearly apparent. (b) Plot of same driver before equalisation. The improved response at 10 Hz is a function of the poor response of the equaliser below 20 Hz. Merely switching the equaliser in cuts the very low frequencies

personal opinion. Some questionnaires were disparate to a degree that would have been unthinkable before the tests began. On a given sample, one person would clearly indicate that the sound of the waterfall was most similar to say Sample D. Another person may indicate the same sound as similar to B. Repeating the test would produce the same results, each listener being very sure of their own opinion. So *did* the sample sound like D, or like B? Analysing the questionnaires from dozens of people could well show a strong body of opinion for D, yet certain people would still, quite unequivocally, opt for a similarity to B. Remember, this was not a test of preference, but of audible similarity. Subjective preference differences we could have expected, but such large objective similarity differences were not anticipated to the degree encountered.

The reactions to the anechoically recorded acoustic guitar chord posed an even greater problem. The change in sound from one loudspeaker to another in certain cases was perceived by some people as a change in tonality. To other people it was perceived as a chord inversion change – as though it had been played further up the fretboard. For example, a chord of C major played as C, E, G, C was perceived by some people as the same chord but with a different timbre, whilst to other people the chord was perceived as having the same overall character, but now becoming a second inversion G, C, E, G. Both chords are C major, but notationally they are not the same. During my years as a recording engineer and producer, I was frequently asking guitarists to be more dynamic, moving about up and down the fretboard. 'Play the F further up the fretboard' I would say. 'It's more comfortable to play it down here,' the guitarist would reply. 'It sounds better up the fretboard,' I would retort. 'It's the same chord,' the guitarist would growl. 'It's not the same chord! Are you going to play it, or am I going to have to come in and play it myself?' I would ask; and so it would go on. It is now very apparent to me that we may well not have been hearing the same effect.

Harking back to the example of the triangle and the bell, relatively minor response irregularities can cause one instrument to sound louder than the other, entirely depending upon the characteristics of the two sets of loudspeakers. If such effects can to some people cause the actual perception of a chord inversion change, then not only the balance but the actual musical arrangement can be influenced by the loudspeakers.

Whilst many people thought the Quad Electrostatic loudspeaker to be excellent, others considered it to be unrepresentative of the other loudspeakers. For the tests, we specifically chose test signals with no musical content in terms of melody; one guitar chord and two flute notes were the most musical of the signals. The purpose of this was to try to avoid drawing people into making judgements based on sounds of which they were familiar with the reproduction on other loudspeakers. Some people considered the flute to be unnatural when reproduced on the Quad Electrostatic, but many people hear flutes more frequently over loudspeakers than in real life. All that I can say is that when we were digitally recording the flute in the anechoic chamber in preparation for the tests, upon checking the recording in the same chamber via the Electrostatic, the similarity to what we had just heard from the flute itself was quite uncanny! The Electrostatic was also by far the most accurate reproducer in terms of impulse/step/square wave response.

5.7 Room differences

When monitoring, it must surely be better to go for sonic neutrality in order to be more fair to all. The true audiophiles will have to learn to spend as much on their domestic *rooms* as they do on their equipment. I can very strongly appreciate the commercial reasons for 'representative' monitoring but I truly cannot see how any useful domestic average can be achieved. It tends to lead to such a smearing of the standard that there are effectively no standards at all. I personally think that the studio industry should once again attempt to lead the hi-fi trends rather than following them; but do the record companies want it, and do the musicians want it? Are we going for excellence and a standardisation of accuracy, or is everything geared to mass consumerism? Somebody somewhere is going to have to decide!

5.8 Drive system differences and limitations

The moving coil loudspeaker was developed by Rice and Kellogg, patented in early 1924 (perhaps they should have called it 'The Kellogg's Rice Loudspeaker') and was the first type of loudspeaker with the potential for 'high fidelity' sound reproduction. In the higher frequency ranges, ribbons, ionic tweeters, piezos and electrostatic units have all since seriously challenged the moving coil loudspeaker, but for reasons of the physics involved, only the electrostatics have been able to challenge over the full range. The electrostatic loudspeaker is to the condenser microphone what the moving coil loudspeaker is to the dynamic microphone. In a similar way, the piezo loudspeaker is to the crystal microphone what the ribbon tweeter is to the ribbon microphone. They are the same technologies in reverse. In the case of the microphones, the diaphragms and distances of movement are very small compared to the sound pressure levels and wavelengths involved, but in the loudspeakers, the movements and wavelength comparisons are much greater. The delicate suspension required for ribbons precludes their use as low frequency loudspeakers when large air volumes need to be moved somewhat abruptly. Piezo-electric crystals, which bend or expand under an applied voltage and vice versa, cannot produce enough movement per volt to move any significant amount of air, so also cannot be used for low frequencies.

Electrostatic loudspeakers have achieved probably the highest standards of sound reproduction from any type of loudspeaker, in the same way that condenser microphones (electrostatic microphones) have long reigned supreme in the world of microphones. In studios, however, where the requirement to solo a bass drum at a real life level is commonplace, the electrostatics begin to fall by the wayside. In order to achieve high sound pressure levels, say 110 dB + at 1 metre, either the diaphragm areas would have to become inordinately large, or would have to be spaced so far apart that the polarising voltages would rise into the tens or even hundreds of thousands of volts. At such high voltages, the attendant problems of safety, diaphragm and component insulation, choice of suitable materials, and many other problems begin to impose severe restrictions. Effectively, in practical studio monitoring applications where wide frequency ranges and high sound pressure levels are involved, the moving coil loudspeaker still stands alone!

Whilst electronics manufacturers have long been producing amplifiers with harmonic distortion in the order of fractions of hundredths of one per cent, loudspeaker manufacturers have been pleased to achieve figures in the lower whole numbers of percentages. I must stress that whilst 1% distortion from an amplifier would be deemed intolerable in terms of high fidelity, such a figure would be acceptable in the realms of loudspeakers, tape recorders and other electromagnetic devices. It is not so much the total distortion *per se* which is audible, but in most cases it is the actual harmonics produced, and the side effects of the distortion producing mechanisms. For this reason, the blind pursuit of lower harmonic distortion is not a worthy goal if viewed in exclusion from other design parameters. To emphasise this, the Yamaha NS 1000 contained one of the lowest distortion mid-range units that the world had ever seen, yet it did not become as widely used as the distortion figures alone would have led many people to expect.

The amplitude and phase distortions which are encountered in loudspeakers are not as easily corrected as those in electronic circuitry. Whereas mid-range compression drivers are generally used through their resonant frequency, suitably damped, cones tend to be used mainly above their resonant frequencies. Under differing types of drive and loading conditions, units are variously mass controlled, velocity controlled, stiffness controlled, inductance controlled, and so forth. It is fortuitous indeed that the physics allows us to achieve limited, 'flat' operating ranges, usually where two of the electro-mechanical properties act in equal and opposite manners over a limited portion of their ranges. The drive units also possess power responses which are not directionally uniform with frequency, therefore the energy that may be directed over one area at one frequency may be distributed over a widely differing area at another frequency. If the energy leaving the loudspeaker is constant at both frequencies, then the frequency which is being spread more widely will be lower in amplitude on the common axis than the frequency which is mainly being 'beamed' or concentrated in that direction. Given all of these and many other non-uniform properties of drive unit physics, it should be more easily understood that a loudspeaker is almost anything but an inherently linear device, and why only the subtlety of precisely defined listening tests can be relied upon to make assessments of sonic neutrality. There are just far too many variables to be able to specify neutrality in terms of figures alone.

In Gilbert Brigg's book *Loudspeakers*, published in 1958, on page 42 he stated that they had just tested every tweeter upon which they could lay their hands. He went on: 'But I must say that when these tweeters were tested on white noise, I am amazed at the differences which were exposed; no two speakers emitted identical sounds. So it must unfortunately be admitted that we are still some distance away from general perfection.'

Had he published his book in 1995, he would have had no cause to change his opinion.

5.9 The real sting in the tail

Despite all of the problems in the electro-mechanical devices, technology is advancing; and as it moves on, some of these very real problems may be overcome. The real brick wall that faces us, however, makes the Great Wall of China seem like a garden fence. The seemingly truly intractable problem is a human one, and we now have excellent evidence of just how great it is. The problem is in the degree of differences in individual human perception, and I fear that to change human beings is way beyond our abilities.

5.9.1 Degrees of variability of human perception

It would to most people seem reasonable to expect two or more human beings to agree in general terms as to when two things sound alike or not. Indeed, when people buy hi-fi magazines to read the equipment reviews, they are expecting this to largely be the case. If the reviewer hears that it is good, then the same should apply to the reader when he or she hears it. A spark of doubt soon rears its head though when one realises that one has 'favourite' reviewers, whose opinions usually seem 'true' to one's own over a period of some time. The question is, when we agree or disagree with reviewers, are we merely dealing with matters of taste and preference, or are we dealing with fundamental differences in perception? Two highly controlled experiments carried out at the ISVR in 1989 and 1990 would indicate that actual perception differences play a large part in human assessments of accuracy, neutrality, correctness, or whatever one may wish to call it. Reference has already been made in Section 5.6 to the fact that a guitar chord, switched between different loudspeakers, was variously deemed to have either changed in timbre, or in notational structure.

In the same listening tests, carried out by Keith Holland and myself, sixteen test sample loudspeakers were individually referenced to four fixed archetypes labelled A to D. The question was asked for each of nine sounds, 'To which of the four archetypes does the test sample sound most similar?' Let us fix on one individual test of one sound and one sample; say, sample 14 of sound 3, as annotated in Chapter 12. As there was no question of preference but simply similarity, we expected the overwhelming majority of listeners to agree. On some sounds and some samples, there was in fact almost uniform agreement, but on other samples and other sounds, there was a marked degree of difference from listener to listener. Looking back through my notes, after the first eleven listeners had completed the test on sound 3, two people thought sample 14 sounded most similar to archetype A, one person thought it most similar to B, five most similar to C, two to D, and one person thought that it did not sound at all like A, B, C or D.

Let us now take the individual case of the person who thought that sample 14 on sound 3 sounded most similar to archetype B. Let us now also reverse the question, 'Of archetypes A, B, C and D, which one sounds most similar to the test sample number 14 on sound 3?' The answer would, of course, have to be the same, albeit in reverse, that archetype B sounded most similar to the sample. When we structure the question this way however, we could easily imagine a set of circumstances under which sample 14 was not a test sample, but the real, live, source of a sound. Under those circumstances, we would be asking which of the archetypes A, B, C or D was most accurately reproducing the original sound; and whilst this is never going to be perfect due to microphone imperfections, it would at least be a reasonable test of accuracy. We would then have a likely situation from the results above that two people would think A most accurate, one person B, five C, two D, and one none. Obviously the results above are not an actual test of accuracy, as the sample was not a live sound but a sound carrying the imperfections of the sample. However, if eleven people could disagree so widely, then the same could hold true in the hypothetical case stated.

At the other extreme out of eleven listeners auditioning sample 6 on sound 8, all eleven agreed unanimously that it was most similar to archetype B. What we are therefore faced with is not eleven listeners with totally differing viewpoints or perceptions, but listeners with areas of agreement and areas of difference; so some types of musical sounds may elicit strong agreement on what is 'right', but other musical sounds may fall into areas where no such general agreement exists. In these latter areas, who could possibly say which of the listeners were 'correct'? Certainly any two or more of the listeners in general agreement, if talking amongst themselves, would probably be wondering what was wrong with everybody else's ears, which is no doubt what happens when satisfied customers speak to their favourite manufacturers, forming a sort of 'Mutual Appreciation Society'. There is nothing wrong with this – it is a function of human differences and cannot really be avoided – but it does go some way to explaining why certain people have favourite brands of microphones, loudspeakers, amplifiers or whatever. The other aspect of this, of course, is that if a person mainly listens to one type of music, or uses one type of microphone technique, or one type of instrument, there will be a tendency to choose audio products whose characteristics or imperfections are benign or even beneficial on those specific types of music, techniques or instruments. This is true of studio monitor use as much as it is of domestic hi-fi.

To further illustrate the inconsistency of human perception, let us consider the results of a test carried out by Dr Andy McKenzie in 1990. My involvement in these tests was as a 'guinea pig'. I was told very little before the test, as I was merely accosted whilst walking down a corridor at the ISVR and asked if I would take part as a listener. I was seated in the centre of an array of loudspeakers, asked to switch a rotary switch from position one through to position six, and to say in which position I preferred the sound for each of a selection of musical signals and a series of different test procedures. On numerous occasions I mentioned that I thought the switch was not making contact, but each time I was assured that the switch was OK and would I please just get on with the test. The switch box in my hand was not itself switching signal, but was remotely controlling electronic switching in the control cubicle, and I soon became absolutely convinced that there *was* a problem. Upon asking Andy one more time to check things over, he reassured me that there was no problem, and if we stopped to discuss things, it would spoil the 'blindness' of the test, thus negating the results. At this point I had to discipline myself to behave like a test subject should. After all, I must have reassured *my* test subjects countless times with the statement, 'There is no right or wrong answer. You cannot "do well" or "fail" – please just answer the questions as best you can.' So, heeding my own advice to others, I buckled down to completing the tests without further interruption.

As soon as the paper was published, Andy both sent me a copy and telephoned me. The problem was with the 'switch faults' referred to earlier, but they were in fact not switch faults at all. The test involved aspects of 'Time/Intensity Trading' to create spacial effects in wide image stereo, and the switch positions controlled differing amounts of simulated direct and reflected signals, from different directions and with different delays. When I thought there was no change due to a faulty switch, in actual fact a change *was* present, but I was not hearing it. The problem in the results was that whilst the positions were highly repeatable, as on one sound, no matter how many times I operated the switch, I still heard either the same change or no change in the corresponding switch positions, the 'no change' positions were not the same from listener to listener. Effects clearly heard by one listener could be totally undetectable by another listener – for example some people heard entirely phantom images in positions where other people heard nothing.

That things work for some people and not for others is a fact of life in the perfume or fashion industries, but for something as 'technical' as a loudspeaker, this degree of individual perception difference came as something of a surprise. Differences were of course expected in terms of preference, as in the latter experiment the question asked in each case was, 'Which setting do you *prefer*?' This was the original reason for the tests, in order to assess the most preferred sensations before launch of a new product (Canon 'Wide Imaging Stereo'), but the absolute inability of some people to detect any change whatsoever between some of the settings suggested a new set of 'Can you hear a difference? Yes/No?' criteria.

Clearly, the company manufacturing this new product would have to expect very mixed reviews in the press, dependent upon the reviewer's own ability to perceive many of the effects. In turn, this could lead to a somewhat hit and miss marketing philosophy such as, 'If it works for you, buy it.' I am not in any way suggesting here that because of these inconsistencies of perception, the product in question would not be worth marketing, as for those people for whom it does work, it will doubtless give much pleasure, and for them it may also be realistic. All that I can say is that for those for whom it does *not* work, they would probably not purchase such units. In reality, this is true to some degree for all loudspeaker systems. On reflexion, maybe the only difference between the company in question here, and many others before them, was that this company was all too painfully aware of the likely perceptual differences of its intended customers, whereas so many other manufacturers have only been aware of preferences.

In both of the tests described above, very real and highly significant differences were detected in what was heard, even between people with experienced and trained ears. As there are too many variables in loudspeaker 'quality' for any instrumentation readouts to coalesce into a meaningful indication of subjective audible acceptance, then where does that leave us? Precisely where we have been for a long time, and precisely where we are likely to stay for as long as *any* errors exist in loudspeaker performance: different manufacturers and different users will adhere to their own sets of design and performance priorities. Remember, if we do not all perceive the same sensations, we cannot all be expected to prefer the same products.

5.9.2 No commonality in any area

Each of the two tests described dealt with different aspects of auditory perception. In the 1989 test with Keith Holland, the anechoic conditions and relative uniformity of the directions of the different sounds ensured that the differences perceived were in the domains of frequency and phase, with inherent drive unit reflexions contributing to the anomalies. No differences in results were apparent if any two drivers were positionally interchanged. In the 1990 tests with Andy McKenzie, all the loudspeaker units used for the simulated reflexions were nominally identical, so all significant differences perceived were due to the timing and positional differences in the simulated reflexions.

5.9.3 Further variability of perception

The variables and non-linearities in all the links of the monitor chain can interact in such a way that system prediction from individual component parameters is not feasible. If we cannot predict, then we can only listen, but listening in itself must also be carried out in a careful manner. Time and time again, organised listening tests have failed to live up to expectations or to produce conclusive results. Evidence is now coming to light which suggests that when we listen in a relaxed manner or are concentrating on something else, the main brain activity in response to the music is for most people in the right-hand hemisphere. As soon as we begin to concentrate on that music, or begin to listen for differences in sounds such as in A/B testing, the more logical left-hand hemisphere of the brain will dominate the response. When this shift of predominant activity from one half of the brain to the other occurs, we are entering a very different mode of perception. Little wonder then that some of the more subtly perceived differences can be so elusive under more controlled conditions.

That the two halves of the brain are different in terms of perception is easily proven by the classic experiment of giving a person a pair of oscillators, one fixed and one variable; one being fed to each ear via headphones. When asked to tune the variable oscillator to the fixed one, there will frequently be a discrepancy in the tuning when checked electronically. When the headphones are reversed, the discrepancy will be in the other direction, showing that one ear, or rather one half of the brain, is perceiving a different absolute pitch to the other. The difference can easily be in the order of 3 or 4 Hz at 1 kHz. Under normal listening conditions, no such pitch distortions are evident.

5.9.4 Relevance of subtleties

Whereas the audiologists may frequently deem that differences which cannot be readily and repeatably detected under controlled circumstances are generally too small to be of consequence, the artistic environment of the working studio can experience a human response to much more subtle changes. The claim of the audiologists that if a monitor system difference is *very* subtle, then it must be specific to a system and will have no bearing on the mix when played elsewhere, is harder to argue against. However, the studio environment is sometimes such a delicate factor in the creative process that it is hard to argue that there is *anything* too small to have an effect on that creativity. The chaotic interaction with the human factors will almost certainly ensure that such things cannot be dealt with in a purely scientific context.

5.10 Summing up

Loudspeaker systems as we know them cannot reproduce an original sound field from an acoustic source. They all therefore produce 'sound field distortion'; thus when any original sound is reproduced via loudspeakers, sound field distortion will be an inherent part of the reproduction, varying from unit to unit and room to room. Different people with different pinnae, different heads, different auditory systems, and different brains, will not respond uniformly to those sound field distortions. Consequently, if a live sound is compared to the same sound reproduced via loudspeakers, the degree of difference between the two will almost certainly not be the same for any two listeners; though for identical twins without environmental damage, it may be! As we have already established that many degrees of difference are signal-dependent, and we know now just how different people's perception can be, we absolutely *must* have a situation where the degree of accuracy of any loudspeaker system will be individual to each listener, his or her choice of music, and many of the other factors previously mentioned.

Chapter 6

Interfacing complexities

6.1 Art or science?

On the subject of the relative balance of art and science in the context of monitor systems, one area which highlights the relevance of this question is that which deals with the problems of interfacing and matching the individual components of the overall system. As will be seen, the degree of variability in the systems as a whole is enormous. When I refer to a monitor system, I am referring to the entire chain from the monitor outputs of the mixing console to the ears of the listeners. A monitor system includes amplifiers, crossovers, cables, connectors, loudspeaker drive units, loudspeaker cabinets, and the listening room itself. Obviously, there is a great degree of science which is applicable to the design of each individual link in the chain. There are also strong scientific bases to the interconnection of the individual links which form the chain, but unfortunately, as no single link is 'perfect', compromises must be made. Those compromises are highly individual, as they follow the subjective preferences and perception differences of different human beings, and as we have seen in the preceding chapters, few people agree exactly about what is right. Where such individual compromise differences cannot be quantified or qualified, and indeed most cannot be pinned down in any absolute manner, a somewhat more artistic interpretation is almost inevitable.

I am by no means *advocating* an artistic final approval – it is merely that the required science is just not developed to the necessary degree. Reproduction via a monitor system must be considered to be a painting of the original, not a photograph. It is impossible to reproduce an original sound field, so any reproduction *must* be less than perfect. The lack of attainable perfection pervades the entire system: crossovers which do not sum in terms of power and voltage; cables which are not perfect conductors; amplifiers which distort and do not have infinite bandwidth; loudspeakers which suffer similar defects as amplifiers, rooms which have finite, non-uniform reverberation times; *and* the non-linearities of the air itself. The variables in the above imperfections are infinite. Quite clearly, there will be combinations which sum in an undesirable manner, whilst other combinations either cancel or are benign in their manifestation. It is for this reason that I have long advocated an integrated approach to the entire system, as I firmly believe that the individual building blocks cannot be adequately specified in isolation.

6.2 Imperfections of rooms and drive units

It is all but impossible to build two truly identical rooms. Even if a perfect room did exist, the very effect of introducing the listener (and chair, and controls, and whatever else he or she may require) into the room would disturb the performance of the room and hence render it imperfect. The complexities of room responses are discussed at length in Chapter 16, but in practice, no one room can support a sound field identical to that of another room.

In the tests at the Institute of Sound and Vibration Research in late 1989, Keith Holland and I carried out a series of listening tests on the electro-mechanical and human perceptual differences on a whole range of mid-range drive units.[1] No two loudspeaker drivers were deemed to be sonically similar over our entire range of test signals – all were imperfect. So, two of our links in the monitor chain, the rooms and the loudspeaker drive units, have already fallen by the wayside in the attempt to achieve any degrees of consistency and perfection.

6.3 Crossover anomalies

The third link, crossovers, is discussed at some length in Chapter 7. Since that chapter was written, more definitive work has been carried out in an attempt to clarify some of the pros and cons of the different approaches. Tests were conducted in which electronic, low level, active crossovers were tailored to match to a very high degree of accuracy the transfer functions of some passive, high level units. No matter how accurately the overall system performance was matched when measured on steady state or noise signals, the systems showed differing impulse responses, and most certainly did not sound the same. There is little wonder that this outcome prevailed, as the electrical loading and interaction characteristics of the two approaches are entirely different. Whilst the active crossovers are usually terminated by a relatively constant impedance, their high level passive counterparts are rarely so fortunate, as a loudspeaker driver rarely presents anything as simplistic as a constant impedance load. In the active unit, the driver and crossover are buffered very effectively by the intervening power amplifier, which provides an almost ideal termination to the crossover filter, and an excellent drive source for the loudspeaker. In the passive, high level units, the close coupling of the filters to the drive units not only precludes anything approaching an ideal filter termination, but also has the unfortunate effect of decoupling the loudspeaker drive units from the low imped-ance signal source provided by the amplifier.

Although swept sine and noise signals can be shown to perform with an admirable degree of comparability between the two approaches, transient signals react in totally different ways, and when approaching overload, the performances of the two systems are radically different. When the time constants of the resonant filter circuits are close coupled to the electro-mechanical problems of a loudspeaker driver, some very unpredictable events may take place. If a cone is pushed forwards under the influence of the signal from, say, a bass drum, then when the transient has passed, the loudspeaker cone/coil assembly will attempt to return to its rest position under the influence of its suspension elasticity. The resultant back-e.m.f. generated by the coil will be modified by the filter circuits *en route* to their arrival at the amplifier's output terminals, where damping will be effected by the low output impedance of the amplifier. The back-e.m.f.s must pass backwards through the filters in order to reach the feedback circuits of the amplifier.

Tighter control is exercised when the system is actively crossed over and the amplifiers and loudspeakers are directly connected to each other. Some most peculiar currents can be generated when further transient and steady state combination signals arrive via the amplifier at the crossover terminals of a high level, passive system, whilst the back-e.m.f.s are being fed in from the other end by a mechanical restoring force from the loudspeaker suspensions. The number of possibilities of combinations of potential musical drive signals is so enormous that trying to predict the behaviour of such a system is beyond practicability. It is a loose cannon on the gundeck. It is these dynamic parameters which are effectively impossible to model in an active version of a passive crossover.

What sparked this comparison work was the noticeably higher failure rate of some mid-range drivers in systems with passive, high level crossovers, when compared with almost identical systems driven via active crossovers.

Dr Keith Holland suggested to me that, from some of his computer models, dependent upon the drive signal, if the system briefly overloaded on a low frequency signal, the resultant distortion products fed into the mid-range units could be much higher than 'rules of thumb' would lead one to expect. Effectively, there are situations where the time constants of the passive crossover circuits can discharge before the clipping of the low frequency half cycle has ceased. As the signal then swings from the full positive to the full negative rail of the amplifier, a product of short duration having rail-to-rail peak amplitude can be fed to the mid-range driver. Discharging once again, the crossover is then primed to deliver a rail-to-rail negative to positive product to the mid driver. The crossover is acting as a voltage doubler circuit, and hence in these cases, a power quadrupler. Such effects are signal dependent and of very short duration, and whilst there is no risk of thermal overload of the coil, a mechanical overload risk does exist, fatiguing the drivers prematurely when levels approaching amplifier overload are used. Subsequent to Keith Holland's mathematical models, independent consultants undertook a verification test using an analogue modelling system which seemed to confirm the problem. The results are shown in Fig. 6.1.

In order to overcome the above overload problems, the initially obvious conclusion was to build a low level active crossover to drive the system in a multi-amplifier configuration, completely isolating the mid-range driver from low frequency signals, whilst adhering as closely as possible to the parameters of the passive unit. In an experimental version, the slopes were matched on the response plot to within the thickness of a pen line, and the systems were measured on noise such that they were deemed to be to all intents and purposes identical. Unfortunately, they did not sound the same.

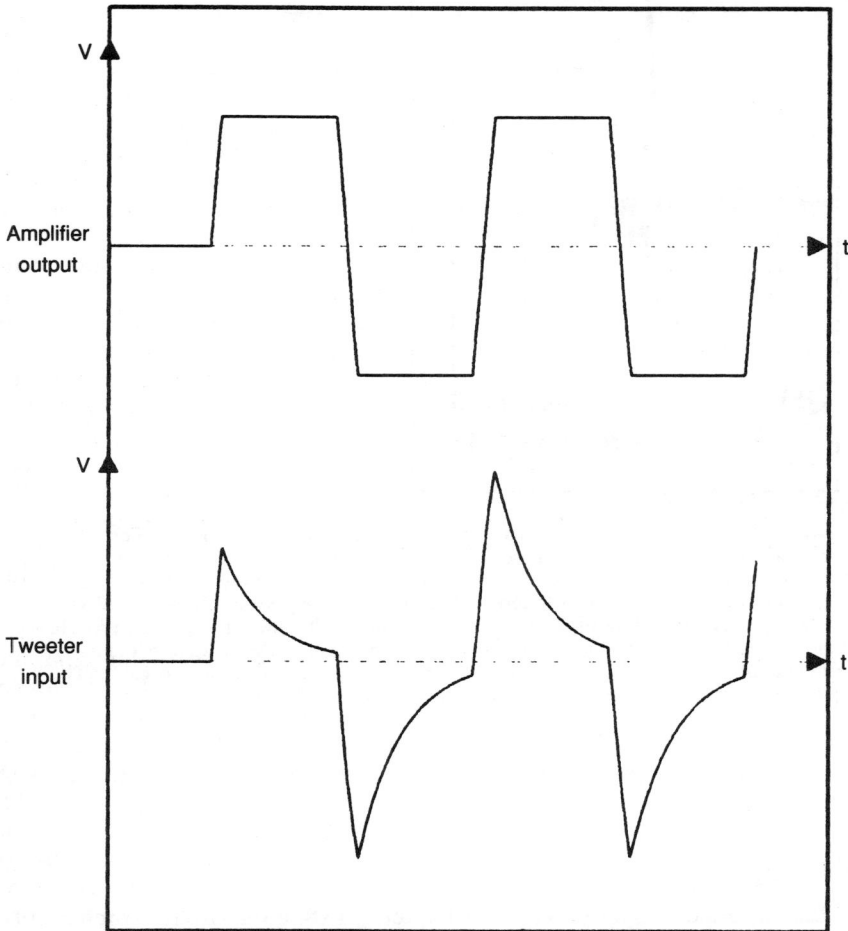

Figure 6.1 A typical example of the effect of a passive crossover on the voltage seen by a tweeter under amplifier clip conditions

Their dynamic characteristics were very different: their step function finger-prints were different. Which was 'better' or 'worse' is not relevant in these circumstances as the object was to reduce the driver fatigue whilst maintaining the well liked, well known, and well appreciated sound of the original system.

Also, the musical drive signal has a great bearing on the nature of these perceived performance differences and distortion products, as shown by Fig. 6.2, taken from Schroeder's *Models of Hearing*. The two plots are composed of the same 31 harmonics of equal amplitude: only the phase relationship of the harmonics is different. At high levels, plot (a) would clearly be more stressful to the driver than plot (b). From the above examples, it will be seen that the crossover format must be taken into account at the outset of the design of the entire system: deciding on the

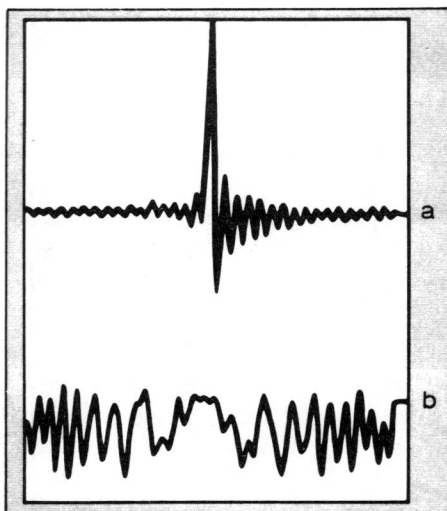

Figure 6.2 One period of waveform made up of 31 equal-amplitude harmonics. Top: all harmonics in approximately zero phase. Bottom: same amplitude spectrum, but phase angles 0 or π rad selected to minimise 'peak factor' (both waveforms are drawn to the same scale). These two waveforms sound very different at fundamental frequencies below 200 Hz, in contradiction to Ohm's Law of Acoustics which maintains that the ear is 'phase deaf'. The author has succeeded in playing simple melodies by changing phase angles of selected harmonics of the top waveform while keeping their amplitudes fixed ('phase organ'). (The figure and description are taken from Schroeder's *Models of Hearing*)

crossover afterwards is not realistic; it is not an add-on, it is of fundamental importance.

6.4 Amplifiers

In the discussion on crossovers in Chapter 7, I state that I had a preferential amplifier type which I tended to use with my own systems; not because I thought that they were 'the best amplifiers *per se*', but because I had found them to be the ones which were generally considered to be most suited to my systems. Since that was written, a most startling series of tests have taken place on the premises of a very well known manufacturer of both hi-fi and monitor loudspeakers. In fact it was Keith-Spencer Allen, the then editor of *Studio Sound*, who drew my attention to their existence and will bear out the findings.

A new range of monitor loudspeakers had been developed, and tests were carried out on four different sizes of systems in order to choose an amplifier to be recommended for use with the range. Of the four loudspeakers, two were of the same generic type, with similar crossover topology and general format. The largest loudspeaker in the range was assessed in listening tests whilst being driven in turn by each of four amplifiers which we shall call A, B, C and D. When it had been generally agreed that amplifier A was preferable, seeming to give the overall character with which the greatest number of listeners were content, the test was repeated on the next size down in the range of loudspeakers. To a certain degree of surprise, amplifier B was chosen to give the best performance on that loudspeaker system. When the tests were repeated on the two other models, amplifier C was chosen. Only amplifier D failed to be nominated first choice on any of the systems.

When the tests were repeated with Keith-Spencer Allen himself in the panel of listeners, the results corresponded with the original tests. The differences were not subtle, or only discernible by 'golden ears'; furthermore, the tests were also eminently repeatable. Not one of the test amplifiers could therefore be deemed to be generally superior to the others. On the other hand, there have been reports and articles published which claim that all high quality amplifiers of better than a given performance are indistinguishable under normal operating circumstances. How come these widely disparate conclusions? If it was possible to produce perfect amplifiers of zero distortion and infinite, uniform bandwidth, then by definition they would all sound the same: they would all be perfect. Choices between makes and models would be solely on output power, physical appearance, weight, reliability and cost. No one would be sonically any better or worse than another. Unfortunately, amplifiers are imperfect, and the discrepancies in the results of listening tests are surely due to drive signal and dynamic loading conditions which play to the strengths and weaknesses of each design.

Given loudspeakers with a generally benign dynamic impedance characteristic, then, on one type of music, it is quite feasible that the performance of a whole range of high quality amplifiers could be deemed to be very close indeed. Changing the drive signal from a mainly smooth to a mainly transient nature, or vice versa, may then produce more distinct differences in the perceived performances. On the other hand, loudspeakers with 'obstacle course' dynamic impedance properties may well highlight some alarmingly noticeable performance differences in the amplifiers under test: instantaneous current delivery capability possibly then becoming very significant between amplifiers of nominally similar steady state output powers. In fact I would go so far as to say that all other factors being within reasonable 'hi-fi' limits, instantaneous current capability and slew rate are probably the prime factors governing sonic performance of amplifiers.

The response from certain loudspeaker manufacturers to the excessive demands which some loudspeaker systems may make on amplifiers has been to attempt to design passive high level crossover circuits which produce relatively uniform load impedance/frequency ratios to the amplifiers. Some very clever and very complex crossover circuits have been designed to achieve this goal, but as we discussed a few paragraphs ago, changing the crossover changes the dynamic loading to the loudspeaker drivers as well as to the amplifiers, and hence almost certainly will change the sonic performance. The drive e.m.f.s go through the crossover one way, the back-e.m.f.s travel in a different way; the dynamic signal coupling between amplifier and loudspeaker drivers can become very complicated indeed.

After hearing of some of the above tests, Tom Hidley contacted me as he was interested in some of the lines of thought. In all of his current installations, he uses monitor loudspeaker systems designed by Shozo Kinoshita. The systems use high level passive crossovers said to be capable of handling 3000 watt peaks before saturation. The impedance drops to around 0.8 ohm at a couple of frequencies, and the systems are rated at around 1 kW of music programme. Kinoshita had intimate knowledge of the drive units, as I believe he was deeply involved in their design at Pioneer/TAD in Japan. These loudspeaker systems were always very demanding on amplifiers, the current systems being installed with JDF amplifiers costing around £8000 ($12 000) per mono amplifier. On a more kindly load, the JDF is probably not readily discernible from another high quality amplifier, yet the options to tame the electrical impedance gymnastics of the Kinoshita crossover, or to electronically cross over and bi-amp the existing Kinoshita systems, are not available. Either of the above changes in the drive system would change the sonic performance. Their only option has so far been to leave this difficult crossover as is, and develop an amplifier which is capable of driving it – the JDF, delivering 3200 watts into $\frac{1}{2}$ ohm. The extra cost does not make the systems any easier to sell, but if that particular sonic performance is what the clients want, then the system as it stands is the only option available, amplifier choice being a critical factor.

6.5 Loudspeaker cables

The cable harness for the above system is now in the order of £2500 ($4000); they are very elaborate affairs indeed. One type of my own monitor system designs is comparable in size, frequency range, output SPL and most other general parameters to the Kinoshitas. The main difference lies in the fact that my systems are 3- and 4-way active. Special esoteric cables have been shown to have no noticeable effects on the performance of these systems, whilst on the Hidley/Kinoshita systems, the effect of esoteric cabling is readily noticeable. Once again, the effect of complex impedances seems to play a significant part in the necessity or otherwise for using expensive cabling. Complex impedances make more demands of amplifiers, and are more likely to produce premature overload and consequent driver fatigue. However, if a passively crossed over system is deemed to work sonically, together with its complex amplifiers and cabling systems, then it must be considered *as* a system. All of our work to date shows that the system can be radically modified in terms of some of its components and structures in such a way that parameters measured against standard test procedures show remarkably little discernible differences. At the same time, another minor change to a system can cause seemingly disproportionately large differences in perceived sonic performance. Prediction is difficult.

The cable issue is discussed further at some length in Chapter 8. Despite much hard work, the correlations are still not definitive. Implications currently appear to relate to effects on back-e.m.f.s and feedback rather than forward current transfer. In many cases, it may well be in some other area that problems lie, rather than purely being in the cable itself: terminations for example.

6.6 Connector reliability

Many of the claims made for the linear crystal and oxygen-free copper cables relate to the electrical imperfections of the inter-crystal boundaries of conventional copper cable. Again, in tests, if these effects manifest themselves at all, it seems to be in the form of very low level signal disturbances. But caution here: as stated previously, whilst some large disturbances are seemingly innocuous, other very minor disturbances can be noticeable. However, even the inter-crystal boundaries of conventional copper are superb conductors when compared with the inter-metal contacts of the connectors. In other words, when a copper cable is soldered via a tin/lead alloy (solder) to a brass (copper/zinc) terminal, in turn screwed to a chrome-plated copper spade terminal soldered to a copper film on a printed circuit board, which leads to a transistor socket which contacts a tinned leg of a transistor and so forth: these things are all thermocouples, producing their own spurious voltages in response to any temperature change. How can the wire alone make that degree of difference? Again, dissimilar metal corrosion between all of these contacts will insidiously reduce the contacts' effectiveness over a period of time, behaving in a very unpredictable manner.

I know that Keith Holland is strongly of the opinion that a person paying hundreds of pounds per metre for cable will make very sure that good quality connectors are used and that all soldered joints are of good standard. Connections will probably be made very carefully indeed, and attention to detail will be rigorously applied. He believes that in many cases, perhaps it is this extra care elsewhere that is the reason for much of the benefit which is subsequently attributed to the cable itself. I know of one classical engineer in particular who re-solders his leads every two or three years to try to ensure good contacts, but what practicability does a mixing console or tape machine offer for the same treatment? Very many loudspeaker drive unit terminals have chromium-plated push connectors which most certainly deteriorate over a period of time as the surface oxides build up. Except under highly controlled and very repeatable conditions, it can be very difficult to isolate entirely some of the above inter-metal contact problems from the effects of the cable itself.

6.7 Geographical and climatic variability of air

So far we have established variables in terms of performance shortfalls in rooms, in the loudspeaker drive units, the crossovers, the amplifiers, and the interconnecting cables. The further link in the monitor chain is the air in the room, which is itself an interconnect between the loudspeaker, the room boundaries, and the ear. To the uninitiated, air is air, but a brief digression into the world of aeronautics may help to emphasise just how variable this medium can be.

During my time at The Manor Studios, I kept one of my aeroplanes across the road at Oxford airport. The aircraft had a service ceiling of around 10 000 ft, at which altitude the rate of climb at full power was virtually zero. I distinctly remember being in Denver, Colorado, high up in the Rocky Mountains, and thinking that on a suitably hot and humid day, the same aeroplane which I kept at Oxford would be entirely unable to leave the ground, no matter what length of runway it had available for take-off. When air is saturated with water vapour, it is much less dense than when dry. That is because water, H_2O, has a molecular weight of 18

(2 + 16) whereas each molecule of oxygen, O_2, has a molecular weight of 32 (2 × 16) and a molecule of nitrogen, N_2, a molecular weight of 28 (2 × 14). The three gases vie for space in any given volume, and displace each other in such a way as to maintain the number of molecules per cubic metre for any given gas pressure. Water, oxygen and nitrogen molecules exist as H_2O, O_2 and N_2 respectively. Given that the atomic weights of hydrogen, oxygen and nitrogen are 1, 16 and 14 respectively, then the above molecular weights can be clearly seen. What is more, the air density reduces by around 1 millibar for every 27 ft increase in altitude. Denver, at over 5000 ft above sea level, would have air approximately some 5000 divided by 27 millibars less pressure than Oxford for the same sea level conditions.

If the barometric pressure on a typical day at The Manor was, say, 1000 millibars, then Denver would be some 185 millibars (5000 divided by 27) less. Add some heat and water vapour to the air in Denver, which both reduce density, and you could well be experiencing pressures of around 750 millibars, enough to completely prevent the aforementioned aircraft, which would have happily risen to 10 000 ft over Oxford, from even being able to take off. Such a pressure change can significantly affect the characteristic impedance of the air, in turn changing the loading on the loudspeakers and their coupling to the ears. With less air to push against, electro-acoustic conversion efficiency will be reduced, and the less dense air can also affect air cavity resonances, both in the loudspeakers, altering the cabinet tuning, and in the room itself. Remember, the difference is enough to prevent an aircraft from lifting itself into the air; it is not trivial! Build yourself a studio in Mexico City at 8000 ft and the differences will be even greater! Climate and weather can be significant factors in system performance and human perception. In a hot, high, humid studio, things may respond very differently compared to a cold, dry, sea level studio.

6.8 Lack of generalisations

If a system works, and a progressive series of modifications or developments are deemed to improve that system, then those improvements are positive steps in the performance of that system. But, modifications which prove beneficial to *that* system do not automatically prove beneficial to other systems. Whilst it is obvious that many of the major prerequisites must be met before a system could really be taken seriously, the more subtle and esoteric concepts which are frequently discussed in the hi-fi press are not always so accommodating in stating their cases for inclusion. Until an awful lot more work is done in some of these areas, cost-effective solutions to the rats' nest of cross-correlations are unlikely to be forthcoming. I do know of one designer having some clients with large cheque books, who operates on a concept of, 'If some people think that it sounds better with X whilst others are not so sure, then as long as nobody thinks that X makes matters worse, we will incorporate X.' That approach does seem to work, but on a process of diminishing returns in the financial area. To be able to define absolutely when X is beneficial, or when it is a waste of time, is hopefully what the future will reveal. To know when to use X to benefit, rather than as a generalisation, will free up the money to also apply Y and Z in their most beneficial roles.

One must be extremely careful about generalising in a way that suggests that 'beneficial' materials or modifications are truly beneficial in a general sense. So many such things are highly system-specific. Hopefully this chapter has highlighted where some of the pitfalls may lie, and stressed the need for caution before building a system from building blocks chosen in isolation.

Chapter 7
Crossovers

'I never realised that a different crossover could change the sound so much!' Such were the words of the owner of a well known studio. Granted, the new crossover was of advanced design, but nonetheless, the crossover which it had replaced was a first class unit.

It is surprising just how many people still specify a crossover for a system, thinking purely in terms of off-the-shelf units, without any idea of the crucial role which they play in the make-up of any monitor system. With conventional crossovers, be they active or passive, what you put in is *not* what you get out. The outputs just will not add up to replicate the input. The only exception to this is the first order, 6 dB/octave Butterworth, but unfortunately 6 dB/octave is far too shallow a slope for most conventional studio monitoring requirements.

Crossover design and technology is enormously complex, and the effect which a crossover has on the audible characteristics of a monitor system is frequently grossly underestimated. Indeed, the problem can be so convoluted that many people seem to choose to use a stock unit in the hope that somebody else has done their homework. Unfortunately, to obtain the best results, the amplifier/loudspeaker/crossover/room combination must really be seen as one unit. This philosophy is brought home by the fact that most specialist monitor manufacturers now use highly dedicated, specific crossovers for their systems, as opposed to using stock items from specialist manufacturers of crossovers, who are now largely aiming their products at a sound reinforcement market. The physical distribution of the many sound reinforcement systems changes from night to night, which, together with the large physical displacement between the individual drivers, renders accurate crossing-over an impossibility given the current knowledge and technology; so crossovers for sound reinforcement and studio monitoring tend to have different orders of priorities.

7.1 Options

There is a very large range of parameters to choose from in crossover design. Active or passive; high level or low level; slope 6, 12, 18, 24 dB/octave or others; slope shapes on initial turnover; power handling (high level)/output voltage (active) – all of these things are crossover variables. Different people, with different philosophies, continue to pursue different paths, but by looking at my own philosophy on the subject, however, we can look at many of the dilemmas in turn.

7.2 Phase leads and phase lags

Somewhat ridiculously in my opinion, amplifiers have usually been tested and rated into resistive loads. As for the truly resistive loudspeakers – I have never seen one! Whenever there is inductance and/or capacitance in a loudspeaker, there is consequently a reactive element where the voltage and current are not in phase with each other, producing the classic 'wattless power' so well known to engineers of heavy electrical machinery. Factories using heavily inductive machinery must install 'power factor correction' capacitors to restore the phase relationship; otherwise the electricity company's wattmeter would not measure the total electricity used, and the subsequent bills would be lower than they should be. A power factor of unity would be resistive, but a reactive element resulting in a power factor of 0.7 would, if uncorrected, result in electricity bills of only 70% of the actual, generated power consumed. The electricity generating people are not too keen on that state of affairs, hence their insistence upon the installation of the correcting capacitors.

Why does this occur? Well, it is probably easiest to explain by an analogy with a familiar piece of everyday plumbing – a WC! Electrical engineering and plumbing really are remarkably similar in their fundamentals. Pipes relate to wires, water to electricity, water pressure to voltage, water flow rate in gallons per minute to current in amperes, and so forth; with pipes or wires, all other things being equal, reducing their diameter will increase their resistance to the flow of either water of electricity. Capacitors have a 'capacity' (the original term for capacitance); and toilet cisterns have a capacity, usually about 2 gallons (10 litres). A resistor/capacitor series circuit has a 'time-constant' – the length of time it will take the capacitor to fully charge through that resistance. The time constant is determined by the multiplication of the resistance in ohms, by the capacitance in Farads. A 10 000 microfarad capacitor (0.01 Farads) fed through a 100 ohm resistor will take 1 second to 'fully' charge. That is, 100 ohm × 0.01 Farads (10 000 microfarads); 100 × 0.01 = 1 (second). A toilet cistern of 2 gallon capacity, when fed through a pipe allowing a flow rate of 1 gallon per minute, will fill to capacity in 2 minutes. Thus, the time constant of the toilet plumbing would be 2 minutes.

Figures 7.1 and 7.2 may help to further the analogy of phase leads, phase lags, and time constants, showing how in reactive systems, pressure and flow are rarely in step.

When a capacitor charges through a resistor, initially current flow is at a maximum as the voltage across the plates is at a minimum, due to the uncharged capacitor effectively being a short circuit. The voltage across the plates eventually rises to the peak value of the charging voltage, and the current ceases to flow. The current flows first, then the voltage rises. The current is thus considered to have a phase lead over the voltage. In the case of an inductor, a voltage must be present across the terminals in order to allow a current to build up a magnetic field within the inductor before current can flow through the circuit. Hence the current in this instance is said to have a phase lag compared with the voltage. Back to the resistor

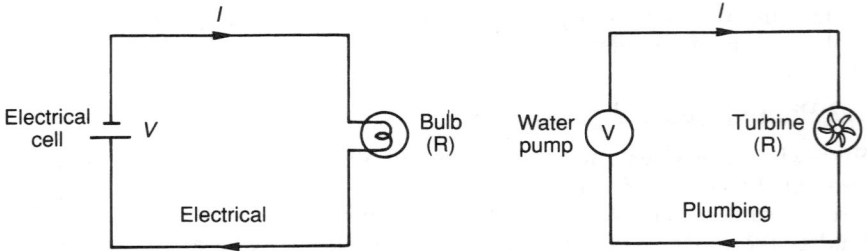

Figure 7.1 In both cases, increasing the voltage/pump pressure would increase bulb brightness/ turbine speed which would always remain in proportion: Increase for increase, decrease for decrease. These are closed loop systems with voltage/current, brightness/speed, in-phase, and are principally resistively loaded

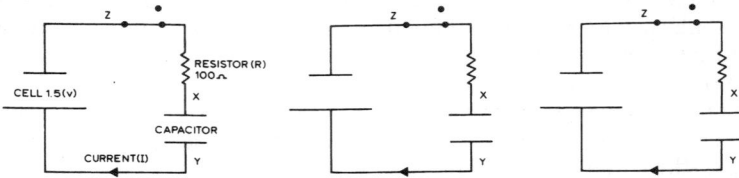

Ohms law V = IR $(I = \frac{V}{R})$

T = 0
Instant of switch being closed
Capacitor begins to charge
Effectively short circuit
Current $(I) = \frac{1.5 (V)}{100 (R)} = 15$ mA
Voltage ZY = 1.5 V
Voltage XY = 0 V
(short circuit)

Voltage across:
Resistor = 1.5 V (max) (ZX)
Capacitor = 0 (min) (XY)
Current through:
Resistor = 15 mA (max)
Capacitor = 15 mA (max)
∴ where V = voltage
and I = current:
Resistor V/I max/max
Capacitor V/I min/max

T+5
Switch recently closed
Capacity charging but not yet fully charged

Current = 7.5 mA
Voltage ZY = 1.5 V
Voltage XY = 0.75 V

Voltage across:
Resistor = 0.75 V (falling) (ZX)
Capacitor = 0.75 V (rising) (XY)
Current through:
Resistor = 7.5 mA (falling)
Capacitor = 7.5mA (falling)

Resistor V/I falling/falling
Capacitor V/I rising/falling

T+10
Some time after closing of switch
Capacitor has charged to full voltage of cell so no further current can flow

Current = 0
Voltage ZY = 1.5 V
Voltage XY = 1.5 V
(effective open circuit)

Voltage across:
Resistor = 0 V (min) (ZX)
Capacitor: 1.5 V (max) (XY)
Current through:
Resistor = 0 mA (min)
Capacitor = 0 mA (min)

Resistor V/I min/min
Capacitor V/I max/min

NOTE: At any given time, the voltage and current for the resistor are either minimum, falling, or maximum together, hence in-phase. The voltage and current for the capacitor are always out of step. The current flows before the voltage rises, so is said to have a phase lead. The voltage has a corresponding phase lag.
   The in-phase current and voltage in the resistor multiply together to produce watts—the resistor heats up—the power is dissipated.
   The out-of-phase current and voltage in the capacitor multiply together to produce volt/amps or wattless power, the capacitor does not heat up, no power is dissipated.

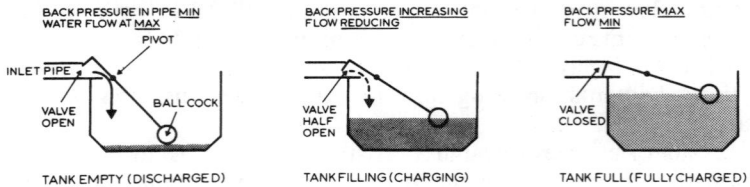

Plumbing analogy of capacity: Flow (I) and back pressure (V) are always out of step at valve.

Figure 7.2 Reactive load

again, current and voltage are always in direct proportion, as increasing the voltage will always increase the current, and are therefore considered to be in phase. Just as with the plumbing system, electrical analogies can be made with the moving systems of loudspeakers.

We shall see in Chapter 10 that the inefficiency of direct radiators is due to the poor loading provided by their coupling to the surrounding air. The loading is reactive, hence the 'power factor' is low, so the transfer from electrical to acoustic energy is poor. Power is measured in watts, which are calculated by multiplying volts by amps. Hence a 220 volt bulb, drawing 1 amp, would yield $220 \times 1 = 220$ watts of heat. On the other hand, if the load is reactive rather than resistive (i.e. if there is a phase lead or phase lag of the voltage or current), watts of heat would not be produced. 220 volts supplying 1 amp through a capacitor would produce no watts (if perfect) so the volts and amps would be multiplied to produce 220 VA or 'volt-amps'. Volt-amps are not representative of power, though the voltage and current must still be supplied. The point is that at any given time, in a reactive circuit, the time of the existence of the maximum voltage is not the same as the time when the maximum current flows. In fact in the previous example, if the reactive circuit exhibited maximum voltage at the time of minimum current, then at that instant no watts would be produced, despite the 220 volt-amps still being drawn. In a resistive circuit, at any time, maximum voltage creates maximum current, and hence maximum watts.

7.3 Importance of phase response

It has long been my opinion that the accuracy of the impulse response is a most important criterion in the performance of any system. In order to create an accurate impulse or transient response, the phase response/phase slope of the system must be as coherent as possible. Only this, together with a smooth 'frequency response' can re-create an accurate impulse. Given the impossible complexity of interaction between the reactive elements of the loudspeaker drivers and the high level, passive crossover components (the inductors and capacitors), I cannot seriously consider such crossovers as contenders in the search for more phase accurate systems. I accept that compensation can sometimes be incorporated in order to present a relatively constant frequency/impedance characteristic at the amplifier terminals, but strange things can still be going on in the crossover/driver interface. All in all, some very strange loads can be presented to amplifiers when coupled to passively crossed-over loudspeakers. The 'wattless power' of phase leads and phase lags can demand enormous current surges from the amplifiers. This cannot be computed on a simple, conventional calculation of power being equal to the resistance multiplied by the square of the current, yet time and again, I hear people saying '100 watts into 8 ohms is about 3.5 amps, so I'll use 5 amp cable.'

The truth is, with reactive loading, 100 watts into a nominal 8 ohms may demand 20 amps under certain loading and drive conditions, which obviously has a bearing on the current carrying/saturation ratings of any inductors used in passive, high level crossover circuitry. It also makes great demands upon the ability of an amplifier, in order to drive some of these absurd loads. An extreme example of this was discussed in Section 6.4 in which transient currents of 60 to 100 amps can be realised into the 0.8 ohm load.

7.4 Amplifier considerations

For many years, the first choice amplifiers for my monitor systems were Crown, as they worked well with the systems on which I used them. I often hear people telling me that one amplifier is better than another because they have made a series of careful listening tests and judged that to be so. These listening tests usually are 'carefully controlled' in one room, with the same programme material and through one set of loudspeakers as a reference standard. They draw up a table in their order of preference and then proclaim to all that amplifier 'F' is definitely superior. It may simply be, though, that amplifier 'F' was more suited to matching the particular complex loading of the loudspeaker which was used. If the test was repeated ten times using ten different loudspeakers and the same series of amplifiers, and if amplifier 'F' was always at the top of the list, then maybe we could be getting close to some more rational generalisation. Even so, it is still quite possible that with a given complex impedance of yet another loudspeaker, amplifier 'F' may run into trouble and amplifier 'D' may prove superior. Rash generalisations that one amplifier, one loudspeaker driver, or one crossover is universally 'the best' are usually out of order.

Amplifier/crossover/loudspeaker combinations have, since around 1980, begun to become increasingly widely commercially available. Sometimes this has allowed the use of designs which could otherwise be problematic. I remember UREI refusing to guarantee their loudspeakers if used with a certain brand of amplifier commonly used in the USA.

The popular amplifier was prone to high frequency instability when driving the UREI 'Time Align' crossover. In the classic 813 and the other members of its range, UREI opted for a passive, high level crossover of the 'Long' design. They circumvented many of the impulse/phase caveats but were still left with the other problem of such crossovers – components in circuit between the amplifier and bass drivers. If the series inductors in the crossover filters feeding the bass drivers have *any* resistance, which they must have, then the damping capability of the drive amplifier is reduced. One ohm of resistance in series with an 8 ohm driver will limit any possible damping factor to around 8:1. This is the case whether the amplifier has a damping factor of 10, 100, or even 1000. The bass cannot be 'punchy' unless the power amplifier has totally authoritative control over the motor system of the loudspeaker. The prime requirement to achieve this at low frequencies is a damping factor somewhere in excess of 40, and a minimum cable/crossover resistance between the amplifier and the bass drivers.

UREI utilised a sensing circuit, terminating on a separate connector on the back of the loudspeaker cabinet, which took a feed from the crossover input, intending that feed to return to a separate control input on their specially designed amplifier. The above system helped to overcome the inherent problems with high level crossovers by taking the amplifier's feedback signal directly from the crossover input, effectively removing the problem of the cable impedance, and at least removing one obstacle to good damping. Certainly in the UK, I rarely saw these amplifiers in use – too many people were still thinking that an amplifier is an amplifier, is an amplifier! They would not pay the price for the expensive, dedicated UREI amplifier. It was to be some years later before fully integrated systems began to be more widely accepted as such, and people got into the habit of buying systems *as* systems.

A word here on MOSFET amplifiers. I have two main concerns. Firstly, MOSFETs are inherently higher in internal resistance than comparable bi-polar transistors. Consequently, when these seemingly ridiculously disproportionate currents are demanded by certain loudspeaker systems, MOSFETs cannot always deliver the goods, unless, that is, the MOSFET amplifier is substantially higher in output power than would be a comparable bi-polar design. Some further limitations on transient ability have resulted from MOSFET designs which have been 'up-rated' merely by the increase in the size of the power supply and the number of output transistors. Paralleling more output devices can result in higher input capacitance to the output stages, which in turn place higher demands on the driver stages. This fact seems to have been neglected in quite a number of designs, where the driver stages have been inadequate in their ability to drive a greater number of output devices after up-rating.

Obviously, this is most apparent on low frequency, high power applications, but, secondly, MOSFETs have inherently higher distortion than their bipolar counterparts. Whilst this is often compensated for on steady state signals by increased negative feedback, under certain drive conditions transients can pass through the system before the feedback systems can respond. There can be time delays in the application of feedback, due to the phase lead, phase lag properties as already discussed, and also group delays, as will be dealt with later in Sections 7.6 and 7.7, into any reactive components such as capacitors or inductors which may be in the feedback circuitry. Usually, the higher the total amount of feedback, the greater the potential problem. As a result of this, transients can sometimes cause the amplifiers to momentarily run wild, with results such as short term instability and transient intermodulation distortion. I believe that this was at the root of the problem with the aforementioned MOSFET amplifier which so disliked being used with the UREI loudspeakers.

Some manufacturers are opting for designs entirely using discrete transistors rather than operational amplifiers, and also using local feedback around individual stages to minimise the distortion of the basic circuit and hence to require a much reduced overall amount of negative feedback. I entirely agree with this principle.

My own preference in amplifiers has long been for DC coupled devices. In these, there are no coupling capacitors and they operate as servo amplifiers. Except for microseconds of group delay, the feedback occurs virtually instantaneously, without any significant phase lead or lag over the range from, say, DC to 100 kHz. With my own design philosophies, wideband amplifiers appear to provide the best overall performance, hence my continuing use of Crowns, amongst others, though they are typical of an entire genre of amplifier. They give *me* the performance which *I* need from a specific philosophy of monitor system design.

7.5 Active and passive crossovers

Notwithstanding UREI's valiant efforts at system design around a passive, high level crossover, I still consider active, low level crossovers and multiple drive amplifiers to be the only truly viable approach to high level, high quality monitoring. There are many reasons for this, some of which are discussed further in Chapter 20, but four reasons in particular are discussed here. These are component tolerance and ageing, power loss and component design problems at high power levels, response slope shaping – particularly in high order (high slope) networks – and the ability to select drive components for audible compatibility, free from restrictions imposed by impedance mismatches or sensitivity discrepancies.

On the first point, component tolerance and ageing, electrolytic capacitors are the major culprits. The non-electrolytic option is not always open, especially when low frequency, low impedance and high power handling are the requirements. These can necessitate large capacitances and high working voltages. Non-electrolytic capacitors may be so physically large that their own inherent inductance may become a hindrance to correct circuit design. These inductances can create undesirable elements in the circuit and cause many hidden problems. It must not be forgotten that the voice coils of the drivers themselves are also electro-mechanical components of the crossover circuits, and that drivers age. Whilst this can usually be dealt with quite simply in active, multi-amplifier systems, their delicate and complex interrelationship with passive, high level crossover components may produce performance drifts which are very difficult to remedy.

The second point is power loss and component design at high power levels. Especially with the higher order 18 dB and 24 dB/octave networks, the increased number of components can produce significant power losses. At high power levels, these losses can produce considerable heating of the components, which can change their value, and hence the carefully chosen parameters can be subject to level-dependent drift. Heating in the voice coils will also change the driver impedances, and once again we move further away from our original design criteria. We have already discussed the problem of finding suitable high value, high voltage capacitors, but inductors can also be a problem. The difficult phase relationship of current and voltage can produce remarkably large transient current surges at high power levels. Designing inductors which will not saturate at these high current levels can be difficult to achieve whilst keeping them sufficiently small to avoid stray leakage inductances. This can once again upset the intended choice of component values. Furthermore, when the inductors are large it becomes increasingly difficult to site them such that their magnetic fields do not interact one with another, causing even further complexities in an already very complex system.

The third restriction is on response shapes. These problems apply to passive crossovers, be they high level or low level. Passive crossovers can of course be used ahead of the amplifiers as with active crossovers, but they are only infrequently used in this way. One advantage of active crossovers, utilising the gain of an amplifier stage, either simply or cascaded, is to provide precise control over the filter shapes. Whilst there is gain still available in the amplifier stage, feedback can be used to provide contours which could not be achieved within the conventional circuitry of passive components.

The fourth drawback of the passive, high level approach is in the choice of drive units. Should it be decided on the grounds of subjective audible characteristics that mid-range unit 'A' with a sensitivity of 92 dB for 1 watt at 1 metre matched very well sonically with bass driver 'B' of 97 dB sensitivity, a problem arises. Accepting the undesirability of introducing components between the bass driver and the power amplifier, and especially considering the far from constant impedance of the voice coil, the introduction of an accurate 5 dB pad into the circuit of the bass driver would be neither desirable nor easily realisable. The only option would be to transformer couple the mid-range driver with a 5 dB step up. Once again, at high power levels, such a transformer, quite apart from weight, size, and expense, could introduce even more phase shifts and inaccuracies into the system. It would not be desirable, and would be difficult to put into practice.

The use of individual power amplifiers provides almost ideal, low impedance, constant voltage signal sources, which damp resonances, are tolerant of widely varying impedance irregularities, and provide a stable drive which can be altered easily should the need arise. They also accommodate the use of drivers of very differing sensitivities, compensation being made with a simple adjustment of a gain control. Once again, back to the UREIs. They overcame the power level restraint of problem two with the use of very high sensitivity drive units, but this *must* restrict the choice in problem four, that is, the choices of drivers on purely sonic grounds. I have been in studios which have replaced the original Eminence bass units with Gauss and JBL drivers, blissfully unaware of the consequence of a sensitivity loss of at least 3 or 4 dB, and even up to 7 dB at the lower frequencies. Without a great deal of experience and knowledge, the UREI really should not be disturbed.

Passive, high level systems are not in my opinion suited to the wider frequency range, higher dynamic range, faster transients and more accurate phase/impulse response of modern digital recording; and especially of computer-generated signals.

As with so many other things, a law of diminishing returns applies to monitor system improvements. More amplifiers are more expensive, but I believe that it is the price which must be paid for more accurate monitoring. In the late 1970s, I gave up the high level, passive option, and whilst keeping abreast of developments in the technology, I still see no likelihood of changing from my decision to opt for the active low level crossovers. Another advantage of the use of low level active filters is the ease of providing adequate transient overload headroom. To achieve a 20 dB voltage margin over a 1.23 volt line level is relatively simple. In terms of power, 20 dB headroom represents a hundred-fold increase. Were a high level passive crossover to be designed with a 20 dB headroom margin, then if the system was rated at 100 watts continuous, the reactive components would be required to saturate at equivalent current and voltage ratings of not below 100×100 watts – 10 000 watts! The achievement of the best transient performance at high levels more or less dictates the choice of the active, low level option, unless unbelievably complex loads can be tolerated by huge amplifiers.

7.6 Slopes and shapes

The next question to be faced is that of the slope, together with the shape of the curve at the turnover point. The most commonly used rates of roll-off are 6, 12, 18 and 24 dB/octave. This is no arbitrary choice of figures; conventional filters, be they electrical, mechanical or electro-mechanical, tend to produce slopes in multiples of 6 dB per octave. Cascading two simple 6 dB/octave filters produces a 12 dB/octave filter, three produce 18 dB/octave and so forth. Unfortunately, with the exception of the 6 dB/octave, or first order filter, they not only create a roll-off, they play havoc with the phase response. The only conventional crossover filters whose outputs can recombine to produce a square wave are the 6 dB/octave filters. The outputs from the high-pass and low-pass sections can be electrically recombined to produce an exact replica of the input signal (Fig. 7.3), as the 3 dB hump at the crossover point is exactly cancelled by the 3 dB loss due to the +45° phase shifts.

Unfortunately, a slope of 6 dB per octave is too shallow for most monitoring purposes, and not only from the point of view of the prevention of low frequencies from entering the tweeters. The individual drive units would need to have excellent response characteristics for two or three octaves above and below the crossover frequency, as their output would only be around 15 dB down even two octaves away from the crossover point; assuming each filter section was 3 dB down at the crossover point itself. Any irregularities in their individual responses within two octaves of the crossover frequency would be clearly audible. Furthermore, with such a range of audible overlap, two octaves either side of the crossover frequency, that would mean four octaves over which two, dissimilar, spacially separated drivers would be significantly contributing to the audible output of the system, losing any hope of achieving a point source. This would inevitably lead to lobing of the polar pattern and time smearing of impulses. Anyhow, drivers capable of responding three octaves beyond the desired crossover point would be difficult to find when designing for high power levels. I do sometimes use 6 dB/octave filters to bring in a tweeter at high frequencies, say 5 or 6 kHz, but this is usually only to compensate for the mechanical 6 dB/octave roll-off above such frequencies, characteristic of certain mid-range units. In the mainstream of things, however, I do not consider them practical propositions.

Second order filters of 12 dB/octave are very widely used in commercial crossovers, the slope becoming sufficiently steep to reduce the problems of driver overlap. The cost and power loss are also acceptable in the high level, passive versions. Unfortunately, however, as discussed further in Chapter 13, electrically they just will not sum (Fig. 7.4). Connecting the outputs of the high-pass and low-pass sections in phase, a dip is produced in the amplitude response graph at the crossover point. As the frequencies approach that crossover point, phase shifts in the filter sections cause one output to develop a phase lead of 90°, whilst the other section develops a phase lag of 90°. The differential between +90° and −90° is 180°, or anti-phase, which is the cause of the cancellation at the crossover point. The common 'fix' for this is to invert the phase of one of the drivers to achieve an in-phase, summed output at the crossover point, removing the amplitude dip and straightening out the 'frequency response'. On steady state signals this is fine, as away from the crossover point (where the drivers are once again out of phase), due to the steepness of the filter slopes, only one driver will be contributing to the total output. As one

VOLTS

+

−

DC INPUT
STEP

LOW FREQUENCY
OUTPUT

+

HIGH FREQUENCY
OUTPUT

=

SUMMED
OUTPUT

TIME →

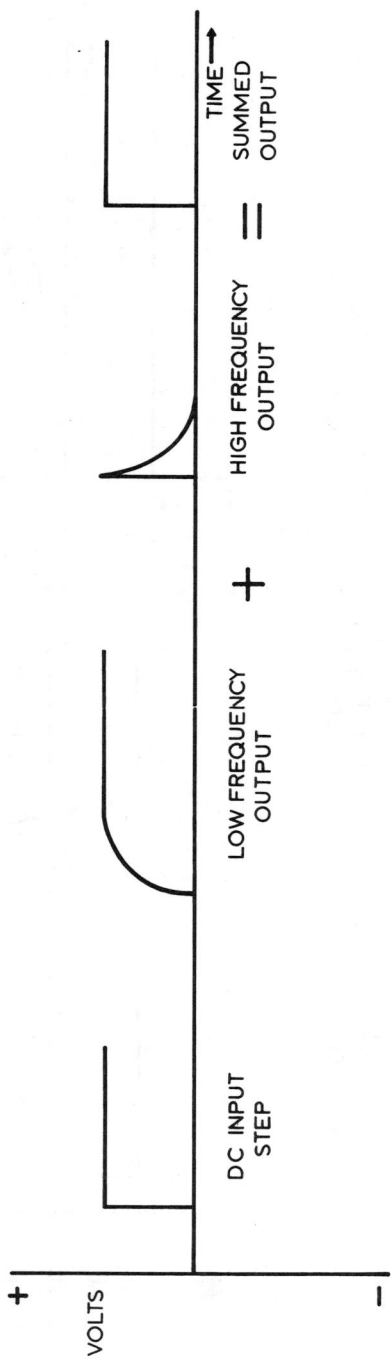

Figure 7.3 6 dB/octave crossover impulse (step function) summing

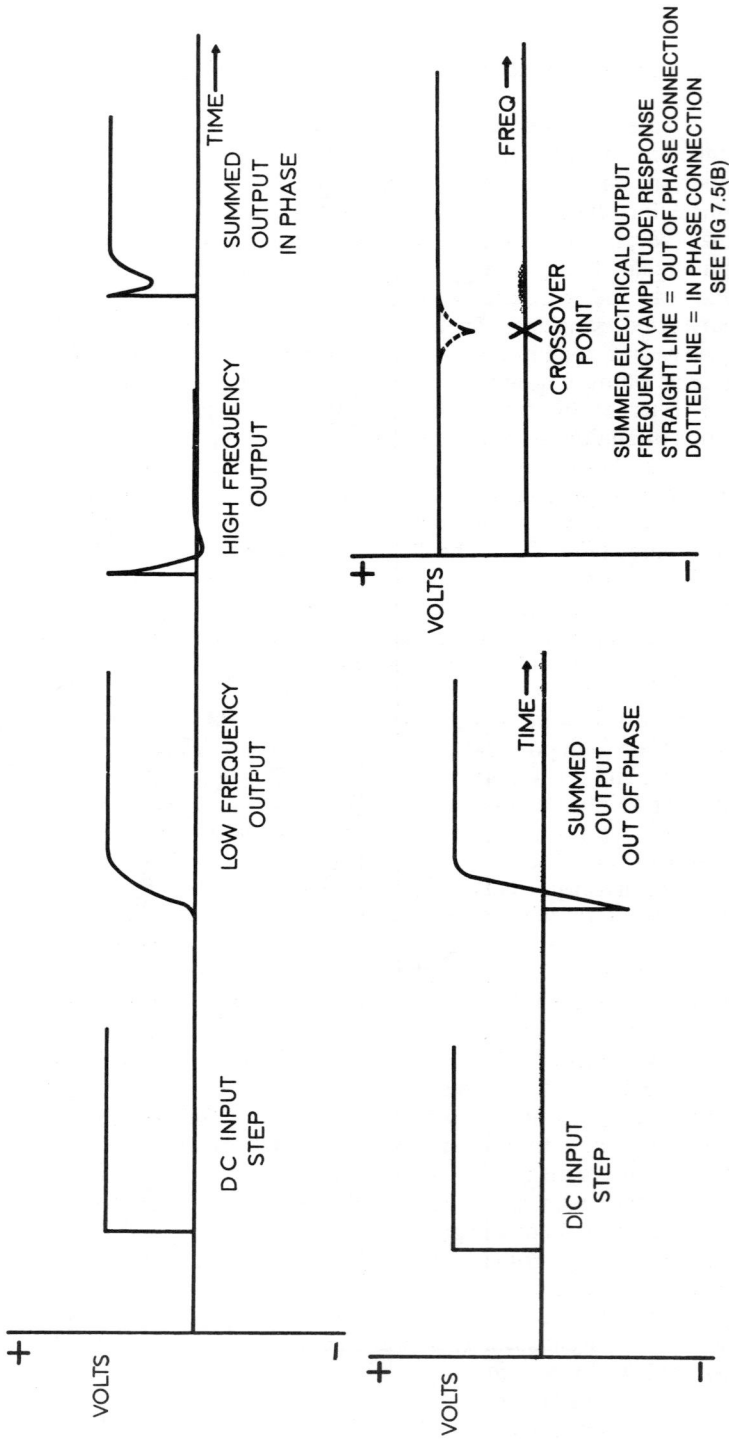

VOLTS

DC INPUT
STEP

LOW FREQUENCY
OUTPUT

HIGH FREQUENCY
OUTPUT

SUMMED
OUTPUT
IN PHASE

TIME →

VOLTS

D|C INPUT
STEP

SUMMED
OUTPUT
OUT OF PHASE

TIME →

VOLTS

CROSSOVER
POINT

FREQ →

SUMMED ELECTRICAL OUTPUT
FREQUENCY (AMPLITUDE) RESPONSE
STRAIGHT LINE = OUT OF PHASE CONNECTION
DOTTED LINE = IN PHASE CONNECTION
SEE FIG 7.5(B)

Figure 7.4 Typical electrical impulse (step) responses for 12 dB/octave crossover. Note that in-phase, or out-of-phase, the output is not a true replication of the input

driver cannot be out of phase with itself, this is usually deemed to be acceptable.

Unfortunately, music is rarely steady state – it is full of crashes, bangs and sharp leading edge transients. A uniform, positive-going impulse (and remember, an impulse contains all frequencies simultaneously) if applied to the input will cause the bass driver to move forward and the treble driver to move backward. This clearly cannot be construed as being representative of the input signal. The choice is clear, either to have phase coherent transients with a dip in the frequency response, or a 'flat' frequency response with phase distorted transients! There effectively is no choice: it is the proverbial choice between a rock and a hard place – neither is desirable, they are both wrong. This point is now being brought home rapidly with the availability of phase coherent, digitally recorded material, free from the inherent phase distortions of analogue recording techniques.

Before somebody says 'filler drivers' – yes, there is a technique for overcoming these problems. A system is connected in phase with the response dip being 'filled in' with a filler driver, contributing the missing output at the crossover point (Fig. 7.5). The filler driver must of course be fed from a suitable band-pass filter, and its audible characteristics must be carefully chosen to match the other drivers in the system. On-axis response is accurate, but the spacial displacement of the drive units create off-axis problems. A four-way system using this technique would become a seven-way, with seven crossover sections – the four main outputs along with three 'fillers' at the crossover points. In other words, hyper complexity! Anyhow, 12 dB/octave is still marginal on high power systems, both in terms of out-of-range protection of the drivers, and the necessity for smooth drive unit responses either side of the crossover points. Personally I reject 12 dB/octave crossovers as a practical solution, and rarely, if ever, have heard them to sound natural.

With third order filters, 18 dB/octave, the slope is generally adequate in terms of both driver protection and overlap. The Butterworth, or maximally flat filter design, frequently used with third order filters, produces slopes which are 3 dB down at the crossover points. The voltage outputs are in phase quadrature (90° out) at all frequencies; in actual fact they are 270° out of phase, being shifted by 135° in each direction, but in steady state terms +90° or -270° are effectively one and the same thing. As the filter slopes become steeper, the areas of overlap become much narrower, so phase discrepancies at the crossover points become of less overall importance, being confined to only narrow bands of frequencies. As we saw earlier with the inductive (reactive) machinery, power factor (phase) problems do not always mean that summing voltages mean summing power. To achieve a voltage sum at the electrical outputs of the crossover does not always guarantee an acoustic power sum from the loudspeakers.

Another summing problem with higher order filters are the group delays which exist in each filter section. It takes a finite time for the signals to pass through a filter, a time which is a function of the filter slope and the upper frequency limit. Because of this property, we have a phase preference on impulses which do not apply to steady state signals. In-phase or out-of-phase connection of third order crossover outputs render little difference to steady state signals, +90° or -270°. On transient signals, however, the out-of-phase connection is usually preferred as the group delay between the

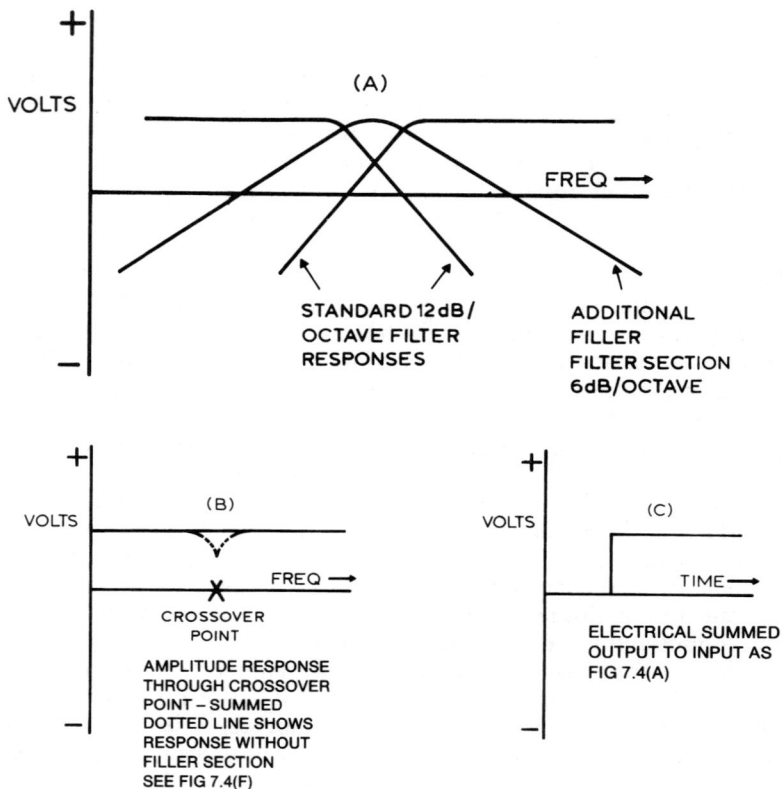

Figure 7.5 Typical 12 dB/octave crossover with additional 'filler driver' filter section. All filter section outputs to be connected in-phase

filter section outputs is much closer at the crossover point than the in-phase delay. 18 dB/octave crossovers are popular and effective. They are capable of excellent audible characteristics, and the Butterworth slope shapes are probably optimum in the smooth transfer of sound from one driver to another. We still, however, have the problem of the outputs being in phase quadrature – 90° out – with the attendant implications for the impulse/ phase/transient response.

Fourth order, 24 dB/octave crossovers go full circle: 360° phase shifts. Effectively, in steady state terms, 360° = 0° = in phase! In so many ways they seem ideal, but I had initially disliked them under certain circumstances on the grounds that the transfer from one driver to another could be too abrupt, audible quite clearly on certain demanding signals, such as sections of strings rising through the crossover point from a paper cone to a metal diaphragmed driver. The maximally flat Butterworth 24 dB/octave filters, however, do not sum to unity directly at the crossover outputs. Although each filter output is in phase with the input, the group delays associated with the individual filter sections have the effect of becoming a little out of step at the crossover points. The Linkwitz–Riley fourth order filters are produced from cascaded second order Butterworths and are also in phase in all frequency bands. The slopes are 6 dB down at the crossover points, so there is a corresponding 3 dB dip in the power response. Whilst they thus do not power sum to exact unity, the notches in the overall frequency response are so narrow as to be virtually inaudible on normal musical programme. For me, they provide the best practical working solution for most of my designs.

7.7 Group delays

As mentioned in the previous section, group delays are the results of the finite time which a signal takes to pass through any filter sections. There are two ways of addressing this problem. One way is to use digital delays on the crossover outputs, delaying the faster outputs to come in line with the slower, more delayed outputs. A second way is to physically realign the drive units with respect to their position in front of, or behind, the baffle. There are two drawbacks to this second approach. Firstly, causing drivers to be recessed or projected from the baffle plane can cause diffractive discontinuities, which break up the simple, flat plane of the baffle; axial response anomalies can also result from such uneven surfaces. Secondly, due to the loudspeaker cone materials having finite speeds of sound transmission, together with propagation delays relating to the reactive couplings to the air, and also any phase dispersion which may exist in horns, no specific point can be regarded as the frequency-independent centre of propagation of the sound. The phantom propagation sources can move backwards and forwards with frequency. One of the advantages of the digital delay approach is that it can be used to optimally compensate for driver misalignment, without repositioning them with respect to the front baffle, and thus avoiding the diffraction problems. The delays can be given compromise settings, to take the best, overall, mean values to compensate for group delays and driver alignment. Although this can never be absolutely precise, especially due to the impulse smearing within the drivers themselves, it does seem to be the best way to overcome the problems using relatively conventional technology.

There is, however, a further consideration about using digital crossovers and/or delays. I do not personally consider a 20 kHz bandwidth sufficient, especially due to the fact that the near future is likely to bring new digital recording media with 40, 50 or 100 kHz bandwidth. As modern analogue tape recorders can respond to the 30 kHz region, I would push strongly for any digital devices in the monitor chain to be of the widest bandwidth technically and financially feasible. 20 kHz brick wall filters have no future on monitor systems.

Taking account of the delays in the drivers is very important. Sound does not emanate from them immediately upon the application of an input signal. There are finite propagation delay times inherent in any mechanical system: a certain amount of springiness and elasticity. Even the molecules are not totally rigid – you get a shunting effect rather like the clatter of railway wagons as the train starts off and takes up the slack. There is a lot of free space in molecules and atoms. If the outer electron orbits of an atom were the size of Wembley Stadium, the total 'solid' matter in that atom would be about the size of a football. There is a certain compressibility in that space, which is why neutron stars can weigh thousands of tons per cubic inch. They are very, very dense. Gravity of enormous strength has squeezed out most of the space to render them 'solid' in a rather more true sense of the word. It is these spaces that X-rays use as they travel through the body. Nothing on this Earth is truly solid, and nothing happens instantaneously!

7.8 Choice of optimum turnover shapes

Returning to the problem of smoothly and gradually crossing over from one driver to another, I have looked at many filter shapes. The Bessel filters are very abrupt, peaking slightly before they steeply roll off. These can make the problem of sound transfer from driver to driver very noticeable. The Butterworth and Linkwitz–Riley filters are better in these terms, but still produce certain audible anomalies. In solving my specific problem of the abrupt transfer from driver to driver, I am indebted to Colin Clarke at Acoustic Services Ltd. Utilising state variable filters, he modified a Linkwitz–Riley design slugging the entry into the slope, to replicate the initial shape of a Butterworth, 18 dB/octave filter. The design gradually assumes the 24 dB/octave slope, but initially it is remarkably similar to the third order Butterworth (Fig. 7.6).

An advantage of state variable filters is that only one slope is generated, being mirror imaged to provide its counterpart on the adjacent filter section. Conventional filters require very tight component tolerances and long-term stability, as any drift or component error can make the overlapping filter sections misalign, causing dips or peaks in the summed response. As the overlapping filter sections are derived from the same filter slope in state variable designs, any drift in one section will automatically be tracked by the adjacent section, thus maintaining a uniform overall response with excellent long-term stability.

There are commercial crossovers available which incorporate variable boost controls to attempt to compensate for crossover point response dips. Other units use variable analogue delay control – though not in the sense of broadband digital delays. Both of these systems are intended to compensate for mislocated drivers, especially in the sound reinforcement field, where placement of widely spaced drivers can never be ideal for accurate phase/impulse responses. The use of any such systems of adjustment must be carefully auditioned in every specific instance of use before they can be deemed to be beneficial or otherwise.

7.9 Electromechanical considerations

Another element of the monitor system, which strictly speaking is in the crossover domain, is the physical, three-dimensional placement of the individual drive units in the system. Effectively, there are two options: co-axial systems, or discrete drive units. The loudspeakers themselves are the mechanical components of the crossovers, both in terms of their relative positions and their inherent electro-acoustic roll-offs. Should a crossover have a 12 dB/octave roll-off at, say, 500 Hz, and be connected to a bass driver which begins naturally to roll-off at 6 dB/octave above 500 Hz, then

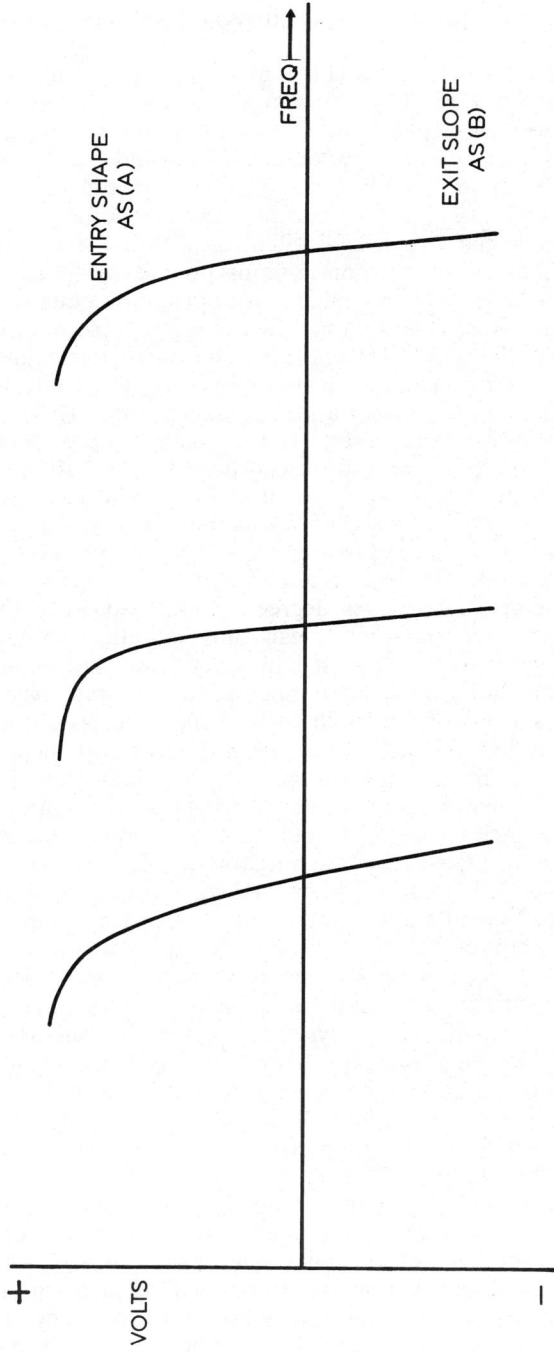

Figure 7.6 Comparison of slope entry shapes and final slope rates of Butterworth 18 dB/octave (a), standard Linkwitz–Riley 24 dB/octave (b) and Clarke 24 dB/octave hybrid (c)

it is no use crossing over into a mid-range driver which begins to rise below 500 Hz at 3 dB/octave.

The resultant would be an 18 dB (12 + 6)/octave crossover on one side of the crossover, and a 9 dB (12 − 3)/octave crossover on the other side. This is another reason, except in special instances, for drivers having to maintain a smooth, relatively flat response, well beyond the desired crossover point. What is more, erratic, high Q peaks and dips can not only disrupt the crossover slopes, they can also introduce responses which are of a non-minimum phase character, which will affect transient accuracy.

Responses which are said to be non-minimum phase are those for which restoration to an even response by means of appropriate equalisation will compensate in terms of amplitude, but will not restore the original phase relationship. Once again, it is of little concern to steady state signals, but it is most destructive to transients. Such non-minimum phase effects can be produced in the off-axis response of a loudspeaker system by the physical, wide separation of drive units. Here, the co-axial designs do have an advantage in that they can be made to be excellent representations of three-dimensional point sources. In the horizontal and vertical planes, the drivers of a co-axial share common axes. By judicious use of delay, the drivers can be synchronised to create the impression of sharing a common point on their front to back, time axis. Horizontal and vertical directivity is maintained by such systems to a greater degree of phase accuracy than their discretely located counterparts, which can only maintain such accuracy over a relatively narrow angle. Almost inevitably, however, for high level applications, co-axial units must be augmented by separate bass drivers. The reasons for this are twofold. It would be difficult to produce a single, say, 15″ driver which had the desired characteristics for both clear low-mid *and* deepest bass at high levels; and, secondly, the very high level bass excursions, if the LF cone is being used as a mid-frequency horn (as in the Tannoy Dual Concentric designs), would tend to modulate those higher frequencies, producing inter-modulation distortion. If the crossover frequency is kept below 200 Hz or so, where the wavelengths are around 5 ft long, the coincident source should not be too disrupted by the physical separation of any additional bass drivers from the common axes of the main co-axial/Dual Concentric units. Such a system is used by UREI.

The other main restriction caused by the co-axial option, however, is the limited choice of units available to a system designer. The 'mix and match' option is not available, which on the higher power systems is sometimes a necessity, as no single manufacturer always produces the entire range of drivers desired by a system designer. Back to the point of the non co-location below 200 Hz. I am not prepared to throw myself into the 'bass is omni-directional below 300 Hz' camp. Strong evidence is now available showing that whilst this may be true for steady state signals, on transient sounds the low frequencies are both directional *and* critical on arrival times. This places a great question mark over the 'common sub-woofer' philosophy, especially where that sub-woofer is located in some out of the way place. I believe that if sub-woofer systems are used, they should be located as near as possible to their satellites, and be in stereo. It may sound good with them remote and in mono, but I doubt that it will be accurate!

Furthermore, many so-called 'linear phase' loudspeaker systems have

Path lengths unequal
accurate impulse arrival
times cannot be
maintained with
horizontal movement

◄——— High ———►
◄——— Mid ———►
◄——— Low ———►

Path lengths equal
impulse response can
be maintained with
horizontal movement

Drivers without
common vertical axis

Drivers sharing
vertical axis

X   Y   Z   X   Y   Z

Relative distances to the three drivers vary
between points X, Y and Z in the horizontal
plane. Impulse arrival times set to be in
unison at 'Y' will show lead of top and lag of
mid at 'X' and *vice-versa* at 'Z'

Distances to the three drivers remain the
same when moving across horizontally in
front of the system. Impulse arrival times
can be set and will remain in unison.

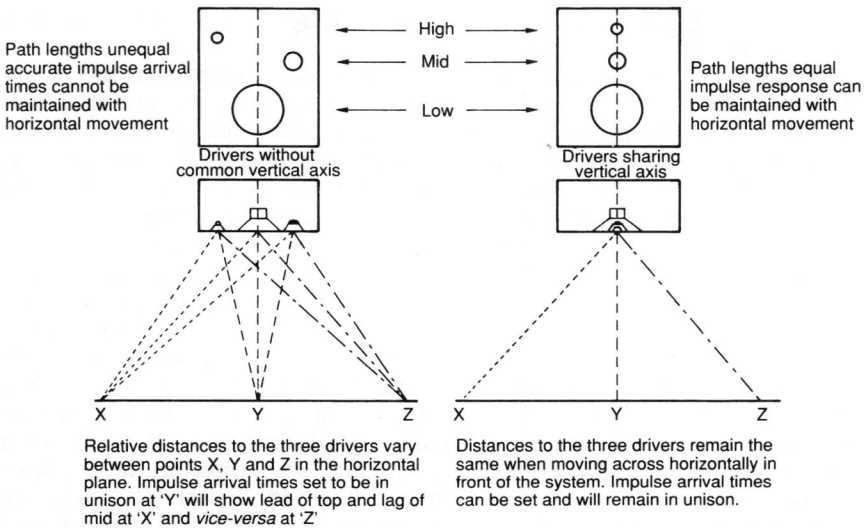

Figure 7.7 Effect of shared vertical driver axis on impulse arrival times. *Note:* Neither system can maintain accurate impulse with vertical movement of listening position. Only dual-concentric units can maintain response in both planes

fallen short of the mark because of undue attention to the prevention of steady state directivity lobing problems, rather than the acoustic reconstitution of the impulse response at the listening position. 'Linear Phase' is also a misnomer. What is actually meant is minimum phase – 'linear phase' actually means a time delay. You get a 'linear phase shift' when you walk, say, a foot nearer to or further away from a sound source.

Systems which use discrete, physically non co-located drive units can only really achieve such impulse accuracy over a more limited directivity field than can the co-axial systems, especially as the wavelengths shorten as the frequency rises. Given that head height in a room is relatively constant, the main variation in the position of listeners' ears is in the lateral plane as they move around the room. Locating the individual drivers, or pairs, around a common vertical axis will minimise these problems, though lobing and non-minimum phase cancellation will be more apparent than from a concentric design (Fig. 7.7). It does not quite require the fixing of one's head in a clamp to hear phase accurate monitoring from non co-located drive units, but in this area concentric/co-axial systems do have an advantage. The option of separately mounted drivers has the advantages that the system designer is able to make highly specific choices of drivers in terms of audible matching of characteristics, power handling, sensitivity, and the ability to split the system in as many ways as may be deemed necessary. It also offers a potentially higher, total system, acoustic output. Further aspects of driver mounting are discussed in Chapters 19 and 20.

7.10 Active control

A means of solving all of these problems is on its way. It is complicated, but it does offer enormous hope for the near future. Although the individual filter sections of a practical crossover may be minimum phase in themselves, they will not sum in a minimum phase manner. This is largely because the group delays in each section increase in proportion to the steepness of the filter slope and in inverse proportion to the frequency, and so are not all the same. In the centre section of the passband, the group delay is more or less constant, but towards the roll-off point, the delay shifts. Any attempt to use, say, a digital delay line to bring the sections into line will succeed in a general manner, certainly improving things, but will not compensate for the changing delays towards the crossover points.

Corrections to this cannot be made by conventional means, as where a 'phase lead' was incurred, correction would need to be applied before the signal appeared. As no effect can be experienced before its cause, this is clearly out of the question by all conventional means. The answer lies in 'Acausal Adaptive Digital Signal Processing and Impulse Reconstitution', using a system of 'Multiple Point, Spectral, Phase, and Transient Equalisation'. It does offer the prospect of a reasonably large area in the room, with near-perfect audio reproduction; see Chapter 18. However, in my own opinion, as mentioned earlier with reference to digital electronics in crossovers, I do not think that a 20 kHz digital bandwidth is sufficient for highest fidelity performance. Perhaps in the near future, this problem will be addressed.

Hopefully, some of the points discussed in this chapter will prepare the reader more thoroughly for some of the discussions in the chapters to follow.

Chapter 8

Loudspeaker cables

In late 1990, I made a presentation to the Institute of Acoustics 'Reproduced Sound' conference on the subject of esoteric loudspeaker cabling. The main purpose of the discussion document was to try to test the temperature of the water with regard to the general attitude to esoteric cabling in the recording, broadcast, film, and the 'professional' industry overall. During the presentation, a DAT recording was played which contained the output from a high resolution differential amplifier which had been detecting the signal difference across six feet of thirteen different loudspeaker cables, passing music from a conventional hi-fi amplifier to a typical 3-way loud-speaker system. I think that it is safe to say that there was general surprise at the relative difference between the cables, especially when measured over such short lengths. The test cables ranged from 'bell wire' through mains cables and general purpose loudspeaker cables, to F.M. Acoustics 'Force-lines' multi-strand, and JDF screened, polarised, 8-core LC-OFC (Linear Crystal, Oxygen-Free Copper) costing around £2500 for a 5 metre pair of cables.

Despite the wide range of variations in the losses within the six-foot lengths, the consensus from the one hundred or so delegates present was that the system-specificity of many of the differences dictated that hard and fast orders of best to worst were almost impossible to achieve. In other words, should a given loudspeaker present a difficult load to an amplifier, especially under dynamic drive conditions, then cable 'A' may well sonically outperform cable 'B'. When presented with a more benign load, it is quite conceivable that the two cables would be sonically indistinguishable. In a third instance, should that benign load be driven by two different amplifiers 'C' and 'D', then were amplifier 'C' to be more prone to RF instability or to dislike capacitive loading, then possibly cable 'A', which was deemed to be sonically superior with the difficult loudspeaker load, could well be more prone to RF problems in the amplifier feedback circuits, and hence cable 'B' may be deemed 'sweeter' with this amplifier. Once again, with amplifier 'D', the two cables may well be deemed to be sonically identical. In these instances, neither cables 'A' nor 'B' could be considered generally superior in all cases. The BBC for example were said to have 'No policy' on esoteric cabling for their control monitoring.

As a studio designer, I have no personal favourite amplifiers, no favourite loudspeakers, and no favourite crossovers or cables. I do, however, have favourite systems for specific sets of circumstances such as the type of music or the rooms into which such systems will be installed. It is the overall balance of the system which is the critical factor, not just the individual performance of any individual component part.

Over the past fifteen years or so, the pages of the hi-fi press have been awash with points of view and advertisements about cabling. It was with a thorough search through these publications, together with a digest of papers in *Audio Engineering Society* and similar journals, that Keith Holland and I began our attempts to distil the essence of the principles from the tangled web of different concepts and opinions. We then embarked on an attempt to measure differences under actual working conditions. The recent advances of integrated circuit technology had enabled differential amplifier chips to be made with around 110 dB of common-mode rejection, which in turn enabled the construction of test set-ups with previously unattainable ability to detect differences over short lengths of cable. We commissioned Ian Piper of ICP Electronics to construct a viable comparator, and were indebted to him for help during our initial experiments.

I was in Granada, Spain, preparing the groundwork for my IOA presentation when the first differential amplifier tests were made by Keith Holland and Ian Piper. When these tests were conducted, Keith had been very sceptical about many of the claims for esoteric cabling, and had been quite dismissive of the whole affair. I was therefore most surprised when, upon my return from Granada, he telephoned me in a somewhat excited state after the initial test run, which had been carried out in my absence. I asked him to send me a DAT of the differential amplifier output, to which he replied that he did not have a DAT machine handy but he had made a cassette. 'A cassette?' I said 'Will that be sufficiently subtle for me to hear the differences?' 'Subtle?' he replied. 'I could play it to you over the telephone and you'd still be able to tell one cable from another.' From a sceptic, that was something! The differences heard were not only in level, but also in spectral balance and 'tightness' of sound.

I must emphasise here, however, that the differences measured were differences in the losses along the lengths of the cables. Only by means of the high common-mode rejection of the differential amplifier could these differences in losses be detected. Although the losses were clearly different from each other, they were way down on the actual signal levels being delivered to the loudspeakers. I must also stress that we were not measuring or perceiving the acoustical output from the loudspeakers. No suggestion is being made here that any massive sonic differences exist between the cables when used in systems in general. What the tests did clearly show, however, was that, unquestionably, differences in performance do exist. Whilst this may appear to be a ridiculously obvious statement to most hi-fi buffs, somewhat akin to proving that fish swim, within the professional industry a very great number of entirely reputable people have shown considerable reluctance in accepting firstly that there was any value in esoteric cabling when compared to cables used in accepted good engineering practice; and, secondly, that any subtle benefits which may exist would be cost-effective or worthwhile in practical situations.

Furthermore, much of the professional industry has perceived some cabling claims as being so contradictory, absurd, or confused, that they have been reluctant to stake their professional reputation on such seemingly nebulous concepts. It was largely to address this problem that the IOA conference discussion was initiated – were there any clear cut cabling types and/or practices which could be deemed to be of universal benefit when applied to a system already constructed to accepted good engineering practice?

Amongst the IOA delegates were a few invited people from the world of hi-fi users and enthusiasts, there to provide the function of observers and to make their cases for the clear-cut advantages which they perceived from the use of special cabling. David Swift, a great hi-fi enthusiast, came along from his home in Whitehaven, wielding a length of multi-thousand strand litz wire. He was virtually prepared to stake his all on the fact that it did benefit his system. As things turned out, I now believe him, but it soon became apparent that his system requirements were of the more demanding nature in terms of cable performance: 30 ft cable runs into difficult loads. The same wire in different lengths on different systems would conceivably show no benefit over conventional wiring. Once again, system specificity was evident and it was this lack of the existence of across-the-board applications, over and above accepted 'good engineering practice', which had led to the continuing scepticism of the 'professional' industries.

The following sections are taken from the discussion document which I presented to the Institute of Acoustics conference, followed in turn by a round-up of general conclusions, drawn from the consensus from the discussions, and from the experiments conducted by Holland, Piper and myself.

Esoteric loudspeaker cables – do they really deliver their promises?

8.1 Abstract

Two thousand pounds or more is no longer an unduly rare amount of money to be spent on a pair of loudspeaker cables; yet many people in professional audio circles are strongly resisting being drawn into any consensus that such cables make convincingly significant audible differences to a system. Many people cite secondary or tertiary effects of the cable installation procedures as the main source of subjective improvement: general care and attention to detail such as clean, tight connectors, careful cable routing and well soldered joints being typical examples put forward for the case against any directly attributable cable benefits. 'Show me conclusive evidence in terms of Ohm's Law' is another often heard demand from the non-believers. The over-hype by the hi-fi magazines has probably done nothing to persuade many of the more conservative professionals that there is any significant substance in the case for such esoteric wonders; yet all along the line, too many people, who I personally respect, have been convinced of the benefits of certain special cables for me to have ever been inclined to dismiss their benefits out of hand.

Debate continues on the subjects of oxygen-free and linear crystal copper versus conventional copper; co-axial versus twin cables; optimum strand thickness with relation to skin effects; silver solder versus tin/lead solder used in terminations; insulated versus uninsulated strands in any one bunch; insulating sheath materials; directionality in terms of one specific end to the amplifiers and the other specifically to the loudspeaker, in other words, non-reversible cables; shielding from external magnetic fields, both LF and RF; general transmission-line properties; and many other controversial areas of discord. Such discussions may be working wonders for the sales figures of the hi-fi magazines, but for the industry in general, it can be doing its credibility no good whatsoever.

8.2 General good practice

Obviously, any reasonable cable is likely to sound better when compared to an excessive length of poorly terminated bell flex. As a general rule, good quality, well terminated, 60 amp, multi-strand, conventional copper wire would seem to be adequate for most purposes. At about £1 ($1.50) per metre for a pair of conductors, the price would also seem realistic for most applications. Minimum cable lengths between the amplifier and loudspeaker driver is a virtually self-evident rule of thumb, as obviously at the extreme, zero length can cause zero effect. Conversely, 8 ohms in an absurdly long length of cable will have untold negative repercussions on the performance of a 4 ohm system.

From our findings to date, the noticeability or otherwise of cable deficiencies is least on electronically crossed over, multi-amplified systems, and greatest on systems with high level passive crossovers, particularly those displaying tortuous dynamic load impedances. An extreme case of the latter is the Hidley/Kinoshita system whose crossover input impedance drops to around 0.8 ohms at certain frequencies. Given the system power rating of 1000 watts, on complex musical drive signals it is not unfamiliar to see transient currents in excess of 100 amps when the systems are driven at high SPLs via their JDF 3200 watt amplifiers. With these systems, JDF supply oxygen-free cables, with DC polarised outer screen, directional conductors and overall 'one specific end to the amplifier' directionality. I mention these systems as, due to their extreme demands, they have proved to be useful test beds for the highlighting of more general trends.

The amplifier to crossover cable would seem to be more critical than the cables from the crossover to the loudspeakers. The amplifier/crossover cable should be as short as possible, as it is in the crossover where the highly complex dynamic loads are realised. If any significant distance exists between the amplifier and the loudspeaker, then the crossover should be brought as near to the amplifier as possible, in order to minimise the distance over which the cables are subject to highly reactive loading. It is also over this length of wire that the complex back-e.m.f.s from the entire system will impose themselves on the feedback circuits of the drive amplifiers. Whatever impedance or irregularity occurs in this length of cable will form the top half of a potential divider network, the lower half being the output impedance of the amplifier, across which the overall feedback circuits derive their error signal. Any RF or other spurii which may superimpose themselves on the crossover/amplifier cable, including any non-linear conductivity (as suggested may be caused by inter-crystal boundaries in the conventional copper cabling), will again modify the feedback signals. I would suggest that some of the benefit attributed by some audiophiles to amplifiers without negative feedback may be partly due to their general immunity from the above effects.

If the benefits of good quality short cables are to be realised, however, the potential for non-linear conduction in this area is greatly exacerbated by the other inter-metal contacts which could include a tinned transistor leg soldered to a copper printed circuit track, in turn soldered (tin/lead) to a copper wire, which may be crimped or soldered to a brass eyelet tag, clamped via a steel serrated washer to a brass terminal, in turn connected to a chrome-plated banana plug or spade connector – and that is only at the amplifier end. The chrome-plated spring connectors on the loudspeaker chassis are notorious for suffering oxide build-up over a matter of only a few months. In fact, chrome retains its shine by virtue of a thin film of oxide which forms on its surface immediately upon its contact with air. The effect of a short length of adequately current rated, well terminated good quality copper conductor would appear to be small when compared to the other potentially non-linear conductors in the circuit. There are people who say that ABX testing of such sensitive nuances via a switch or relay system is invalidated by the introduction of a switch contact into the circuit. I cannot find any justification of this neurosis either by experimental measurement, by listening, or by intellectual reasoning, when the circuit contains such interconnections as described above.

8.3 Conductor geometry and skin effect

The overall cross-sectional area of the conductors must be adequately capable of passing the highest dynamic currents likely to occur on any given system, without any instantaneous temperature-induced resistance rises sufficient to be detected audibly. Cross-section will be a function of overall length and inherent resistance per linear unit. The individual number of strands which form that total cross-sectional area are the subject of heated debate. Once again, the problem seems to be aggravated by difficult dynamic loads and minimalised by active crossover/multi-amplifier drive systems. In the latter, in a typical four-way system, it is unlikely that any cable would be handling more than four octaves. In a passively crossed over system, eleven octaves can make more stringent demands on the transmission line linearity of a pair of loudspeaker cables. Indeed, as an extension of this principle, even in a passive, high level system, it is only the amplifier/crossover cable which carries the full, wide-band programme.

Some manufacturers sing the praises of cables composed of hundreds of hair-like strands in order to maximise the ratio of surface area to cross-section – the skin area – whilst others claim that this approach maximises the potential for surface corrosion and inter-strand non-linear conduction, due to the copper oxide rectification principle. It has also been claimed that individual conductors of the size of typical telephone installation wires are the optimum choice for the skin/core balance of current flow. Still other manufacturers insist that such strands must be individually insulated in the bundle, both to reduce the problems of long term corrosion and to prevent the cable impedance from varying due to the randomised inter-strand contact varying as the cable is moved, either by vibration or for relocation.

I have found manufacturers claiming that skin effects become significant at around 10 kHz, whilst other people claim that the effect is evident from 1 kHz. I have also met some experienced and learned people who claim that such effects could not be evident until 100 kHz or more. Such discrepancies in their claims are not comforting.

8.4 Group delays and insulating materials

As with the variability in the claims for skin effects, there are factions who support the concept that transmission line group delays must be considered in monitor system design. Again, others contest that such group delays as do exist are usually only in the region of micro-seconds, and are clearly irrelevant to audio applications. Signals reflected back from the non-ideal terminations at either end of the transmission line, which the loudspeaker cables constitute, do have finite 'lives' within the lines, and can once again, if present to any significant degree, superimpose themselves upon, in particular, the feedback 'error' signals. The question is, just what degree is significant?

Insulation materials are dielectrics which exhibit charge migration under certain high level drive conditions, 'bouncing back' again a finite time after the cessation of the drive signal. As with group delays, to what extent are these effects audible and under what circumstances?

Another aspect of insulating material technology is the effect to which the insulation can inhibit or chemically advance the onset of surface corrosion. While this is not a directly audible effect itself at any given point in time, if such effects *are* evident, over what period of time do these system degradations occur – months, years, tens of years or lifetimes?

8.5 General conclusions

Once a cable has been installed and approved, one very valuable asset for any cable to possess, along with the connectors to which it is attached, is consistency. Insidious changes in resistance or linearity of conductivity are alarming properties for any system to possess. The suspicion of a system deteriorating with time is unnerving both for the conscientious amateur and the professional alike. If one cannot trust one's monitoring to remain relatively constant, then the very foundation of one's judgement is undermined. If skin corrosion is both significant and promoted by certain insulation materials, then cables exhibiting such symptoms are clearly to be avoided. If cables with multiple strands do exhibit perceptible changes due to inter-strand contact irregularity when vibrated or relocated, then these potential variables would also be deemed undesirable, and cables exhibiting such properties should not be specified for any serious system.

Whilst realising that real world applications often make strange demands upon system designers, I doubt that any system could be deemed to be well designed if it had power amplifiers 40 ft away from the loudspeakers, especially if those amplifiers were in turn fed from a plugboard on the end of a similar length of not so heavy flex. May I open the debate, however, by saying that a well designed, well installed system – with amplifiers close coupled to the loudspeakers, if necessary having one amplifier adjacent to each loudspeaker as opposed to a common stereo amplifier necessitating longer loudspeaker cables – will perform optimally using Mk 1 copper cable of adequate cross-section. I submit that the more esoteric cables are a means to an end. They may solve the problems of difficult installations or particularly tortuous dynamic impedances, but, in general, a well selected, conventional copper cable of adequate current rating, short length, and optimum strand configuration for the application, when well terminated will perform equally well.

Pistols at dawn on the French coast anyone?

8.6 Note

Sections 8.1 to 8.5 were the actual paper presented to the Institute of Acoustics before the demonstrations and discussions. What follows is an analysis of many of the highlights of the subsequent discourse.

8.7 The cables or the amplifiers?

If a cable linking a given amplifier and loudspeaker combination was to be changed for another cable, any perceived difference would usually be attributed to the cable. There are instances however where, for example, Zobel networks, frequently installed to reduce the risk of instability, have been dispensed with in amplifier outputs, the manufacturer presuming a certain minimum usual impedance in the cabling, and relying on this for stability. Changing to a low resistance, low inductance, high capacitance cable can have adverse effects upon the operation of such an amplifier, so the perceived effect of such a cable change could well be based more on amplifier loading differences than on the cable *per se*. Some people feel that such a problem should be addressed in the amplifier, but some amplifier manufacturers would argue that they are producing amplifiers for optimum performance on most typical systems. This highlights my long-held view that systems can only be considered *as* systems. Individual component parts cannot be designed in isolation.

8.8 Cable variables

Notwithstanding the above case of an amplifier manufacturer relying on a given minimum of cable impedance, in which case it should be argued that the cable is part of the amplifier circuit, it would seem hard to argue with the fact that the best cable is *no* cable. Richard Lee at Wharfedale stressed most vociferously at the IOA conference that in tests carried out at Wharfedale, the only conductors of significant length which could *not* be detected by a listening panel, when inserted between the amplifier and loudspeaker, were two lengths of enormous lightning conductor. These unwieldy lengths of rod were the only cables tested which, to them, sonically approximated to no cable.

8.9 Oxygen, and charge migration

Cables are, in reality, highly complex networks of resistance, inductance, and capacitance; made even more complex by the diode-like contacts which can exist between individual crystals or strands where oxygen is present in the form of surface oxides. Indeed, the copper oxide rectifier was the first widely used rectifier in the world of electrical experiments. Removing oxygen from the copper removes one source of variability, but the natural affinity of the copper for oxygen will ensure its eventual re-oxidisation, unless steps are taken to use insulators which are as impermeable to oxygen as can be achieved. Unfortunately, though, oxygen permeability alone cannot be the criterion for insulating materials, as dielectric absorption and charge migration are also important factors in insulator choice. When a capacitor is charged then briefly shorted in order to discharge it, it will sometimes be noticed that after a short period of time, a small voltage will have reappeared at the terminals. This is a function of dielectric absorption which, to a lesser degree, is observed in cables, the greatest influence on the degree of absorption being the choice of insulating material. The delayed appearance of a voltage after the short is removed is due to the absorbed charge gradually being released from the insulating material, reappearing as a charge across the plates.

8.10 Skin effect

The much vaunted skin effect is the subject of great debate. Essentially, higher frequencies tend to travel along the skin of a cable rather than through the core. An early approach to address this fact was to use finely stranded cable which would present the greatest surface area to cross-sectional area ratio, but it has been suggested that when such strands are bundled together they behave not as strands but as one conductor. Individual strands in the bundle may be near the outside at one point in the cable, suddenly diving into the centre of the bundle in the twisting process then reappearing elsewhere. The high frequencies thus skip from strand to strand in a way which is both unpredictable and variable as the cable is bent or moved. Obviously, if the cable is made from conventional copper with surface oxides, it can be seen that each inter-strand contact is a potential rectifier, but by-passed by the resistance of the easier route of the individual strands. Most electrons will tend to follow whichever is the line of least resistance.

The use of litz wire, in which each individual strand is insulated with an enamel, attempts to overcome this contact problem by preventing inter-strand conduction. I have seen loudspeaker cables employing the litz principle with up to three thousand micro-fine strands per conductor. When wires are drawn so fine, effectively they begin to approximate to linear crystals, as each of the individual crystals of copper is squashed and stretched. Each individual strand should thus have fewer inter-crystal boundaries, and hence less potential for low-level rectification. Yet here again, there is controversy over whether to draw or whether to extrude. Extrusion is said to produce long crystals which are in less of a state of internal stress than those produced by the drawing process. Not only are stressed crystals said to be more generally brittle, but they are more prone to fracture due to current stress.

8.11 Magnetic considerations

Current stress is induced by the inter-molecular electromagnetic attraction and repulsion which is a function of the magnetic forces resulting from the current flow. Certain materials exhibit the property of magnetostriction, the expansion and contraction in sympathy with the electrical current flow. In multi-strand cables, the individual strands can also be moved by interaction of their external fields. Any such movement in cable geometry will inevitably change the inherent inductance and capacitance, leaving us with a dynamically varying set of electrical parameters. On a larger scale of movement, the inter-cable relationship was clearly demonstrated in the tests which we carried out prior to the IOA conference. A pair of FM Forcelines multi-stranded cables were spaced apart whilst listening to the output from the differential amplifier, with signal flowing between the amplifier and loudspeaker. As the cables were brought together, the sound from the differential amplifier output noticeably changed in its spectral balance, clearly suggesting a change in the series inductance of the pair. I note that FM supply plastic clips to maintain the cables at a constant spacing.

8.12 Beyond power loss

The multi-strand cables based on the skin-effect philosophy are said by some to have failed to take into account that power loss is not the only parameter to be considered. There are factions who claim that larger cables, in the 0.8 mm diameter region, offer a more balanced response. The resistance and linearity changes due to inter-strand leaps, and the frequency-dependent pathways through larger cables, together with the attendant impedance differences of those pathways, and magnetic movement of the cables themselves, are all claimed to be more audibly noticeable with their attendant phase shifts than the volt/amp losses due to the skin effect alone. The intricacies of these inter-reactions were clearly demonstrated by Professor Malcolm Hawksford from the University of Essex, who showed graphs to the IOA conference indicating the behaviour of cables of five metres or over above 100 kHz. Resonances and transmission line effects could clearly be demonstrated in terms of transfer function errors above 100 kHz and into the megahertz region. Here was an area where cables exhibiting good HF transfer could introduce serious levels of RF interference spurii into amplifier circuitry to detrimental effect, whereas a slightly poorer or more lossy cable at HF could be better on a system where RF interference or HF amplifier stability were problematical.

Clearly the potential for a cable to affect performance is established, but the problems still seem to be in the precise balance of parameters, and their relationship to the amplifier and loudspeaker characteristics. It seemed to be generally accepted by the IOA conference that the 'correct' specifications for a cable could not be defined as an engineering exercise, as the inter-reaction potential with the rest of the system was too complex. Cables chosen as a part of an overall system were entirely valid in that system, but were not necessarily superior in another system. Many people seemed to feel that in the case of the more expensive varieties, the money could usually be spent elsewhere in the system to greater subjective effect.

8.13 System susceptibility

Obviously, when all else has been done, the cable may be all that is left for state of the art improvement; however, system design can help to remove the worst problem for cabling. I have now been designing and installing studio monitor systems for twenty-five years; even my first systems were low level crossover/multi-amplifier designs. Adequately sized and well chosen cable has always seemed to render such systems relatively immune from cable problems, whereas at the other end of the scale, the Hidley/Kinoshita, nominally 4 ohm 1000 watt passive systems seem to be very cable-dependent. The explanation appears to be that cabling problems seem to be proportional to length, frequency range, and current. A multi-amplifier system with amplifiers located close to the loudspeaker units requires only short cables, and only a narrow frequency band to be carried by each cable, thus reducing two out of the three critical criteria. The third, current, seems largely to be relevant only when the other two are large. On a large, wide band system, the low frequency signals can easily be 60 dB above the level of the subtle information carried in the high frequencies. When the signal is carried in one cable, the low frequency currents can, by some of the aforementioned processes, modulate, swamp, or mask by distortions the high frequency subtleties which are responsible for clear, open reproduction. Remember, 60 dB expressed in terms of power is 1 000 000 to 1! The override potential is enormous.

The potential for inter-crystal/inter-strand contact distortion is also reduced in multi-amplifier systems. Remember, if the low frequency distortion products are 60 dB down on the signal, they can be on a par with the level of the high frequency information. With the LF distortions restrained within the LF system, they will not be reproduced by the LF drivers, which usually cannot reproduce high frequency distortion products, and hence will leave the high frequency signals clear. Any such similar effects of current-induced problems in the HF sections, if separated from LF currents, will be 60 dB down on the HF signal, and hence causing little problem.

8.14 Bi-wiring

The get-out from this problem for people using high level, passive crossover systems is to avoid any of the cabling in the system having to carry the full range signal. The answer lies in bi-wiring: taking separate cables from the amplifier to the separated high and low frequency filter inputs of the cross-over (Fig. 8.1). Although the cables are paralleled at the amplifier output terminals and – unlike the cable of a multi-amplifier system – receive a full range *voltage* drive, the high frequency cables carry no low frequency drive *currents,* and hence the magnetic problems which are current related, are not excited. The high frequency signals are thus unaffected by currents flowing in the low frequency cables. The principle can be extended still further by using two amplifiers as shown in Fig. 8.2. In this configuration, the passive high level crossover is still utilised, but the cables are buffered by using two separate amplifiers with the common connection being made ahead of the amplifier. The amplifiers still receive full band voltage drive, but are only called upon to deliver currents proportionate to the filter sections which they are driving. Certainly from my own experi-

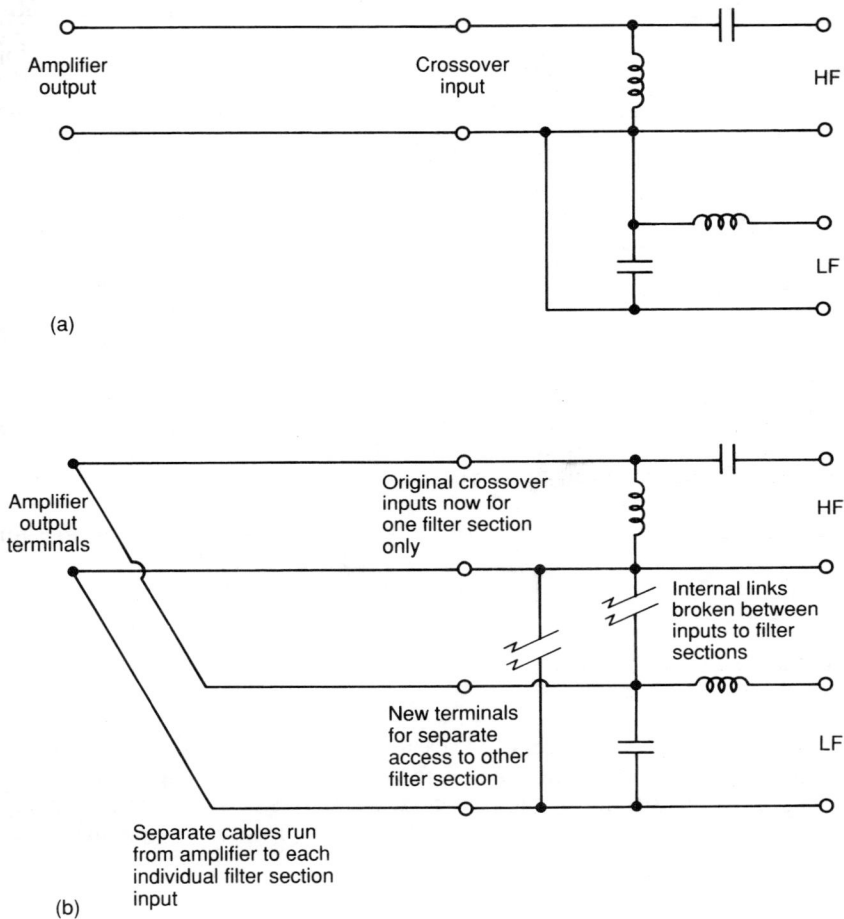

(a)

(b)

Figure 8.1 (a) Conventional amplifier to crossover wiring; (b) bi-wiring

ences, bi-wiring probably has a greater effect than any other single aspect of cabling to passively crossed over, full range loudspeaker systems.

8.15 Magnetic radiation – both ways!

I suppose that I should point out that more than one person of general repute and competence has spoken to me about aspects of direct human pick-up of the electro-magnetic field surrounding loudspeaker cables while being driven. Whilst it is impossible to discount any such concepts, I know of no attempt to qualify or quantify by experimental means any such effects. On the subject of radiation in the reverse direction, however, RF interference, facilitated by the potential of a cable to act as an aerial, can superimpose itself on the feedback circuits causing grittiness, frequently due to precipitating HF problems in the amplifiers. Such problems are entirely

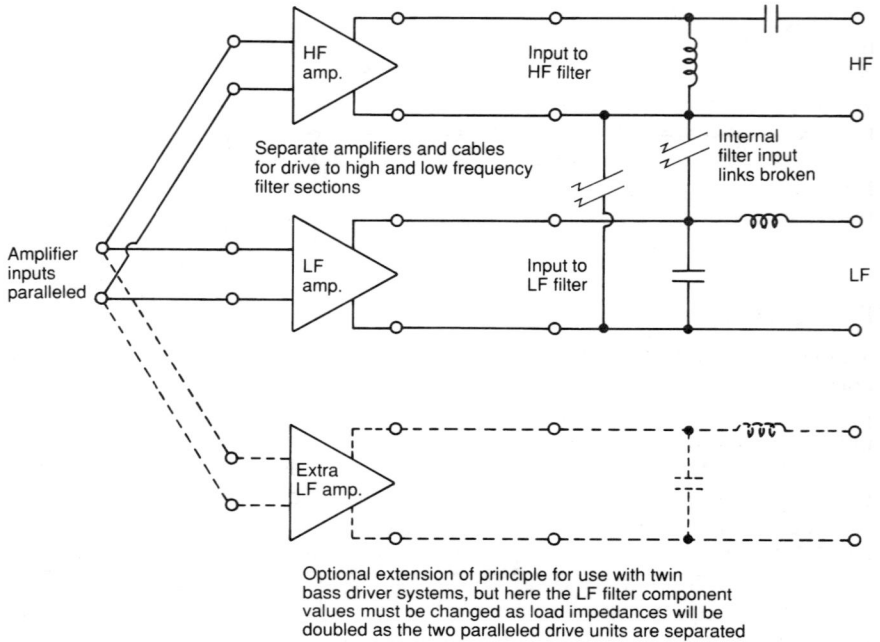

Figure 8.2 Extension of bi-wiring to bi-amplification whilst still utilising passive, high level crossover

dependent upon environment, so a cable is not necessarily 'better' *per se* because of screening. Again, such a requirement is system specific, depending upon the sensitivity of the particular amplifiers to RF interference, and also on the degree of any such interference in the location where the system is to be installed.

8.16 Geometry

Different geometrical layouts of conductors are manifold. All deal in their own particular way with a given designer's optimum desired balance of resistance, inductance, capacitance and RF rejection. High capacitance cables were in vogue some years ago. Much of this philosophy was based on line impedance matching, particularly with valve amplifiers, where a nominal 8 ohm output could well be realistically fed into a nominal 8 ohm loudspeaker input via a cable of nominal 8 ohm impedance. I say 'nominal', as the output and input impedances are often highly frequency dependent. High capacitance cables have been blamed for the failure of certain types of transistor amplifiers, but conversely there are systems on which they still give pleasing results, yet again showing the system-dependent nature of cabling choice.

8.17 Insulators

Geometry can also be influenced by insulator design and material, 'Teflon' (PTFE) seems to be generally considered to be a superior material. Insulator thickness can affect dielectric absorption factors and can also vary the inter-strand magnetic attraction/repulsion strengths. Some cables are now being manufactured with strands of insulators within the bunches of individually insulated conductor strands. This is in order to help each strand to more closely represent a single core, by reducing magnetic inter-reactions and attraction/repulsion forces. Compared to non-litz cables, they are also free from the overall skin effect aspect, reducing the problem of high frequencies skipping from strand to strand, as they treat the whole of a tightly packed bundle as a single conductor.

8.18 Conclusions

The variations on all of the different themes are beyond quantification. The complexity of cable behaviour, especially when passing high currents, precludes absolute prediction of how any particular cable will perform with any given system. Although good engineering principles will usually lead to good results, if a cable, somewhat against the odds, is deemed to improve the subjective quality of a system, then that cable should be given due consideration when specifying that system. Actively crossed over, multi-amplifier systems do appear to be relatively immune from cabling problems as long as good quality, adequately sized cable is used, and the amplifiers are kept within two or three metres of the loudspeakers. The relative immunity is inversely proportional to the width of the frequency band handled by any one cable.

Full range, high power, passively crossed over systems should, where possible, maintain the shortest possible distances between the amplifier and crossover, and crossover and drive units, consistent with any magnetic inter-reaction considerations. In almost all such cases, bi-wiring the amplifier to crossover connection will improve results, separating the current flow into the two filter sections. This prevents the low frequency currents from modulating, and indeed moving, the wires carrying the much lower levels of subtle, high frequency information. The benefit is probably further enhanced by the fact that each cable is feeding a much less demanding load impedance, reducing still further the complex current patterns.

The problem of correlating subjective and objective differences in these areas is largely based on the fact that, in response terms, we are dealing with the bumps on the bumps on the bumps. The combinations of brain and ear have an awesome ability to resolve fine detail, orders of magnitude beyond our best measuring equipment. The non-linearity of some of the cable phenomena preclude our ability to wholly predict system performance from first principles. There are many 'pet theories' which have no sound basis in known fact; there are also many contradictory viewpoints to be reconciled. The ability of the rest of the system and listening environment needs also to be considered in terms of its ability to resolve some of the cable difference aspects and should also be taken into account. It is very, very difficult to dictate absolute, hard and fast rules to cover all situations. Considering the complex inter-reaction of performance criteria, both within and between the components of a monitor chain, I do not think that it is too much of an evasion of firm opinion to say that if you can afford it and it works for you, use it! Clearly this latter point may explain one of the reasons for the lack of consensus or undue interest from much of the professional audio world, where rules of thumb and predictability may often count for more than they do in the 'enthusiastic' world of hi-fi. At the end of a professional installation, endless time and resources are rarely available for in-depth studies into the fine detail of subjective system subtleties.

One final aspect of all this is that it may well be that some people are actually far more susceptive to cabling effects. It is well known that we all hear differently, and I have reports of well controlled circumstances where reputable test subjects have failed to detect differences which others have deemed to be obvious and unmistakable. Should such tests be carried out in groups, the peer pressures and the 'Emperor's New Clothes' syndrome can often carry a consensus when, in reality, none exists. Having said that, however, there is little doubt that some people do exhibit pattern recognition to a far greater degree than others. Once again the human factors are lurking, waiting to confound our best efforts of objective analysis.

8.19 Footnotes

Recent tests on some very high power, low impedance, wide band, 'worst case' systems have borne out many of our initial conclusions. Bi-wiring certainly seems to work. I can see no caveats in the concept. Martin Colloms, in a *Hi-Fi News* article on bi-wiring in June 1986, stated that in his opinion, bi-wiring with conventional loudspeaker cable almost universally would give a system a greater sonic improvement than any change from decent, conventional loudspeaker cable to any of the more esoteric varieties. He also stated that as the conventional copper cables were replaced with more specialist types in bi-wired systems, the overall sonic quality improved, but by no means with as great a differential between the more expensive and cheaper cables as noticed with non bi-wired systems. With these statements I entirely concur. I note also an article by Karl Brown in *Sound Engineer and Producer*, January 1989, again advocating this technique. In independent tests carried out by Tom Hidley in Switzerland and Shozo Kinoshita in Japan on their high power studio monitor systems, they concluded that on any run of over 1.5 metres, bi-wiring produced a noticeable improvement to the sonic performance of the system. Nothing was directly measurable, but that is of no surprise as we do not yet have analytical equipment with anything even approaching the resolving power of the human ear/brain combination.

A reputable manufacturer of specialised cabling has recently faxed to me a statement confirming that the directionality of the cable overall, that is, one end clearly marked for the amplifier and the other for the loudspeaker, was based on theoretical concepts of speed of electron 'flow'. They accept that over short distances, sonic differences would be imperceptible were the cables to be reversed. They confirmed also that the overall screens were for rejection of electro-magnetic interference which could disrupt amplifier performance. Were no high levels of electro-magnetic interference to be experienced, the presence or otherwise of the screen would again be sonically insignificant. They did however stress, and this point I fully accept, that they are producing cables for professional use (by the converted) where system performance criteria could be guaranteed, even when system installation environments could not. A principle of 'belt and braces' was therefore valid and justifiable. It did not, however, mean that, for example, screened cables sound better *per se*. They *may,* in certain circumstances, but that does not mean to say that they *will,* in all circumstances.

To sum up, wherever possible, restrict the bandwidth over which the cables supply current into their load – bi-wire if possible. This of course explains why electronically crossed over, multi-amplifier systems appear to be relatively immune to cable problems. Where full range loudspeakers of low impedance or high peak dynamic current-demand must be fed by single pairs of wires, keep them short. I realise, however, that that may require separate amplifiers placed close to the loudspeakers, thus transferring a problem to the longer, low level signal cables required to couple the amplifiers to the control unit/pre-amplifier – but low level signal cables, well, that's another story.

I am still persistently being asked to make clear-cut statements in response to questions such as, 'Is solid cable superior to stranded cable?' I cannot give yes/no answers as the specificity which we have discussed ensures that we have a dynamic problem. Dynamic problems, by their nature, refuse to stand still; they are therefore reluctant to stay in any pigeon-holes to which we may try to allocate them. Solid cables may well by some be thoroughly and rigorously tested and be shown to be superior. Unfortunately, however, such tests are valid only for the specific circumstances under which they were conducted. Here we face two problems. Firstly we face the false syllogism, 'Cats have tails, Fido has a tail, therefore Fido is a Cat'. We must thus resist at all costs the temptation to equate the superiority of a generic cable type on one set of tests with the general superiority of that cable in all instances.

Secondly, we are also faced with a parallel to the 'five blind men and an elephant' syndrome, in which each man, touching a different part of the elephant, concluded with some certainty that it was a bat (the ear), a wall (the slab side), a snake (the trunk), a tree (the leg) and for the sake of decency we will stop there, but the point is that 'reality' can depend upon aspects of presentation. The fact that so many of us buy so many different types of loudspeaker within any given price range implies most strongly that we perceive different aspects of performance in quite unique and personal orders of priority. Should one group of cables be deemed preferable by a greater or lesser number of the listeners, then such preferences neither validate nor invalidate their specific superiority in all other cases.

If in one specific, fixed system, 85% of listeners show a preference for a given cable, then I think it reasonable to assume a generally accepted superiority, but were I allowed to change any component *in* that system, then the tests would all have to be repeated in order to re-establish any order of cable superiority. Whilst amplifiers exist which either do or do not anticipate a certain amount of resistance, inductance and capacitance in the cabling, then this case will clearly be so. Environmental variability is another such disruptive influence on pigeon-holing.

'Good cabling practices' must take into account all variables, and as such it is entirely misleading to make statements of extreme generalisation. There are even circumstances where 50 feet of bell wire could, whilst delivering

500 watts, 'improve' the performance of an already 'bad' system, but as a 'bad' system is outside the confines of 'high fidelity' I think that we can generalise by saying that 20 feet of a high quality cable will outperform 50 feet of bell wire. In gross generalisation, however, much beyond that statement we cannot go, hence the somewhat lengthy nature of these conclusions. Any other approach would surely only serve to even further muddy the waters of a subject which has all too frequently been confused by those very attempts to turn specifics into generalisations. Whether we like it or not, loudspeaker cabling is a very complex subject. It would be a dis-service to all concerned to suggest anything less.

Chapter 9

Monitor equalisation and measurement

9.1 Introduction

Probably all studio designers are asked from time to time by studio personnel to make one loudspeaker, room, or monitoring system sound like another; at least in terms of certain characteristics. It is all too frequently expected that adjustments to the 'frequency response' will bring two dissimilar units into line. Clearly this cannot be so, for if it were, then by the same token, a cheap violin could be equalised to sound like a Stradivarius, which we all know cannot be done.

If two loudspeakers are to be expected to sound the same, then effectively they must not only *be* the same, but must be auditioned under similar circumstances. The same requirement of similarity also applies to rooms, which again must be of similar construction in order to produce similarly perceived performances. In terms of classical acoustics, the pressure amplitude response, the phase response, and the presence of non-linear distortions can be shown to determine the total performance of any loudspeaker or system, but the key question is to just how close a degree must those performance characteristics match before two units or systems can generally be deemed to be similar?

The response graphs published by many manufacturers are usually highly sanitised versions of reality; in many cases, marketing departments would not relish publication of more detailed response plots. Ironically, if nobody used 'smoothed' plots, then little would change in the order of things, as everybody would just have shifted on to a more sensitive scaling. What such scalings would show, however, would be just how far from perfection we remain.

Along with numerous other studio designers, I ceased using monitor equalisers for 'voicing a room' at the end of the 1970s. There are many 'old wives tales' in society which appear to linger on long after the truth of a situation is known. Judging by the number of people who still ask 'Which graphics do you use?' or 'When are you bringing in the spectrum analyser?' whenever and wherever I am completing a monitor installation, the myth of all 'professional' studios using monitor equalisation obviously is alive and well.

Monitor equalisation, *per se*, is not an absolute taboo, but there are only three common sets of circumstances where it may be used. It must *never* be used to correct for room reverberation time problems, as this would be attempting a cross-domain correction which can only end in tears. Somewhat ironically, this was perhaps the primary use of such equalisation in the 1970s.

The whole subject of monitor equalisation can be a minefield. Even when graphic equalisers are used in allowable circumstances, then unless applied in very smooth and gentle sweeps, third-octave equalisation rarely reflects the true situation. For example, suppose a monitor system has a bump at say 58 Hz which shows at 63 Hz on a third-octave analyser. If the nearest available frequency on the equaliser is 63 Hz, this would be pulled down till the real 58 Hz was flat on the analyser (at 63 Hz). The result of this would be a dip at the next highest frequency which would need to be boosted. This in turn would create a peak at the next frequency, which would need to be cut, and so on. From a single 58 Hz bump, we can end up having a flat picture on the analyser; however, this is achieved by the drawing of a roller coaster on the graphics. All we started off with in reality was a minor bump at 58 Hz. In no way is the picture painted by the graphics the inverse of our original situation, and as such it has no justification in being there. After all, when the music is playing, we are intending to listen to the speakers, not look at the relatively crude analyser.

Despite the reading on the analyser, acoustically we would quite categorically not have a flat response in the room. I have never yet seen a monitor graphic which was truly reflecting the inverse of the room/loudspeaker combination. Switching out the equalisation can often allow you to hear just how much more natural and clear things sound, except possibly in extreme problem cases. Without the monitor equalisers, you can actually hear if you are over-equalising something on the console – it is not masked by unnatural, equalised monitors. Equalisers also tend to introduce unwelcome phase shifts, especially when exhibiting the alternating up/down pattern. This makes a mockery of achieving minimum phase shifts in crossovers or any other parts of the system.

When a large studio in London was completed in 1978, the rooms were fitted with 27-band graphic equalisers and set up from scratch. These were rechecked every few weeks, but after a few months some strange things were noticed about the sounds. The technicians checked and double checked, but the prescribed curve was still visible on the analyser. It ultimately transpired that with gradual adjustment, compensation made every few weeks had resulted in a totally different set of equaliser settings to those noted upon first installation. The upshot of all this is that two or more entirely different settings of the graphic equalisers can achieve a flat response on the analysers. Clearly they cannot all be right, and in all probability none of them are! They never accurately correspond with what is really happening. To cap it all, even the different makes of analysers and microphones do not always correspond, and there are even discrepancies in the microphones used – grazing, free-field, omni and cardioid. It all depends what you are measuring as to which ones *should* be used, but they do seem to get transposed rather a lot.

9.2 Suitability

The three sets of circumstances where monitor equalisation *can* be used are:

1 To correct for driver discrepancies where such deviations from the desired norm are of a minimum phase nature.
2 To compensate for 'room gain' where a loudspeaker is placed somewhat near to a corner, or is mounted in a wall, thus producing a bass build-up.
3 To apply a 'desired curve' where a studio operator requests, say, a roll-off of a shallow slope above a certain frequency.

All of the above cases are amplitude/frequency aberrations, so a frequency domain correction is in order. Reverberation time discrepancies are in the time domain and hence cannot be remedied by frequency domain fixes. My practical solutions to the first three problems have been to:

1 design wide band systems of sufficient linearity, consistent with sonic acceptability, that high or low frequency boosts or cuts were not required to linearise the systems
2 site the loudspeakers such that any 'room gains' were allowed for at the design stage
3 adjust to 'desired curves' as much as possible by level changes of the tweeters.

My approach to the irregular or undesirable reverberation time problems have been to either fix them acoustically within the rooms, minimise their effects by use of loudspeaker directivity control, or to learn to live with them.

The problems are based on the correlations between what first passes the ear, and what frequency balance is then left in the reverberant hang-over. In our first three cases, we are dealing with adjustments to be made to the output from the loudspeaker. A minimum phase roll-off (*or* boost) is one which can be equalised in a way which will restore phase accuracy with amplitude accuracy. Should that roll-off (or boost) be non-minimum phase, then equalisation of the amplitude response will not restore the phase response, and hence will distort transient waveforms. If the loudspeaker output is rolling-off with a minimum phase characteristic, then application of equalising boost will restore both amplitude and phase response. Unfortunately, however, each 3 dB of boost will double the power delivered into the loudspeakers when signals arrive at those boosted frequencies. It is hence all too easy to run into headroom problems, either in the amplifiers or loudspeakers, when such boost is present. Furthermore, it is almost unheard of for an equaliser to exactly mimic the inverse of the problem, so one frequently merely changes the problem. One-third or even one-sixth octave analysis is far too crude a tool with which to make any judgements of similarity of response. Somehow, the regularity of the bumps and dips in an applied corrective equalisation always appears to be more noticeable *as* equalisation, as compared to the equally up and down, but less regular, response performance from a loudspeaker with a natural response more akin to what was originally desired.

I believe that it is the regularity of split octave equalisation which tends to make equalised monitor systems sound equalised, even when equalised for 'allowable' reasons. In the case of response irregularities where the problem is of a non-minimum phase nature, such as found at loudspeaker crossover points, any attempt to correct for amplitude response irregularities will inevitably cause aberrations in the phase response, which in turn will almost certainly impart an unnatural character to the overall sound. Such problems are to some degree inevitable in a large system, but must be addressed as far as possible at source. They cannot and must not be dealt with by split-octave equalisation.

When a loudspeaker generates low frequencies in a largely omni-directional pattern, constraining the angle of radiation by means of walls, floors and ceilings will concentrate the power in a forward direction. Most complete loudspeaker systems for use in rooms, to some degree or other, incorporate this property in their design philosophy to augment bass output. Again, when loudspeakers are built into walls, the power concentration is usually included in the design criteria. From time to time, however, loudspeakers are positioned less than optimally in rooms not designed for them, and here a rising or falling amplitude response can become apparent. Such a rise in the overall response of the room should not be confused with the room reverberation time response anomalies. The room gain by constraint is both minimum phase and possesses the same time decay characteristics as the loudspeaker system. It is therefore a frequency domain problem, affecting all relevant frequencies in the same sense; it can therefore be equalised back to a linear overall response. Room reverberation on the other hand is a function of the wavelengths of reflected energy in the room modes. Such reflexions are of a non-minimum phase characteristic; they also cause both peaks and dips in the overall response which will not coincide with the discrete frequencies of conventional equalisers. See also Note 2 at the end of Chapter 4.

The major problem with attempting to make electronic equalisations compensate for reverberation time problems is that, unlike an omni-directional microphone measuring a steady white or pink noise within the room, our ears can easily discriminate between the first pass of the sound and the reflected after-sound. When an over-long reverberation time exists at certain frequencies, the reverberant energy will make the room seem subjectively louder at those frequencies. Unfortunately, if the response of the monitor system is attenuated at those frequencies, then as a transient first passes the ears, energy will have been removed from that initial wavefront at those frequencies. The amplitude and phase characteristics of the transient will thus have been distorted and hence a 'natural' sound is almost certainly unattainable as the transient integrity has been lost.

After a period of acclimatisation in a room, the ear and brain soon learn the general character of the room from speech and other noises. A person soon becomes able to make automatic compensation for room reverberation time. Unless this reverberation is gross, when it can mask much of the fine detail in the sound, it is quite remarkable just how well we can adjust our perception to take the room into account. Were we to attempt to equalise the monitors to the steady state performance of the room, then the leading edges of the sounds can sound very unnatural. Not only are they amplitude and phase distorted, but our ears/brains are still adding our automatic compensation. The legacy of the 1970s, when room equalisation was the norm, still has a lot to answer for.

Bear in mind that I am discussing analysis of studios. There are situations where it is often necessary to compensate for time problems by frequency adjustment, such as in concert halls. When a band performs a one night stand in an auditorium with a five-second reverberation time, clearly the *acoustics* cannot be dealt with either in terms of cost or time. Here we may have a gross situation where if at certain frequencies the hall rings badly, and masks much of the detail for seconds afterwards, then something drastic must be done. In a live performance it is more desirable to lose transient integrity and maintain some definition, rather than the other way round. It is therefore often necessary to reduce the offending drive frequencies, but such gross situations should never occur in a studio, and even if they did, there is time to fix them acoustically. The whole show will not be moving to a different location the next day. The old techniques of pink noise, a measuring microphone, and a spectrum analyser are clearly outmoded, and are potentially dangerous in the wrong hands, but we will look later at what alternatives are available.

9.3 Time/frequency reciprocity

Time and frequency are linked in the definition of frequency, 'the rate of change of phase with time'. Were we to look at a single frequency on a spectrum analyser, we would see a single column display. The spectrum analyser represents a plot of amplitude against *frequency*. A remarkably similar picture would be displayed if we were to look at an impulse on an oscilloscope, which displays amplitude against *time*. A sine wave is a point in frequency, an impulse is a point in time. Conversely, were we to look at a display of a sine wave on an oscilloscope, we would see that it is a continuum in *time*, a full display across the screen. Again, were we to look at an impulse on a spectrum analyser, then, albeit briefly, we would see all columns energised – the impulse would represent a continuum in *frequency* (Fig. 9.1). Strictly speaking, a pure sine wave can never be turned on or off. It exists in perpetuity – for all of time.

The interrelationship between the spectra of the impulse and the sine wave is the Fourier transform, named after the French mathematician who first described the mathematical relationship between the time and frequency domains. A Fourier transform must be carried out frequency by frequency, and hence for a wide-band signal it is a singularly tedious exercise. The advent of computers brought the possibility of the FFT or fast Fourier transform. By use of such FFT analysers, impulse responses can be generated from noise signals such as white noise, pink noise or a pseudo random binary signal (PRBS). PRBS contains discrete frequencies of, say, 20 Hz intervals, which allow the analysers to 'lock' on to the signal with greater speed than waiting for each frequency to eventually, randomly, appear in the white or pink noise, which, as in the case of the impulse, contain all frequencies.

A further function of the reciprocity of impulsive and steady state signals is their characteristic relationship in terms of their phase properties, which is one reason why the phase integrity of a system is so important where a 'natural' sounding reproduction is sought.

As the definition of frequency is the rate of change of phase with time,

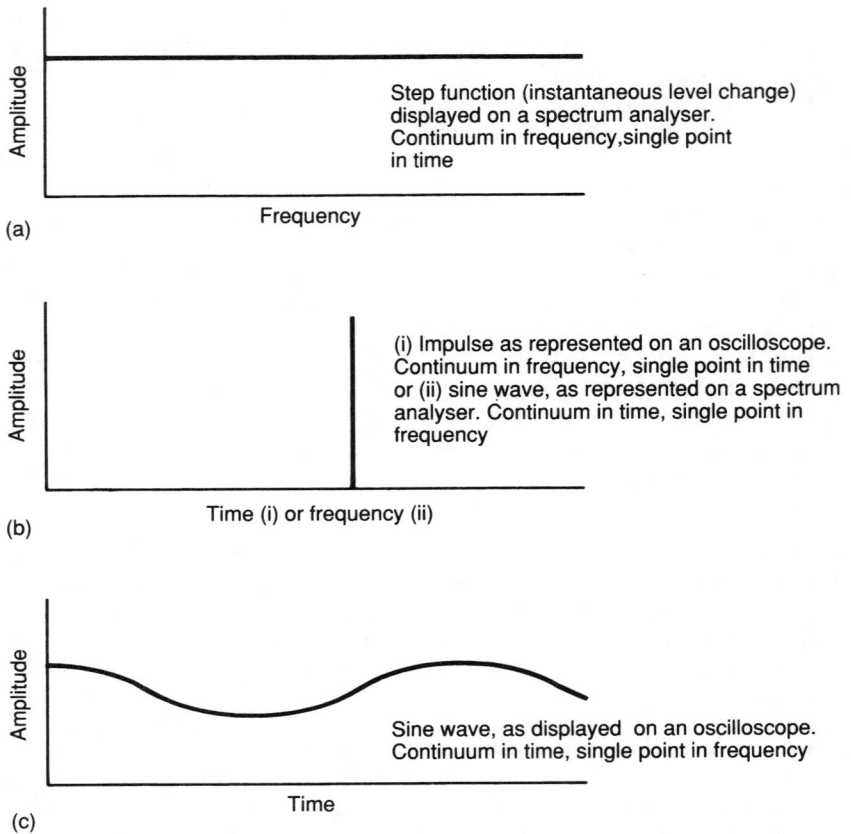

(a)

Step function (instantaneous level change) displayed on a spectrum analyser. Continuum in frequency,single point in time

Frequency

(b)

(i) Impulse as represented on an oscilloscope. Continuum in frequency, single point in time or (ii) sine wave, as represented on a spectrum analyser. Continuum in time, single point in frequency

Time (i) or frequency (ii)

(c)

Sine wave, as displayed on an oscilloscope. Continuum in time, single point in frequency

Time

Figure 9.1 (a) Step function; (b) impulse/sine; (c) sine wave

only an individual frequency can have a phase shift. A sine wave, phase shifted, will show a displacement of its pattern along its time axis (Fig. 9.2(a)). A transient impulse cannot have a phase shift as it has no single frequency, but contains all frequencies simultaneously. A movement of the display along the time axis would represent another point in time when that impulse occurred (Fig. 9.2(b)). That movement in time would represent not a phase shift but a phase slope.

Whenever we meddle with graphic equalisers to correct such non-minimum phase distortions as room reverberation (where to improve amplitude response, we destroy phase accuracy), we tend to get further away from reality, not closer to it. So if we are going to attempt to assess what is really happening in a monitor response, we must be able to look simultaneously yet separately at both amplitude and phase with respect to frequency.

By means of fast Fourier analysis and the related MLSSA technique (Maximum Length Sequence System Analysis), in which suitably long sequences of computer-generated, repeatable, pseudo-random noise are used as test signals, the phase and amplitude information can be assessed *in situ*. They can either be looked at in terms of loudspeaker performance,

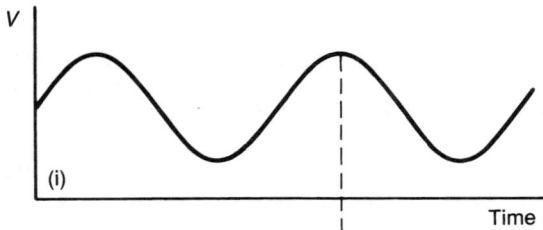

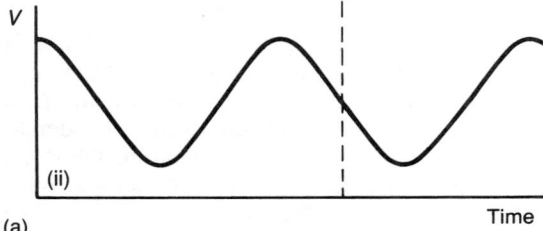

(a)

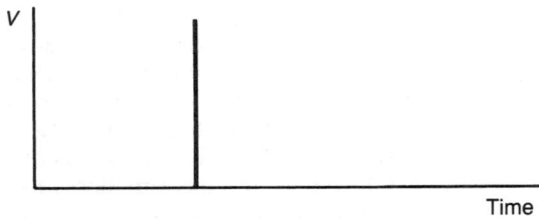

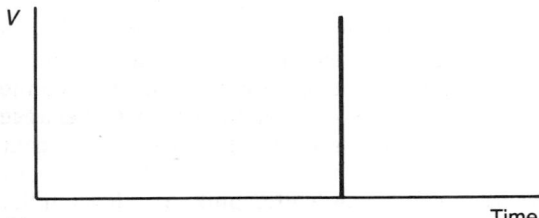

(b)

Figure 9.2 (a) − 90° phase shift causes crest of (ii) to move with respect to (i) but the same overall time frame is still filled. Only the pattern moves along the axis. (b) Time shift of impulse causes movement along axis. A different part of the time-frame is occupied

room performance, or the performance of the combination. For FFT analysis, either our familiar random noise sources may be used, or, alternatively, impulsive or noise sources. Each have their advantages and disadvantages.

9.4 Practical test signals

Any half cycle of a square wave contains a very great number of frequencies in very specific phase and amplitude relationships. As any waveform distortions will upset the time/frequency balance of a system, then obviously no accurate square wave reproduction would be possible if distortions were present. In order to accurately reproduce a square wave, therefore, an accurate pressure amplitude response, commonly referred to as 'frequency response', and an accurate phase response must be prerequisites. In view of this, accurate square wave responses of loudspeakers should be the ultimate goal, since if a square wave can be reproduced accurately, *all* waveforms should be reproduced accurately.

For testing purposes, the two extremes of square waves are frequently more convenient signals to use, and indeed are superior signals to use as they contain *all* frequencies. At one extreme, the delta function, or pulse, is a square wave with an extremely short duration, and is usually accompanied by a short mark/space ratio. At the other extreme is a step function, in practice a half of a square wave with long duration 'on' cycle. In order to prevent comb filtering of the measured results due to truncating the data, as a conventional square wave rises then falls, the test signal should have either an infinitesimally small 'on' time, or an infinitely large 'on' time. The short, spike, delta function and the long step-function are practical realisations of these requirements. During the subsequent signal processing, either can be regenerated from the other by a process of differentiation or integration, so the choice of which one to use is down to convenience.

The pulse, or delta, function does have certain practical drawbacks. Firstly, it is difficult to generate accurately without overshoots or ringing. Secondly, as the burst of power is so brief, it is difficult to either hear or see by means of an oscilloscope whether any part of the system is clipping. If mechanical clipping of any drive unit did occur, non-linear distortions, whilst possibly imperceptible aurally, would render the measurements useless. Thirdly, the low frequency content of a delta function is so low that the bass drivers really do not get a chance to be put through their paces, nor are they given the power to excite any resonances to any measurable degree. A step function is easier to generate, works the bass drivers much harder, and is generally much more easily judged in terms of the 'loudness' of the signal, giving the person performing the test a good indication of the overall power level being fed to the loudspeakers. There is thus less chance of clipping the system or encountering accidental, gross, non-linear distortions.

By means of a fast Fourier transform, a white, pink or PRBS noise source can also be used to produce an impulse or step function fingerprint, but unless complex gating techniques are applied, an anechoic chamber is required as the effects of the rooms on the steady state signal cannot be separated from the responses of the loudspeaker systems themselves. The use of simple gating techniques allow a step to produce its characteristic time history fingerprint *in situ*, along with both pressure amplitude and phase responses via FFT analysis. Except in the smallest of rooms, where the first reflexions appear before the low frequency component of the step function can be fully integrated into the FFT, there is no requirement for the monitor systems under test to be removed from their daily working surroundings, and hence from their actual loading conditions.

Remember, the room is itself a part of a loudspeaker's air loading. Changing the position of a loudspeaker cabinet from being flush mounted in a wall to being free standing in a room, or even to being placed in a corner, may significantly affect the character of the air loading. In a location where the low frequency drivers have more air to push on, more work can be done. Where this extra work manifests itself as a change in acoustic output, effectively the performance of the loudspeaker has been altered. This is our previously mentioned 'room gain' which *is* correctable by appropriate equalisation. Irrespective of room reflexions, a loudspeaker is not the same under all conditions. It is necessary to measure a loudspeaker in its particular working environment for accurate assessment, especially of the low frequencies. It is for this reason that the impulse/step function fingerprints are more practically achievable from an impulse source, rather than by FFT transformation from noise sources, conventional forms of which would require the loudspeakers to be removed to anechoic conditions if the room effects were to be entirely separated out.

9.5 Application to perceived drive unit differences

In a representative sample of around thirty high quality mid-range systems, whilst most sounded very similar on signals with a highly tuned content, such as resonant drums, sine waves or smoothly enveloped tonebursts, not one like-sounding pair could be found when listening to white noise or a recording of a waterfall – not even some identical drivers from the same batch of the same production line. Ironically, less expensive, lower quality drivers could be matched more easily, as inherent flaws tend to cause an easily recognisable pronouncement of a certain response characteristic. Such characteristics, together with other response limitations, often tend towards a matching process of considerably greater simplicity. Somewhat ironically, the more accurate mid-range units become, the more the differences appear to stand out.

People often mention that recordings of a certain well known band always seem to sound good, irrespective of on what, or where, they are played. By chance, having been looking at an EPQ, band splitting visual monitor, consisting of a series of picture tubes, each monitoring the waveforms of a narrow frequency band, it was noticed just how strong the fundamentals appeared to be on those recordings. There was a great predominance of 'clean' pseudo-sinusoidal waveforms. Such narrow band or sine wave orientated signals produce a distinctive fingerprint which is very difficult to upset. The opposite extreme is the tendency for a wide band linear system to allow differences in wide band signal response anomalies to predominate over the similarities. The better the system, the more the discrepancies will show.

The point to be made here is that the present conventions of using pink noise or swept sine waves is of little relevance beyond a certain point, when very high quality drive units begin to produce very similar response graphs which have no bearing upon the humanly perceived discrepancies in their sonic character. From the work done to date, I believe that the step function fingerprints do relate to perceived sonic performance, with the units which visually produce a recognisably more accurate output response sounding more 'natural' under representative listening conditions. The lower three examples in Fig. 12.9 (see Chapter 12) show the step function

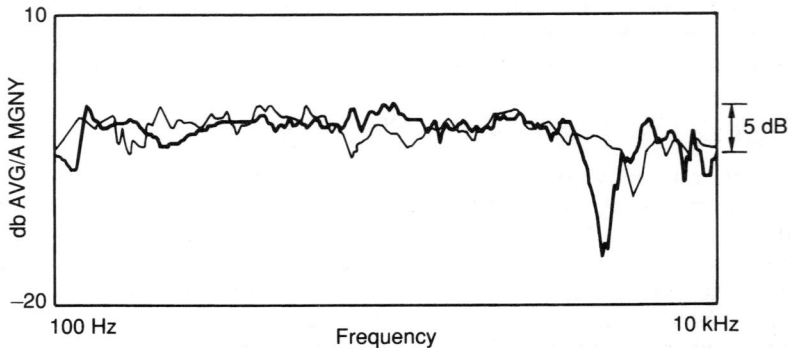

Figure 9.3 (a) ▬ Electrostatic; (b) ▬ 2-way monitor system. Both respectably performing within ± 2½ dB over most of their range

time history responses of three widely used monitoring loudspeakers. Although they all have acceptably flat 'frequency responses', the three systems sound very different. From these time history waveforms of their step responses, it can clearly be seen why they sound so different.

It must be added here, however, that although the manufacturers' published data on the above three loudspeakers is commendably 'flat' in terms of 'frequency response', a certain amount of averaging has been performed, which is generally considered to be acceptable. If one really decides to look at an unsmoothed spectrum as derived from an impulse, the actual differences in the amplitude and phase responses are more readily apparent. These are the differences which conspire to produce the gross waveform differences seen in Fig. 12.9. The amplitude response anomalies between Figs 12.9(b) and 12.9(d) are shown in Fig. 9.3. Third-octave analysis just cannot show this degree of detail. It is often quoted that the ear, in general, perceives one-third octave power bands, and that minor discrepancies in two signals would not be perceived if the average power level in those one-third octave bands remained the same. Unfortunately, the compounding of these discrepancies, not only in terms of amplitude, but also in terms of phase and hence waveform, conspire to make performance differences which are clearly audible, but do not show on one-third octave analysis.

Further to this, some manufacturers publish idealised performance plots with accompanying statements of permitted production deviations. Whilst these specifications may appear to be commendable, production differences at the tolerance extremes may represent 4 dB difference in the output of two drivers at any one frequency for units with a ±2 dB tolerance. Such specifications are of great use to system designers and users, but they do not represent any clear statement on sonic compatibility between different drivers with similar specifications. Albeit largely in narrow band disturbances, it has been my experience that impulse testing has been exposing performances more in the order of ±5 dB rather than the ±2 dB often quoted by manufacturers. However, I am not accusing the manufacturers of 'sharp pencil' practices, as they have been using accepted techniques. I *am* implying though that the currently accepted techniques do not relate to sonic performance similarities. None of these problems are addressed by conventional split-octave analysis and equalisation. Such equalisation may in some circumstances produce the result of a more pleasing sound, but such adjustments rarely produce any greater degree of 'accuracy'.

9.6 Loudspeaker/room combination responses

When given a highly linear monitor system, a room will probably dominate in the perception of the overall sonic performance, hence the reverberation time characteristics of the room will assume very great importance. The complication here, however, is that the knowledge of the reverberation time of a room, when calculated from an omni-directional source such as a starting pistol, is of little consequence unless the loudspeakers were to energise the room in a similar manner. Without any fixed standard of directivity, it is impossible to say precisely what would be the 'best' reverberation time characteristic for any given room. Only when considered as part and parcel of the monitor system can the RT requirements of any room be specified or judged (q.v. the 'non-environment' rooms discussed in other chapters).

Impulse testing of the loudspeaker/room combination can, by fast Fourier transform, give a graphic representation of the actual, overall steady state response of the whole monitor combination. Only when the steady state response and the gated, on-axis impulse response produce one and the same overall linear response graph can the response then be deemed to be in any way 'accurate'. Tailoring of the response by means of equalisation may then be possible to suit individual requirements based on average listening levels, fatigue problems, or other considerations, but at least any adjustments will actually be heard to 'track' the applied electrical equalisation, without introducing other anomalies. Where an off-axis response becomes irregular, or lobed, excitement of the room modes will not be uniform, even for minor differences in the position of the loudspeaker. Again, by the monitoring of the step function responses of the loudspeakers at various positions, it can be determined to what degree a room is being driven in each direction, and reverberation times at problem frequencies can be adjusted, either acoustically in the room or by repositioning of the loudspeakers to drive the problem modes to a lesser degree.

Whilst many people have put forward proposals for omni-directional loudspeakers to mimic more closely the characteristics of the original recording space, such polar distribution would cause an inordinately high reflected to direct ratio in the listening room, blurring the stereo imaging. About 60° to 80° would seem optimum for the horizontal directivity of the mid and top of a monitor system. Having said this, however, the impulse/step function fingerprint at a 30° off-axis position should be as close as possible to the on-axis fingerprint for an even response while moving around the room.

Figure 12.9(a) shows a practical application of the technique. The trace shows the step function response of a large, two-way monitor system, in a suitably designed control room, measured at a distance of around two metres. The rapid rise time, well behaved decay and conspicuous lack of resonant overshoot are all reminiscent of an electrostatic. The system is in fact capable of producing over 120 dB behind the mixing console and is

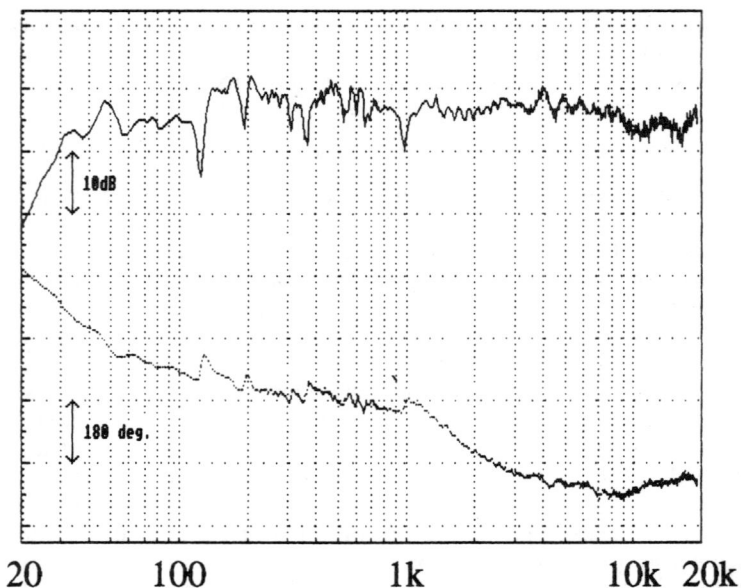

20   100   1k   10k 20k

Figure 9.4 LMH unsmoothed, uncorrected, 2000 point FFT plot of amplitude and phase at 2 metres

horn loaded above 1 kHz. Many experienced listeners have deemed this system to be audibly very similar to typical electrostatics, but of much greater potential SPL. It is certainly by far the closest to the electrostatic of Fig. 12.9 in terms of its step function fingerprint.

When I first put forward these proposals for the publication of step function plots in loudspeaker specifications,[1] many people asked why I should propose again something which had been largely rejected in the past. It had been generally considered that although the response plots were all dissimilar, they showed little information in terms of what those differences actually were. My feelings at the time were that far too many loudspeakers were too wrong to produce fingerprints close enough to display much meaningful information: the loudspeakers were failing to deliver, not the technique!

As Fig. 12.9(a) shows, when one does begin to close in on the target, the picture suddenly begins to make sense. The amplitude and phase response plots of the system are shown in Fig. 9.4. Obviously, these must be commendably flat in order to have any hope of producing a step response such as Fig. 12.9(a). The point is, however, that such a response, in the room at the listening position, was not achieved by any electronic equalisation. The horn, shown in Chapter 11, was specifically acoustically matched to the desired drive unit, and also to the properties of the intended room. Room reverberation time problems were treated entirely acoustically within the room. The only significant kinks in the curves lie at the crossover frequency, but as this is of a non-minimum phase nature, correction by conventional means could not restore both amplitude and phase to linearity. It was considered that attempts to linearise any part of this region would ultimately be detrimental to the sonic neutrality and natural musicality of the system. In any event, the band of disturbance is so narrow as to be deemed virtually inaudible and is best left as is for the time being. The drive units themselves were selected for their inherent wide frequency response range, and their subjective audible neutrality. The two by no means necessarily go hand in hand.

I still strongly recommend that step function fingerprints and their fast Fourier transform derived phase and frequency/amplitude graphs should become the accepted reference standard for all loudspeaker and loudspeaker/room combinations which are intended for accurate studio monitoring. The general drift of technological advance should be towards more faithful reproduction of such step functions, after which everything else should begin to fall into line.

One interesting outcome of the aforementioned listening tests carried out on the mid-range units at the ISVR in Autumn 1989, was that the perception of steady state signals appeared to be more sensitive to amplitude/frequency discrepancies, whilst the transient sounds were most perceptibly different when smeared in terms of the amplitude/time plots by poor phase accuracy in the systems. In the frequency domain, a phase lead or lag can be corrected by relatively simple, conventional electronic means, but in the time domain a time lead, if required, would necessitate an acausal correction.

As discussed in the previous chapter, and to a further extent in Chapter 18, digital signal processing technology has provided a practical reality to the concept of active control of the entire time, phase, amplitude convolution. By the introduction of a modelling delay between the signal source and the loudspeaker system, then by allowing feed forward signal paths around that modelling delay, response errors can be detected, inverse modelled, fed-forward, then recombined with the original signal, before the total delay time of the error correction system has elapsed. Such active control may well be the way forward in impulse reproduction accuracy. But one thing is certain, however, given the complexities of the interrelationships, a spectrum analyser, measuring microphone and a graphic equaliser is a grossly over-simplistic assessment and correction system. Except in a very few specialised circumstances, over the years they have probably done far more harm than good. Indeed, where a general high or low lift is all that is required, few equalisers can sound more natural than a simple Baxendall-type 'bass and treble' control!

Reference

1 P. R. Newell, K. R. Holland, 'Impulse Testing of Monitor Loudspeakers'. Proceedings of the Institute of Acoustics, Reproduced Sound 5 Conference, Windermere UK. Vol 1, Part 7, pp. 269–275 (1989).

Chapter 10

Mid-range horns vs direct radiators

There has been a great tendency amongst recording studio personnel to tolerate horns, rather than to wholeheartedly accept them. They seem to have been deemed to be a 'necessary evil' used under certain circumstances, really belonging to a different world of public address and cinemas. There is an irony in this: most horns in use in recording studios were indeed primarily designed for public address and/or cinemas, whereas almost all direct radiator units used in studio monitor systems were specifically designed for studio monitors. Some of the bad press for horns almost certainly came from their 'secondhand' application. Horns suffer from a general lack of widespread understanding, and from many charges against them which are simply not true. Possibly more than anything else in the monitor chain, if some misapplied examples possessed certain problems, then they were *all* deemed to possess the same problems.

There are six major differences between horns and direct radiators:

1 Sensitivities
2 Phase dispersions
3 Mouth reflexions
4 Non-linearities
5 Mechanical alignment
6 Directivity patterns

10.1 Sensitivity

The high-frequency horns of co-axial types of drivers, Tannoys and Altec 604s for example, are probably the only horns in common use which have not been used primarily on the grounds of sensitivity. In these designs, the concept of a horn HF unit was the most practical means of achieving the closest approximation to a three-dimensional point source. In almost all other instances, horns have been used in studio monitor systems for their sensitivity and efficiency – the acoustical power out for the electrical power in. Just why we achieve this extra efficiency does not appear to be too widely appreciated, but it is more a case of the very poor efficiency achieved by direct radiators (the non-horn designs), rather than any special magic being attributable to the horns themselves.

Think of the silencer on the exhaust of a car (Fig. 10.1). If we look at the

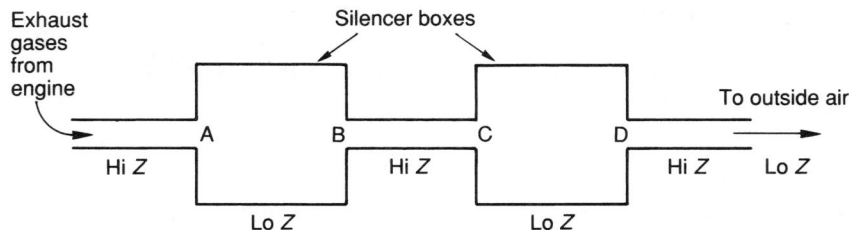

Figure 10.1 Cross-section discontinuities at A, B, C and D severely attenuate the exhaust noise by poor acoustic coupling between sections. The system is analogous to an electrical transmission line of alternate high and low impedance sections with no matching circuitry such as transformers

exhaust system in profile, we see a series of narrow pipes, interspaced with larger diameter pipes – the silencers. Within the silencer boxes are heat-proof absorbent materials, but the prime effect of the silencer boxes is to create a series of abrupt changes in the cross-sectional area of the exhaust system. Each time that we encounter an abrupt expansion in cross-sectional area, we encounter a power loss in the transfer of sound from one section to the next. This is due largely to the reflexions from the acoustic impedance changes at the cross-sectional area discontinuities. Resistive and reactive loading can also be considered rather like the effect of the springs on a motor car. Were the wheels attached rigidly to the body, each and every bump on the road would be transmitted to the seat of the driver with very little loss. This 'resistive coupling' – that is, in phase – would transmit each bump in the road almost instantly and most uncomfortably to the seats of the occupants. By placing springs in the coupling from the wheels to the body, much of the energy is not transmitted by this 'reactive', springy coupling. Another effect of this 'reactive' energy absorbing system is that the near instantaneous coupling is lost. A shock impact at the wheel will be time delayed by a small amount before reaching the seat of the driver. The result of this loss of intimacy between the road and the driver nevertheless provides the occupants with a generally more comfortable ride, as the reactive coupling absorbs and reflects back much of the energy of the impacts from the bumpy roads.

In each of the above cases – the silencer and the spring – the 'wave' of input energy meets a coupling which gives it less rigidity to push against. The coupling absorbs the shock. In a conventional room, a direct radiating loudspeaker suffers a sudden change in cross-section – say from a 5″ cone to the cross-section of the room. The match is very poor, and as a result the efficiency of the power transfer suffers. Many direct radiator loudspeakers are lucky to achieve a transfer efficiency of even 1% between the electrical input power and the acoustic power delivered into the room. Ninety-nine per cent of the output power of the amplifiers turns to heat within the loudspeakers themselves. Imagine a cyclist pedalling down a steep hill in a low gear. He or she could pedal furiously, expending an enormous amount of energy and working up an incredible sweat, but contributing a pitifully small amount towards increasing the speed of the vehicle. In low gear, the cyclist has little pedal resistance to push against in order to transfer the power from his or her legs to the machine.

Horns are acoustic transformers – impedance matchers – gearboxes! If our cyclist could change to a higher gear, a new situation could be achieved whereby the pedals present more of a resistance to the furious pedalling. In this instance, more power would be transferred from the legs to the wheels, via the pedals, and thus more work would be done. Despite the downhill roll, significant further acceleration could then be achieved. A horn provides a gradual taper, from the small size of the driver diaphragm to the large mass of air in the room. It gives the diaphragm more to push against – a more resistive, or 'solid' coupling, less springy and energy absorbing. Horn loading can achieve practical energy transfer efficiencies of the order of 50%. At the extremes of both cases, a direct radiator could require over 100 times the input power of an equivalent compression driver/horn combination to produce the same sound pressure level in a room. In practice, the differential between the efficiencies of studio quality horn/driver combinations and direct radiators is more likely to be in the order of 15 dB – but even this represents a power differential of 32 watts into the direct radiator for every watt into the horn. The importance of this in current studio design is that many direct radiators are having to be operated at the very limits of their capabilities, whereas horns offer a substantial power reserve.

The sound source of any loudspeaker driver must take into account the wavelength of the frequencies which that driver will be expected to handle. With the wavelength at 1 kHz being just over one foot, the sound source diameter should be significantly less, and certainly no more than one foot in order to achieve coherent, in phase, wide directivity, 'point source' radiation. Conventional cones and domes, with diameters of 6″ or less, are usually employed in non-horn studio monitoring systems. With drivers of this size, 200 watts give or take 2 or 3 dBs is the maximum input power which can currently be handled without severe thermal overload problems. Most of us are aware of just how hot a 100 watt light bulb can become after a few minutes' use. From this it will be appreciated that twice this power into the small voice coil of a mid-range driver will produce considerable heat within that driver. Remember, a 1% efficient driver would convert 198 watts of the 200 watts input into heat within the driver itself.

Confining such heat into a small space will have certain predictable results. Obviously, sufficient heat input will cause the whole thing to burst into flames – unless the voice coil melts before the onset of combustion! These are the absolute upper limits, the self-evident catastrophic failures; but before such failures occur, other problems can manifest themselves. Drivers experiencing repeated cycles of heating and cooling are more prone to deformation, premature failure, and impaired long-term consistency of performance. Large temperature changes in the voice coil also cause increases in the voice coil impedance. A voice coil of nominal 8 ohms can become 16 ohms when running very hot. This impedance rise reduces the available power which can be drawn from the power amplifier, resulting in a power compression, with the input to output no longer having a linear relationship. Such compression at high drive levels can be one of the reasons why certain direct radiator monitors can be considered 'smooth' at high levels. They compress! Whilst the bass, and thus the apparent power, rises linearly, the mids wind themselves down. One 'feels' more power, yet it doesn't seem to hurt; at least until the onset of gross distortions!

If we still insist on a capability, of 120 dB at the listening position in a control room, then, as we have seen, the larger rooms are already on the upper drive limits of mid-range direct radiators. Hence the frequent use of horn systems.

The mass of the moving diaphragm assembly of a typical studio compression driver is in the order of a mere few grams. In certain circumstances, it can even be less than a gram. Although these drivers are frequently rated at 100 watts or more of continuous music, typically they would be asked to handle only 5 to 10 watts peak in a studio environment. In a conventional 4-way studio monitor system utilising horn loaded upper-mid and high frequency drivers, the main power requirements would be largely at the bottom end. With crossover frequencies at 300 Hz, 1200 Hz and 6 kHz, to produce 110 dB in the room would require a power split in the order of 200 watts low frequency, and 35 watts in the lower mid. A mere 4 or 5 watts peak would be required for the upper-mid horn, and less than a single watt into a horn loaded tweeter. Given the massive magnet assemblies of the most frequently used mid-range compression drivers, the temperature rise of the voice coil under all such normal operating conditions would be insignificant.

One obvious, first advantage of this increased sensitivity is from the largely isothermal operation. A long, consistent working life is assured, together with an absence, in practice, of power compression due to voice coil heating. The light weight and largely resistive loading of the diaphragm allows for very fast and accurate response to transients, whilst maintaining excellent sensitivity. The other significant advantage of the higher sensitivity is headroom, so for any given, desired SPL, the headroom will be higher for a horn, compared to equally power-rated direct radiators, by the margin of the sensitivity advantage. Once again, this bodes well for cleaner, faster, transients at high sound pressure levels.

10.2 Phase dispersion and cut-off

The second major area of difference between horns and direct radiators is in the characteristic phase dispersions. All frequencies do not travel down horns at equal speeds. Below cut-off, sound does still travel down the horn, but only at such an efficiency as would occur if the compression driver was acting as a direct radiator. In other words, the horn is no longer loading the driver, it is no longer an acoustic transformer.

The approach towards this cut-off point has great ramifications in respect to phase response. Strictly speaking, the speed of sound in horns does not directly correspond to the speed of sound in air. There is a direct, finite relationship in air between wavelength and frequency. Within a horn this does not quite apply in the same way. Phase speeds are frequency dependent and are affected by two major aspects of the horns. The first is the cut-off effect, and the second is due to reflexions from the impedance change at the mouth. The separation and individual analysis of these two sources of dispersion has been the subject of much intense research to determine the precise, audible effects of each source. We do know that in the case of the dispersion related to the cut-off, difficult, high Q phase shifts can begin to take place in the octave above cut-off. For high linearity

systems, unless a horn can exhibit an exceptionally smooth throat impedance response through cut-off, it would seem wise to avoid using a horn in
this area by using a high slope crossover, at least an octave above the cut-
off. There is nothing, however, in this range restriction which places horns
at any disadvantage to direct radiators. They too can exhibit wild irregularity at their range extremes, which in terms of phase and amplitude cannot
in practice be accurately corrected by analogue electronic means. The general
philosophy of ensuring that drivers have an octave of smooth response on
either side of the crossover frequencies is still as valid today as ever it was.

Phase speeds are also influenced by the reflexions from the mouth
discontinuity, continuing to occur at higher frequencies than the cut-off
related dispersions. As with any other type of loudspeaker, it is difficult to
make horns to cover, phase-accurately, more than three or four octaves.
Within the central design range, however, a section usually exists which
exhibits an acceptably tight range of dispersion. One crucial factor which
relates to the range of spread is the length of the horn. The greater the
distance from the diaphragm to the mouth, the greater the length of time
which the waves spend in the dispersive section of the horn. There is a
measure of practical compromise here, however, as too short a horn can
increase the mouth-related dispersions, due to a less desirable termination
from too abrupt a change from the mouth to the room.

With careful design, phase dispersion in horns need not be significantly
greater than related problems in direct radiators. In many cases, however,
horns have been used over too great a range of frequencies, pushing too far
down towards cut-off and too high up into the beaming range. Designing
to published 'frequency response' limits has been all too frequent an
occurrence, without giving due regard to the full implications of dispersive
tendencies. Phase dispersions will show up in an impulse response, but are
unlikely to be seen on a spectrum analyser. Too many horn loudspeaker
designs have been used far too close to their response limits. This is just one
example of horns gaining a poor reputation from uninformed misapplication, often by companies from whom we would be entitled to expect better!

Phase dispersion also occurs in direct radiators, especially around the
cone break-up modes. The speed of sound through the cone is greater than
the speed of sound in air. Sounds travel up the cone and are reflected back
from the edges, back towards the coil. These break-up modes are dispersive
when they cancel or reinforce the airborne vibrations. One reason for
complex, soft cone surrounds is to absorb these waves as they reach the
edge of the cone, and hence prevent their reflexion back towards the coil.
Cones used in uncontrolled break-up are not of minimum phase character.
So once again, a horn, carefully designed and equally carefully applied, is
not necessarily at a disadvantage to direct radiators. Horns are perhaps
more subject to misapplication, but do not blame the horns for that!

There is another area in which direct radiators and horns may have
differing phase characteristics. Richard Small published a paper in 1970 on
'Constant-Voltage Crossover Network Design'. He stated that whilst direct
radiator diaphragm motion is mass controlled, horn driver diaphragms are
resistance controlled. The result is a constant phase difference of 90°
between the transfer characteristics of the two types of drivers. Indeed, as
previously stated in the section on relative sensitivities, the reactive and

resistive loading are the prime reasons for the differing sensitivities of direct radiators, as compared to horns. Dr Keith Holland and I have been reassessing this situation, especially in the area of reflected impedances. These have possible implications for differing phase characteristics, dependent upon whether the units are crossed over passively, with any resistance in circuit, or actively, where the units are effectively driven from a constant voltage source – the output terminals of a power amplifier. There are obvious implications here for the overall phase response of any system of which they form a part (Fig. 10.2(a)–(e)).

10.2.1 The phenomenon of horn cut-off

The following description of horn acoustics began with a request to Keith Holland to answer the frequently asked question: 'Why do horns have a cut-off frequency?' A thorough literature search had failed to reveal a physical explanation, and the subsequent formulation of a physical theory of horn behaviour has subsequently led to useful extension of the one-parameter modelling exercise.

Horns are waveguides which have a cross-sectional area which increases, steadily or otherwise, from a small throat at one end to a large mouth at the other. An acoustic wave within a horn therefore has to expand as it propagates from the throat to the mouth at a rate dependent upon the local flare rate of the horn. A comparison between the propagation of waves in two simple acoustic systems may help to explain the physics of horn behaviour; one in which the wave does not expand as it propagates, and one in which it does. In the first instance, consider the propagation of a one-dimensional free progressive plane wave, such as a low frequency sound in an infinite pipe. A wavefront, defined here as an iso-phase surface, undergoes no change in cross-sectional area as it propagates, and the normalised acoustic impedance at any point along or across the pipe, ignoring losses, is purely resistive and equal to one.

A plane velocity source placed anywhere along the pipe thus has no reactive acoustic loading on it at any frequency, either as added mass or as stiffness.

In the second instance, consider the propagation of a three-dimensional free progressive spherical wave radiating from a monopole point source. In this case, a wavefront continually expands as it propagates, and the normalised acoustic impedance at any point is dependent upon both the distance from the source and the frequency. The impedance approaches unity, and hence is largely resistive at large radii and at high frequencies, but is reactively dominated at small radii and low frequencies.

When the product of the radius and the frequency is equal to the speed of sound divided by 2π, the reactive and resistive parts of the impedance will be equal in magnitude. A spherical velocity source of finite size will therefore be subjected to either resistively or reactively dominated acoustic loading, depending on the size of the source and the frequency of vibration. The only physical difference between the propagation of waves in these two systems is the expansion or 'stretching' of the spherical wave as it propagates – similar to the stretching of a balloon skin as the balloon is inflated. In the case of the plane wave, a forward or positive particle velocity is accompanied

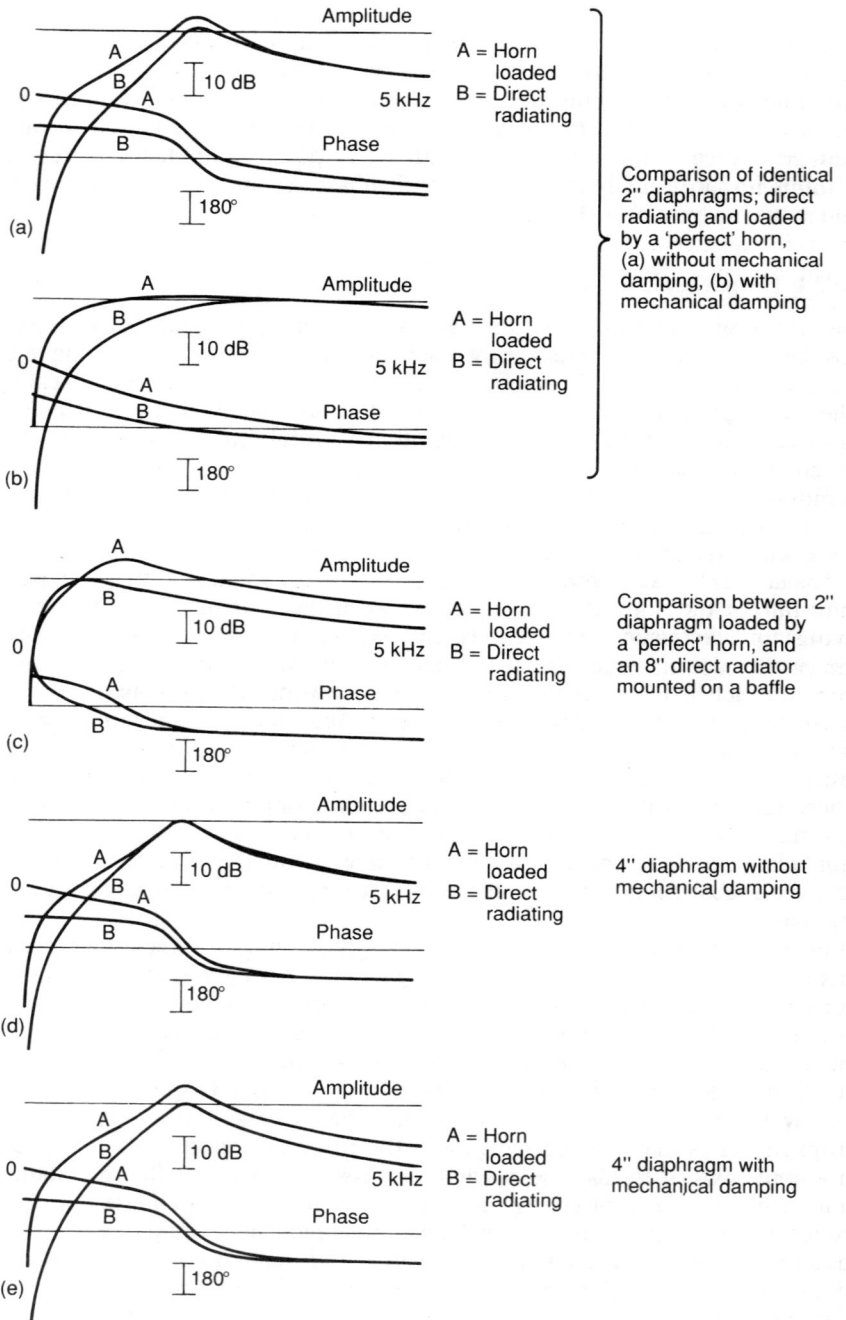

Figure 10.2 Diaphragm loading comparisons (courtesy of Keith Holland)

by a positive increase in pressure at the same point, due to the parallel motion of all of the adjacent particles, and sound is thus radiated. In the case of the spherical wave, however, because an outward positive velocity causes adjacent particles to move apart, the positive, plane wave-like radiating pressure is accompanied by a negative 'stretching' pressure due to the expansion. It can be shown that the plane wave-like propagating pressure is proportional to and in phase with the velocity, and independent of frequency and radius. The non-propagating stretching pressure is proportional to displacement, inversely proportional to the frequency and radius, and is in phase quadrature with the acoustic particle velocity.

The expansion of the wave, and hence the stretching pressure, thus has the effect of reducing the resistive part of the impedance at low values of the distance–frequency product, and increasing the reactive part. Such positive reactance is usually associated with added mass, but in spherical waves there is no extra inertia involved when compared to plane waves. The usual description is therefore inadequate to explain this property. In this case, the positive reactance is clearly due to 'negative stiffness' and not added mass. The region near to a small source where this reactance dominates is known as the hydrodynamic near field of the source; its extent is frequency dependent. The region outside this where resistance dominates is known as the far-field.

The concept of stretching pressure can be applied to horns by considering their flare rates. The flare rate can be defined as the ratio of the rate of change of wavefront area with distance to the area of the wavefront. A conical horn has a flare rate that is inversely proportional to the distance from the apex of the cone. Indeed, a spherically radiating source can be considered to be a special case of a conical horn, and thus shares the same expression for flare rate. At any given frequency, the radius at which the resistive and reactive parts of the impedance are equal in magnitude occur where the above-mentioned frequency–distance product is equal to unity. The flare rate at this radius is equal to twice the free-field acoustic wave number (2π divided by the wavelength) and is identical to the flare rate in an exponential horn having this cut-off frequency. With flare rates below this value, resistive, far-field type propagation takes place, and with flare rates above this value, reactive, near-field type propagation occurs. Using this physical concept, the difference in behaviour between different types of horn can be explained.

The radial dependence of the flare rate in a conical horn, which shares the propagation properties of a spherical wave, gives rise to a gradual transition from the reactive, near-field dominated propagation associated with the stretching pressure, to the resistive radiating, far-field dominated propagation as any wave propagates from throat to mouth. This transition in a conical horn from near to far-field, reactive to resistive dominance, is gradual with increasing frequency and/or distance from the apex; distinct 'zones' of propagation are thus not clearly evident. As it is the property of resistive loading which gives to horns their greater efficiency when compared to direct radiators, then as the resistive loading rises, so the radiating efficiency rises. From the above, the characteristic gradual throat impedance cut-off slope of a conical horn should be readily appreciated (Fig. 10.3).

On the other hand, an exponential horn has a flare rate which is constant

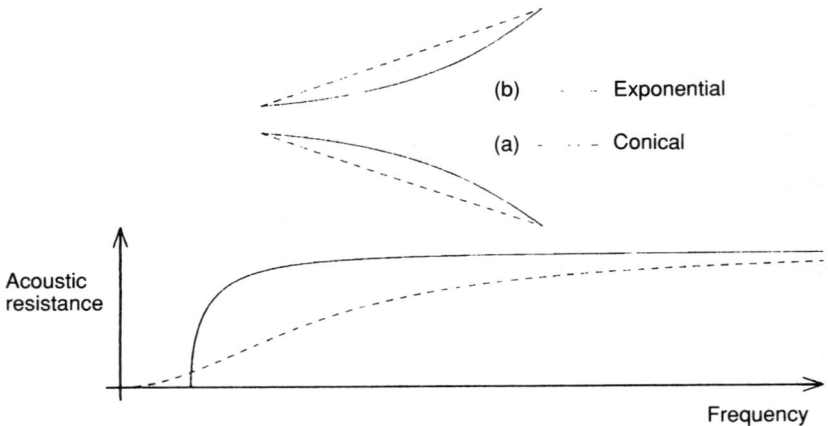

Figure 10.3 Comparison between the normalised throat resistance of an exponential and a conical horn of comparable size (ignoring reflexion from mouth). The acoustic resistance curves are typical of the 'frequency responses' to be expected from such horns

with distance along the horn. Therefore at frequencies below cut-off, the reactive or near-field type of propagation dominates throughout the entire length of the horn. Below cut-off, if the horn is sufficiently long, an almost totally reactive impedance exists everywhere within the horn. Conversely, above cut-off, the resistive, far-field type of propagation dominates, again almost everywhere throughout the entire length of the horn. Above cut-off, propagation within an exponential horn is physically similar to a spherical wave of large radius, with minimal stretching pressure. Below cut-off, propagation is dominated by the stretching pressure and is thus similar to a spherical wave of small radius. Clearly, the abrupt cut-off phenomenon of an exponential horn occurs because the transition from resistive to reactive propagation takes place simultaneously throughout the entire length of the horn as the frequency is reduced below cut-off, producing the typical throat impedance characteristics of the exponential horn in Fig. 10.3.

The use of conical horns in audio is thus limited because for practical sizes of horn, the resistive loading, and hence high radiating efficiency, rises only gradually, and thus the useful high efficiency radiation only begins to develop into a uniform response at much higher frequencies than in an exponential horn of comparable size. At the frequencies below cut-off in any horn, the reactive propagation is somewhat similar in nature to the reactive loading of direct radiators, and thus falls to the comparably low radiating efficiency of a direct radiator of the same throat size. The behaviour of other types of horns, such as hyperbolic, hypex (hyperbolic/exponential hybrid) and sinusoidal can be explained in a similar manner using the concept of stretching pressure.

10.3 Mouth reflexions

Here we have another area which is almost exclusively in the domain of horns. The problem should occur with direct radiators, but as we have already seen, the coupling from the cone (the mouth) to the air is so poor that reflected energies are too weak, and travel over too short a distance, to have any noticeable effect on the overall output.

In the case of horns, the distances for mouth reflexions to travel back to the throat are sufficient to be of consequence, and the power in the reflexions can be proportionately quite high. Poor mouth design will cause improper termination and significant reflexion. The waves reflecting back down the horn will eventually arrive back at the throat, where they can add to or subtract from the subsequent direct waves. These compressions and rarefactions in the negative direction superimpose themselves upon the throat pressures, and hence modify the throat impedance and the loading on the diaphragm.

As we saw from our friend on the bicycle, impedance (gearing) and effective loading and coupling (the pedals and chain) have a great bearing (no mechanical pun intended!) upon the power delivered to the wheels. In the same way, a varying throat impedance will have a great effect upon the diaphragm loading, and hence the output power. In turn, the output power from the driver, modified by the directivity of the horn, will translate into the pressure amplitude response, or 'frequency response' of the horn/driver combination. Reflexions from the mouth have a bearing therefore on the frequency response and on problems of phase dispersion. The size of the mouth dictates the lowest usable frequency for a 'flat' response, whilst the shape of the mouth has a bearing on the directivity. These facts must obviously be borne in mind by the designer whilst attempting to build-in characteristics for minimising the reflexions.

It is worth pointing out here that the low frequency coupling limit, which is determined by the mouth size, is not to be confused with the cut-off frequency. The former determines the coupling from the *mouth* of the horn to the room; the cut-off is determined by the loading presented to the diaphragm by the *throat* of the horn. The cut-off frequency is entirely a function of the rate of flare of the horn itself, irrespective of the length or the mouth size. Saw through a horn half-way down its length, and you will not affect the cut-off frequency. The smaller mouth will no longer couple the lower frequencies to the room, so the 'frequency response' will not go so far down, but otherwise the horn will remain substantially the same – minus any effects of lips or other mouth paraphernalia. The only other significant effect of this truncation would be that the different mouth of the short horn would inevitably cause different mouth reflexions, as compared with the longer horn. These different reflexions would inevitably produce a different pattern of waves to affect the throat impedance, and, hence, would subsequently manifest themselves as a different series of ripples upon the pressure amplitude or 'frequency response' plot (Fig. 10.4).

Only with an infinitely long horn can mouth reflexions be totally avoided, but with careful design the disturbances caused over the preferred frequency range can be reduced to relatively insignificant proportions. Poorly designed or obstructed mouths can develop cross modes which in turn have a bearing on reflexions back down the horn. Once again, however, this is a function of bad or inappropriate design, so should not be a criticism aimed at horns *per se*!

Automatic impedance measurement

Date: 19-7-88 φ_0 = 345.9 m/s
Sample type: 2A   ⎯⎯⎯ Resistance
MIC. positions: 30 & 55 mm   ⎯⎯⎯ Reactance

(a)   Frequency Hz

Automatic impedance measurement

Date: 18-8-88 C_0 = 347.6 m/s
Sample type: 2HL   ⎯⎯⎯ Resistance
MIC. positions: 30 & 55 mm   ⎯⎯⎯ Reactance

(b)   Frequency Hz

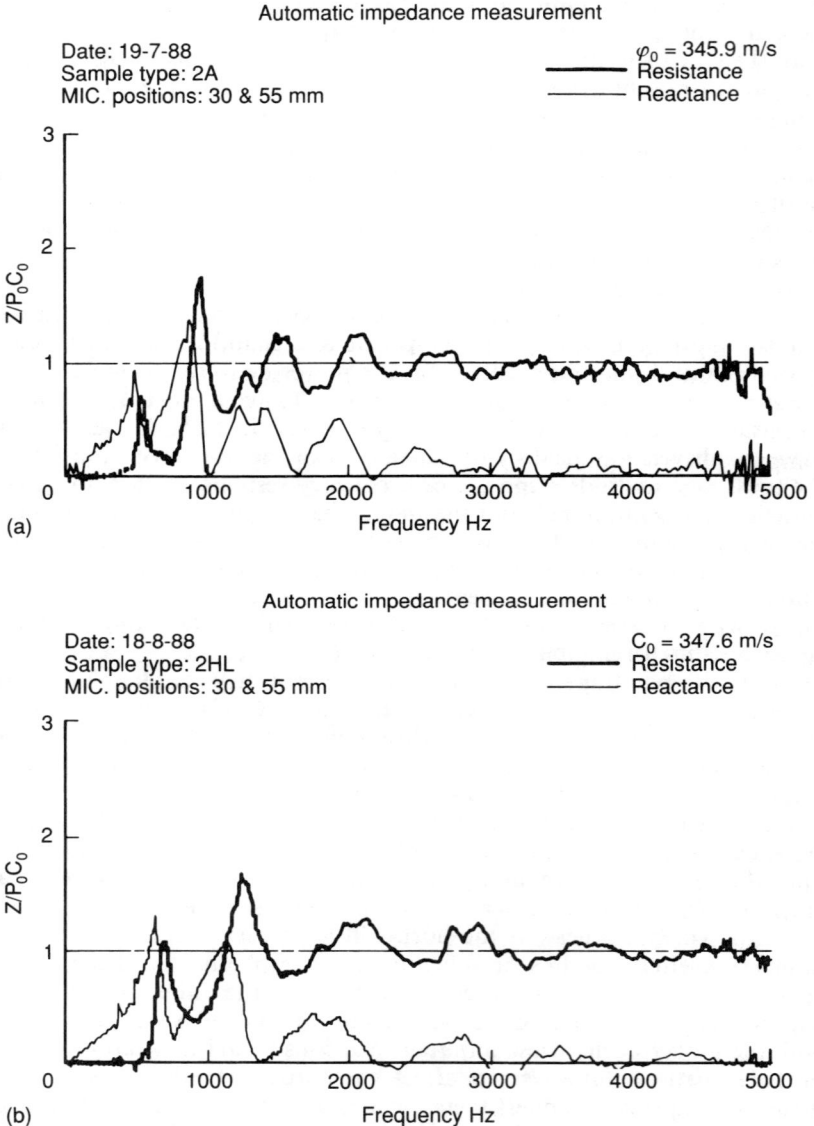

Figure 10.4 Throat impedances, hence implied 'frequency response' of the same horn, (a) with and (b) without lips. Note very different reflexion patterns from the altered shape of the mouth. Clearly the two horns will sound quite different (both examples baffled)

10.4 Non-linearities

In other words – distortions! 'They get harder as you wind them up!' Many of you, at some time or other, have probably heard that comment with reference to horns. True, to some degree they do, but the effect is sometimes noticed well out of reasonable context. Horn systems are often capable of much higher acoustic output powers than comparable direct radiator systems. The extra 'cleanliness' at relatively high levels often allows a horn system to exceed the levels at which direct radiator systems would be incurring severe overload distortions. This can occur without the horn systems necessarily appearing subjectively louder. In other words, the absence of gross distortions often tends to fool the ear into being unaware of just how loud something actually is. When horns apparently become 'harder' in or above this region, comparative criticism *vis-à-vis* direct radiators can be somewhat unfair, as the latter may well have failed or grossly distorted 10 dBs earlier!

Notwithstanding this, there are certain characteristics of horns which can tend towards non-linearities at high levels. Electro-thermal effects of voice coil heating should not be a problem in studio situations. The thermal capacities and radiation loses of the magnet systems are large when compared to the heating effect of only a few watts in the voice coil. Heating, however, can play a part in non-linearities due to the heat caused by air compression adjacent to the diaphragm.

Classic 'air overload distortion' has long been known to be a problem of horns when operating at high SPLs. The cavity between the diaphragm and the phasing plug is of a relatively small volume. At high sound pressures, the very rapid movement of the diaphragm compresses the air in this cavity when forward-going, and rarefies the air in the cavity when moving backwards. Any given unit force of compression will cause a smaller change in air volume than that same unit force in rarefaction. For example, a given volume of air, say 1 cc, acted upon in compression by force X, would compress to, say, 0.82 cc. In rarefaction, that same force would expand the air to a volume of, say, 1.22 cc.

The volume change on compression would be $1 - 0.82$ cc $= 0.18$ cc; the volume change in rarefaction would be 0.22 cc (Fig. 10.5). As a result of this, the positive-going half of the output sine wave would be smaller than the negative-going half-cycle, producing non-linear distortion in the output. This begins to happen at levels above which the air can no longer 'move out of the way quickly enough' as the diaphragm moves back and forth.

If you are still having problems visualising just why the compression and rarefaction half-cycles should produce different volume changes, then think of it this way. The compression half-cycle has only 1 cc available, even for infinite compression into an infinitesimally small 'singularity'. The rarefaction half-cycle has the whole universe into which to expand. The resistance to the pressure build-up on compression is therefore much more 'urgent' than the corresponding pressure reduction on rarefaction. As a loose analogy, connecting two identical voltage sources in-phase will effect a total voltage increase of only 6 dB. Out-of-phase connection would produce a voltage reduction of minus infinity – total cancellation.

Going back to the point on heating, there is another form of distortion which we had been looking at at the ISVR (Institute of Sound and Vibration Research). These are the results of the very high sound pressure levels encountered close to the diaphragm. When producing 120 dB behind the mixing console, a monitor system can be producing, say, 130 dB at the mouth of a horn. In turn, this can be up to 150 dB at the throat of the horn, and around 170 dB at the diaphragm. 195 dB is one atmosphere, so

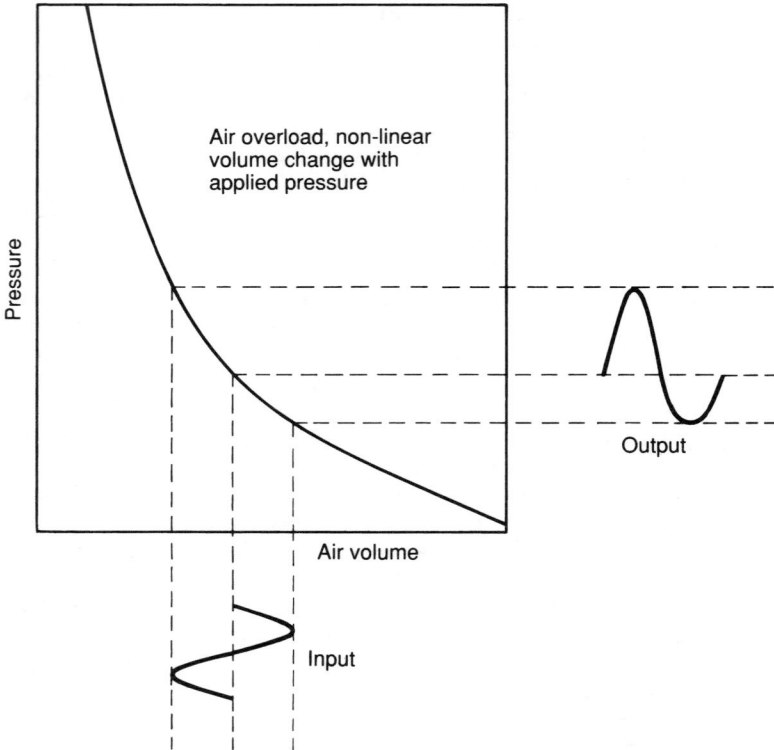

Figure 10.5 Conventional, air overload distortion

as can be appreciated we are dealing with very large pressure changes in the phasing plug region. At these SPLs, air is not a linear fluid.

Some years ago, whilst doing some background research at Cape Canaveral, I was very surprised upon witnessing the launch of an Atlas-Centaur rocket to hear all of the 'distortions' which I had heard whilst watching launches on television. I had always put the crackling sounds down to microphone or amplifier overload – certainly originating somewhere in the recording/reproducing system. Believe me, those searing, tearing, crackling sounds are as apparent near to the rocket as they are on television. They are also far more evident than one would expect, even at a range of five or ten miles from the rocket. The significance of this was pointed out to me by Dr Chris Morfey, at Southampton University. Dr Morfey had studied the noise footprints of rockets, and had found that the high frequency components of the sound, relative to the low frequencies, were not falling off with distance according to general expectations. Non-linear air acoustics were thought to be at the root of the problem, and as the blunt end of a Saturn V can generate a good 200 dB on lift-off, we are certainly well into the non-linear regions.

At these SPLs, the compression half-cycle generates such localised increases in the temperature and pressure of the adjacent air that the localised speed of sound is higher than in the ambient air. On the rarefaction half-

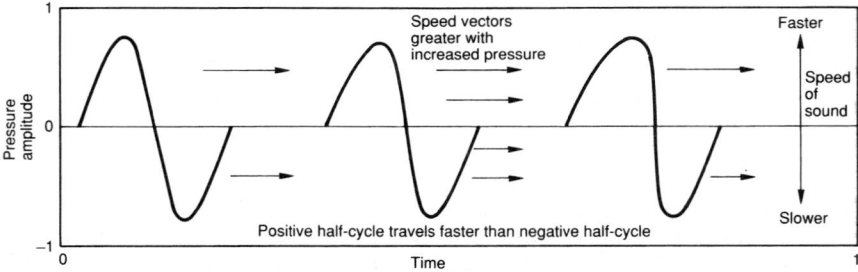

Figure 10.6 Onset of non-linear shock distortion at high SPL

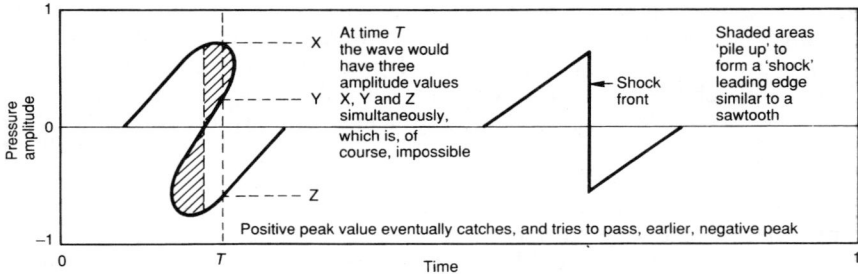

Figure 10.7 Full development of non-linear shock distortion at high SPL

cycles, the localised temperature drop creates the condition for a lower than ambient speed of sound. The speed of sound in air is a function of temperature. The speed of sound at sea level is higher than the speed of sound at 35 000 feet, where the air temperature is usually around $-56°C$, which is why aircraft flying fast and high carry Machmeters as opposed to air-speed indicators scaled in knots. Mach 1 is the local speed of sound, irrespective of the actual value in knots, or miles per hour.

The result of this (Fig. 10.6) is that at SPLs of, say, 155 dB +, the positive-going half-cycle will travel faster than the negative (rarefying) half-cycle for as long as the sound wave propagates above the non-linear threshold. The positive half-cycle of the wave will catch up on the next negative half-cycle, till the sine wave tends towards becoming a saw-tooth. It is this 'sharp edge' on the low frequency saw-tooth wave-form which keeps the rocket noise from 'dulling' even at great distances. The sine wave can never fold-over, as in Fig. 10.7, as that would mean that it could have three values at one point in time, which is clearly impossible. Instead, the waves tend towards the production of shock-waves. This is similar to an aircraft 'breaking the sound barrier', the waves 'piling up' on top of one another, creating the shock, or 'sonic boom'.

Even if direct radiator loudspeakers were capable of producing 170 dB at the cone, the sudden expansion, and hence the pressure reduction into the room, would be so abrupt that the wave would have no effective distance to propagate at high levels of intensity. The positive wave cannot catch up on the negative wave if it has no distance in which to catch up. One cannot have an Olympic, Zero metres sprint, for with no distance, there can be no race! In horns, however, the pressure reduction is much more gradual. As we have already discussed, it is the gradual taper from the diaphragm to the outside air, which enables horns to achieve a better acoustic impedance match, and consequently a significantly greater conversion efficiency. Effectively, over whatever distance exists from the diaphragm to the point where the SPL drops to the 150 dB region, non-linear propagation distortion can be considered to be a potential problem.

To maintain a smooth response, it is necessary to maintain a steady flare all the way from the mouth to the diaphragm. Any parallel section of the flare, or abrupt changes in cross-sectional area, can produce wild irregularities in the response. Maintaining the loading all the way to the diaphragm frequently requires the tiny flared tubes of a phasing plug, which adjacent to the diaphragm often causes the phasing plug to look rather like the top of a pepper pot. With 120 dB in the room, all squeezing through the top of a pepper pot, it is no wonder that things become a little 'loud' inside those holes. In the case of phasing plug design for high-level horns, to achieve a compromise between efficiency and any tendencies towards the above effects, slightly reducing compression ratios can have quite a beneficial effect, whilst only incurring a tolerably small penalty in terms of reduced sensitivity.

Once again, at reasonable, or even quite loud, studio levels, these non-linearities should not be very likely to occur. They are, however, most relevant in cinema and public address systems, and, yes – they can get harder as you turn them up, but not at studio monitoring levels. (See Chapter 12 for a more in-depth assessment.)

Another aspect of these tiny holes or slots in the phasing plug is that in attempting to transmit such high energy through such a confined space, the wave motion can become a turbulent flow in the small tubes. This can produce noise (or non-harmonically related distortion) in the output, which is a function of the air flow, which is in turn a function of sound pressure levels. Dr Wolf at NASA's Ames Research Center in California tells me that there are formulae to calculate length/diameter relationships for minimum turbulent noise flow generation. With the tapered holes required in phasing plugs, this is somewhat difficult to calculate, but Keith Holland worked on this at the ISVR. This is also tied in with venturi effects, where the flow speed changes with the change in cross-section. This is the effect which causes the lift on an aeroplane wing and the atomisation of petrol within a carburettor – as the tube diameter decreases, the flow speed increases and the pressure drops.

There are problems uniquely associated with horns and compression drivers, but they are usually only encountered at SPLs where equivalent direct radiators would have long since expired.

Many manufacturers seem to have been reluctant to publish any detailed harmonic distortion figures for their drive units, direct radiating or horn loaded. A table of some comparative results is shown in Fig. 10.8. Whilst certain electrostatic and dome mid-range units can produce outputs with no distortion products above − 60 dB, or 0.1%, most direct radiators operate in the − 50 dB to − 60 dB range of distortion products. It is true that some compression driver/horn combinations may be more inclined towards the − 40 dB or 1% distortion level, but, nonetheless, many better examples exhibit no distortion products at studio monitoring SPLs above a level of

75 dB @ 3 m		dB HARMONICS			
LS	Input (V)	2nd	3rd	4th	5th
1 kHz					
QUAD ESL	2.04	−59	−60	−69	−66
Son Audax 8″ PR17/HR100/1AK7	1.01	−59	−54	−68	—
AX2 Horn/Emilar EK175	0.24	−61	−66	—	—
2.8 kHz					
QUAD	5.6	−62	−49	—	—
Audax	0.75	−61	−60	—	—
Emilar/AX2	0.17	−58	−73	−70	—
5 kHz					
QUAD	3.5	−66	−51	—	—
Audax	0.64	−62	−59	—	—
Emilar/AX2	0.3	−47	−57	—	—

(−70 dB = 0.03%, −60 dB = 0.1%, −50 dB = 0.3%, −40 dB = 1%)

Figure 10.8 Harmonic distortion comparisons of electrostatic, direct radiator, and horn/compression driver. Measured via B + K 4134/2032

−50 dB or 0.3%. It is generally considered that harmonic distortion levels of below 0.5% are not audible from loudspeakers, due to masking effects and other perception problems. Once below this level, I believe that other factors become more important than harmonic distortions. Cones, domes, electrostatics, ribbons and horns are all capable of achieving acceptably low harmonic distortion products.

10.5 Mechanical alignment

In the mid-1970s, there was a sudden surge of interest in so-called 'time domain alignment' or 'linear phase alignment'. The precise application of the principle was somewhat convoluted, and I believe that many misconceptions were erroneously applied to some designs. The fundamental principle was to synchronise at the listeners' ears the arrival times of the impulses from the individual drivers. This was usually attempted by vertical, depth alignment of the individual voice coils (Fig. 10.9). Somewhat arbitrarily in many cases, the voice coils were deemed to be the location of the sound source, putting horn mid-range units at something of a disadvantage (Fig.

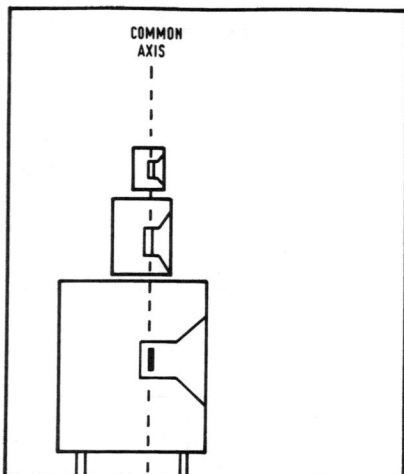

Figure 10.9 Conventional voice coil alignment intended to produce phase-coherent wavefront, obviously ignoring effects of mechanical propagation delays or electrical group delays

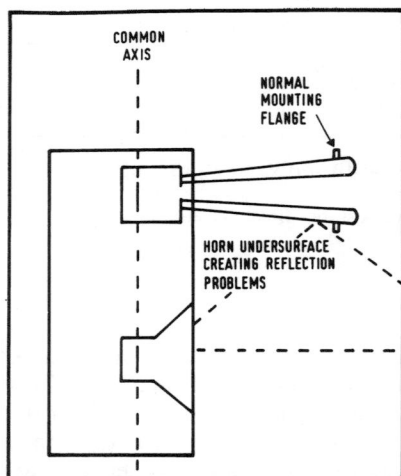

Figure 10.10 Obvious absurdity of alignment similar to Fig. 10.9 when using long horns. Conventional mounting with the normal flange aligned with the baffle would likely produce a delay of around 1 ms for every foot of voice coil misalignment

10.10). As can be seen, aligning the voice coil of the compression driver with the voice coil of the bass drivers could wind up with the horn protruding two feet from the front of the baffle. Certain companies utilising 'long horn' designs have resorted to delay lines to electro-acoustically realign the system. In my experience, however, there has sometimes been over-compensation. This has been due to not taking into account the group delay effects in the crossovers, and the propagation delays in the drivers themselves.

Taking these things into account, and considering the group delays affecting the lower frequency drivers, the shorter horns may in many cases actually be more desirable physically than direct radiators. The necessity for pushing back the mid and high units into separate enclosures is, in the case of short horns, dispensed with. The horn would also not suffer the diffraction problems of some awkwardly mounted cone and dome units.

10.6 Directivity

One advantage which mid-range horns definitely possess over direct radiators is the available range of polar pattern control, or directivity. Whilst the overall expansion of the area of the horn must closely follow the desired mathematical curve – for example, exponential, hyperbolic, catenoidal, hypex – the relative dimensions of width and height can be varied; though obviously the axisymmetrics offer only a circular pattern, akin to that from direct radiators. This variation in mouth shape can be used to cause the output to spread out over a considerable range of areas. Conventional horns are likely to cover 60° to 120° in the horizontal plane whilst narrowing the pattern to between 30° and 60° in the vertical. Such a capability can allow horn systems to direct the sound to cover a desired area, rather than being restricted to the symmetrical polar patterns of a cone or dome. As frequencies rise, pattern control is eventually lost as the high frequencies beam out in a narrow pattern as though the horn were not there.

The beaming problem can be overcome by the use of 'constant directivity' horns which are usually exponential/conical hybrids, maintaining their polar pattern with relative disregard for frequency. The price paid for this advantage is that the total available amount of high frequency power falls off as the frequency rises above the point where the diaphragm mass begins to take effect. Constant directivity is achieved by the appropriate use of diffractive angles in the geometry of the horn. In a conventional horn driver, the total power at higher frequencies, where beaming begins, reduces. However, the tightening pattern of the beam concentrates the energy on axis, thus the on-axis response can remain 'flat' with a reducing high-frequency response off axis. The total power of the two systems remains the same, so if the constant directivity horns spread a reducing power over the whole area, then in order to achieve a 'flat' response on axis, the electrical drive at those frequencies must be increased in order to compensate. If the roll-offs are still in the minimum-phase domain, the electrical equalisation will restore the phase response as it restores the amplitude response. Should the roll-offs not be minimum phase, a price for constant directivity would also be paid for in terms of phase accuracy. Purely personally, I prefer to use the conventional type of horn, due to my inherent wariness of the audibility of equalisation circuits in monitor systems. Once the pattern control has fallen below acceptable limits, I prefer to cross over into a matching, complementary high-frequency driver. Further aspects of the sonic limitations of constant directivity designs are discussed in the Conclusions section of Chapter 12.

10.7 Conclusions

Whilst the properties of direct radiators are well covered in other books, I have concentrated in this chapter on many of the properties of horns

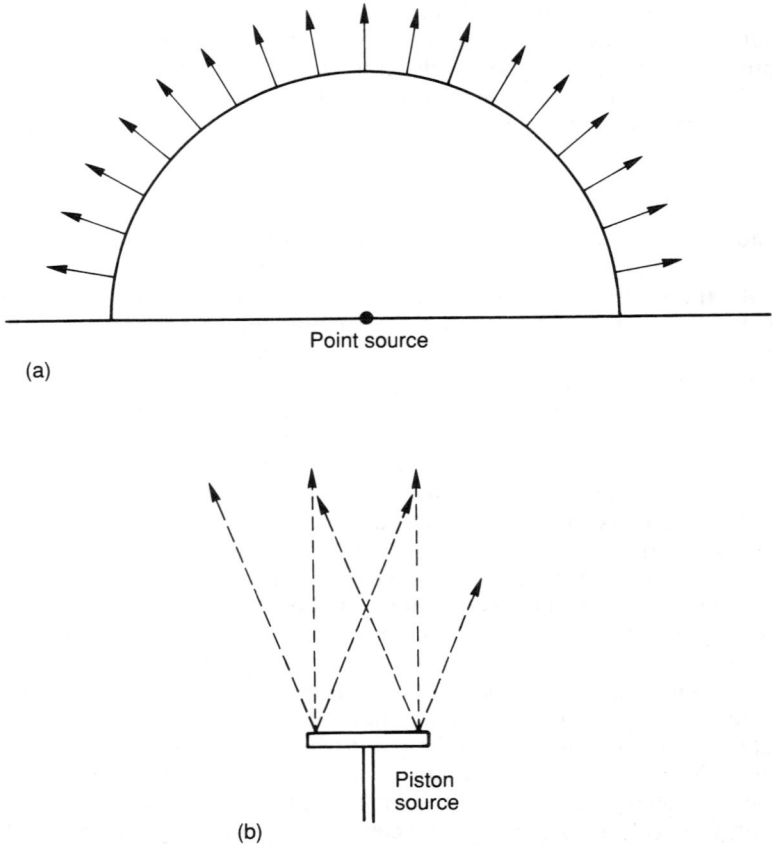

(a)

Point source

(b)

Piston
source

Figure 10.11 (a) Spherical expanding wave with all particles moving away from each other;
(b) radiating pistons produce particle motions which can interfere with each other, producing
interference patterns and subsequent lobing on the polar pattern plots

precisely because such information is not so widely promulgated. I use
direct radiating mid-range drivers just as frequently as I use horns, but in
the upper SPL ranges I almost exclusively use horns. In the lower SPL
regions, horns *per se* need not be at any significant disadvantage if specifically designed for their end use. Indeed, the spherical/spheroid expanding
wave propagation of a horn can be more akin to a point source than the
piston action of a direct radiator. In the case of a spherical expanding
wave, only one portion of the sound source is ever pointing directly at the
listener, as opposed to the multiple path lengths which give rise to the
lobing effects of direct radiator polar patterns (Fig. 10.11). I am not trying
to actively promote the use of horns, but I *am* endeavouring to point out
that they *can* have advantages, often little realised, in certain critical
applications.

Horns used in studios have frequently been 'borrowed' directly from
public address and cinema applications. In many cases, this has been done
without due regard for the different requirements of the different disciplines.

In cinema, sound reinforcement and studio usage, the compromises lie in different areas because the problems and priorities lie in different areas. Studios do not have to contend with the reverberant conditions of cinemas, or with the problem of projecting the sound to a balcony 200 ft away. Similarly, in the largely reverberant field of that balcony, less importance is placed on the absolute transient integrity which is so necessary in studios. Just because a horn has been used for twenty years from 800 Hz to 20 kHz in one particular situation does not automatically mean that it is a suitable horn for the same range in a different application. There are very many factors which must be taken into account.

There is tremendous potential in studio mid-range horn/driver combinations: very low distortion, very fast transients, high output potential, high reliability and consistency relative to acoustic output, facilitated pattern control, and many other advantages. These potential benefits have often been abused by attempting to use the horns either to cover too great a frequency range, or in an inappropriate frequency range. They have been employed with a degree of misunderstanding, and a lack of a comprehensive awareness of their strengths and weaknesses. They must be interfaced carefully, both with the other drivers in the system, and in their relationship with the room.

Only a very small percentage of horns have been specifically designed for studio monitoring. Horns can also be costly, both to develop and to manufacture. Studios account for less than 1% of professional loudspeaker sales, so it is probably understandable that large manufacturers apply more of their resources and expertise in other directions. I have felt for some time that too little attention has been paid to the potential of horns, and that they have often been maligned unfairly, frequently due to a lack of understanding of many important aspects of their behaviour. The previous discussion has largely been on what the text books either do not tell you, or fail to explain lucidly.

In a 1988 magazine interview, David Hawkins of Eastlake Audio commented that he believed that the hi-fi industry was now leading the studio industry in innovation and research. I tend to agree with him! The commercial recording industry has not effectively reinvested in the development of its most basic tool. It would be unreasonable to compare too directly the worlds of studio monitoring and hi-fi, since the requirements are different: domestic systems are not designed to cope with a solo bass guitar at 115 dB, nor the concentrated high frequency output of a solo synthesiser. It is also impossible for the studio industry to produce systems which mimic, in every way, the wide variation of what is accepted as 'right' in hi-fi. Historically, however, studio systems were looked up to as the basic standards of reference. What the advances in domestic hi-fi have done is to make people more critical and aware of just what they are listening to. The demands for accurate monitoring are now greater than ever.

Carefully designed mid-range and high-frequency horns, correctly and carefully applied, have much to offer in precision monitoring above 1 kHz. Contrary to much popular belief, clean hi-fi in this range is not the exclusive territory of direct radiators. New generations of horns are a force to be reckoned with!

Chapter 11

Background to the search for a new, high definition mid-range horn

The previous chapter discussed some of the physical strengths and weaknesses of horns when compared to direct radiators. The chapter following this one deals with the listening tests which were carried out in order to determine some of the things which contributed to those strengths and weaknesses. In this chapter, we can see the thinking concurrent with the start of the listening tests; by such means, much of the relevance of the outcome of the tests will be seen in a clearer perspective.

11.1 Horn properties

Horns have been designed mainly for efficiency and directivity. Relatively little has been done to research what actually happens to the sound waves inside the horn, and the relevance of this motion to what the horn actually sounds like. From a vast array of available horns, it has been necessary to make some attempt to correlate physical characteristics, mathematical and geometrical properties, materials, and mounting methods, with particular attention to desirable or undesirable sound qualities. In order to design a horn specifically for studio monitoring, the following headings require further detailed discussion:

1 Directivity – beaming, fingering, constant-directivity
2 Geometry – diffraction, radial, multicellular, compound designs
3 Distortion
4 Colouration – reflexions, eddy problems, obstructions in wave path, wave propagations
5 Cross-sectional relevance – rate of flare, square to round throats
6 Construction material – practicability, effect on sound
7 Physical to acoustical relationships – diaphragm to mouth distance
8 Effects of abrupt cross-sectional discontinuities, velocity modifications

11.1.1 Directivity and geometry

Directivity control has long been a prime consideration in horn design. The first objective is to establish the desired horizontal and vertical directivity pattern, then to approximate as closely as possible to this ideal over the entire intended frequency range. One flexible approach to achieving the

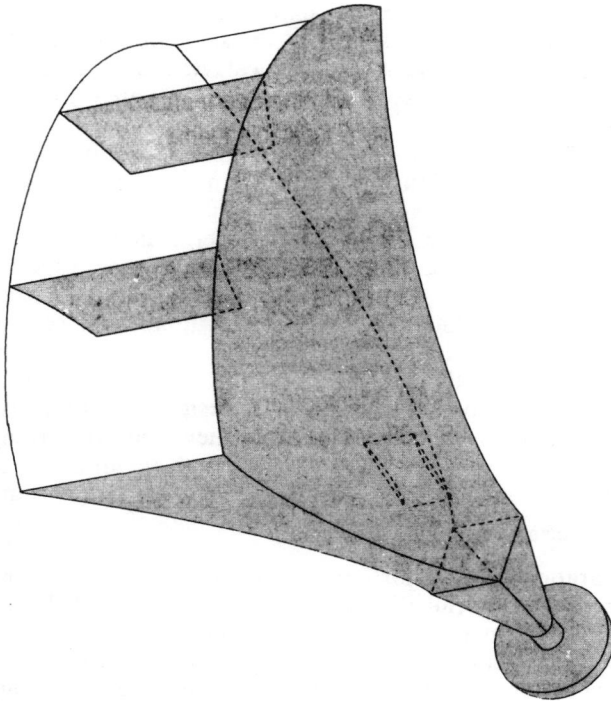

Figure 11.1 Archetypal mid-range horn

above is the multicellular method. Multicellular horns are clusters of similar horns, bunched together so that each component horn can be arranged in such a direction that the overall coverage area is served by the entire cluster. Drawbacks can include 'fingering' at high frequencies, when 'beaming' begins to occur as the polar pattern of each small horn narrows. A cluster of, say, eight small horns may well produce a relatively coherent low frequency directivity pattern, as at those frequencies the mouth areas sum. At high frequencies, however, eight narrow beams or 'fingers' are produced, far more noticeable at close quarters than in longer throw, public address use. A further problem with multicellular devices is that it is very difficult to find an ideal way of mating a large number of individual throats to the single, circular output of the compression driver.

A variation on the multicellular theme is the use of dividers to modify a basic radial horn into smaller sections (Fig. 11.1). These can be either near the throat, carefully shaped to maintain the flare, or at the mouth to break up transverse standing waves. This method helps retain the low frequency coupling of the larger section of the horn, whilst spreading more evenly the distribution of the higher frequencies. There is still something of a tendency towards high frequency 'fingering' but somewhat less than the multicellular method. Rarely, however, are all the sectors perfectly symmetrical and equal. Once again, perfect mating to a simple, circular throat can present difficulties. The radial horn gets its name from the slab sides, like the radial cuts which form a slice of cake. Radial horns are also sometimes known as 'sectoral' horns, as they are again like a 'sector' or 'slice' of cake. In the horizontal plane, the flare is therefore always conical, with straight sides. The geometry necessary for achieving an overall exponential flare (or other) is in the vertical plane only.

Constant-directivity designs have attempted to use cross-sectional geometrical changes to maintain a relatively accurate and equal directivity pattern across the entire design frequency range. A significant problem with these designs is that maintaining a constant polar pattern, independent of frequency, is achieved at the expense of a flat, on-axis amplitude response with frequency. The resultant necessity for equalisation circuits to flatten the frequency response can preclude the use of these units with certain studio monitoring designs philosophies. Implications for constant-directivity concepts are further discussed at the end of Chapter 12. Briefly, though, it is difficult to avoid reflexion problems with CD designs, as they use reflexion creating abrupt cross-sectional changes to achieve their directivity control.

11.1.2 Distortion and colouration

The sources of colouration and distortion are manifold, and it is the reduction of these two elements which has presented the greatest challenge to horn designers. Most assessment of horns seems to have revolved around pressure amplitude measurements, but in most instances the main sources of colouration are various forms of phase distortion. Once these distortions in particular can be brought within the range of high quality cone and dome drivers, the compression horn can then be reasonably expected to be capable of exceeding in many ways the performance of most other high output mid-range devices.

In the highly sensitive throat region, any abrupt changes in cross-section can be deemed more akin to concepts of pistol silencer design than to achieving the objective of a smooth transfer of sound; yet many well used studio horns have major discontinuities in this very region.

In order to achieve a horn of the highest fidelity, a single compression of the driver diaphragm must produce a single, coherent pressure wave at the mouth of the horn. This wavefront must then be allowed to leave the mouth and disperse into the room with as smooth a transfer as can be accomplished. The uniformity of this wavefront is in many instances shattered by absurd cross-sectional changes, both in the throat of the horn and also at the mouth. This can be further exacerbated by any obstructions in the horn itself, sectional dividers, dampers, wave guides and other similar devices. In one example of 'reputable' manufacture, a circular strengthening pillar was positioned exactly on axis, and I have even seen serious designs with bolt-holes penetrating the throat! Reflexions from these obstructions, together with further cross-sectional changes which they may impose, create multiple wave paths to the mouth of the horn. Some waves may even be turned through ninety degrees, eventually dissipating their energy in the walls of the horn and never reaching the listening room. The resultant of the reflexions, absorption, and path length variations can be phase chaos when the sound eventually reaches the ear. Colouration and intermodulation products are only to be expected from these above aberrations. The shapes of the flares themselves would also appear to have a dramatic effect on the tonality of the different horns. Once again, however, prime consideration in the design of so many horns appears to have been given to directivity pattern criteria, rather than to what the horns actually sound like.

Many expensive, high quality horns are mated to drivers with two-inch throats. Advances in diaphragm materials and designs have enabled responses to be pushed, usually with the aid of equalisation, to 20 kHz and beyond, which has led to a situation of diverging interests. The commercial tendency is to utilise this driver response, by attempting to enable the horn itself to retain its pattern control to ever higher frequencies. In reality, a 2″ horn throat can accept approximately three entire wavelengths at 20 kHz across its diameter.

When achieving 125 dB at 1 metre, the sound pressure level in the horn throat reaches levels where air compressibility effects can lead to non-linearities in the sound propagation down the horn. Wave motion at such high levels in a wide diameter throat results in poorly controlled wave motion within that throat. I personally feel that a 1″ (25 mm) throat at 20 kHz – roughly a ¾″ (19 mm) wavelength – would be a maximum in both diameter and frequency for well controlled 'natural' uncoloured sound.

The throat should contour very smoothly into the geometry of the flare, and the mouth should then smoothly release the pressure wave into the listening room, with minimum interference and the most gradual transition. Obstructions in the sound path should be discouraged, to preclude the possibility of reflexions, eddies, and abrupt disturbances in the flare rate. Preliminary investigation suggested that, all other things being equal within practical limits, the shorter the distance from the diaphragm to the horn mouth, the less coloured the sound. This would agree with a 'common sense' approach in that one would expect a larger horn to impart more of its particular characteristics upon a sound travelling 'down the tunnel' for a longer period of time.

11.1.3 Cross-sections and materials

A definite link exists between flare shape and colouration, and despite sound being a wave motion as opposed to an air flow, I had personally often contemplated the relevance of some aspects of transonic aerodynamics. Aircraft designed for transonic speeds comply with the 'area rule' by making only very gradual changes in overall cross-section, slice by slice from nose to tail. Thin wings may taper, and gradually fatten in section at the roots, smoothly blending into a streamlined fuselage cross-section. A well known example of this type of aerodynamic principle is the Lockheed SR71 Blackbird (Fig. 11.2). Even though wave motion in a horn does not flow in a linear way, it does move, albeit over very short distances, and that movement is by definition at the speed of sound. Turbulence as such is a function of a flow, and would not be applicable in wave motion. Eddies, however, could possibly be caused by non-adherence to the 'area rule'. I am using the term 'area rule' here, not as a definitive term, but as the closest approximation of a known rule to an intuitive feeling. In the phasing plug, however, at very high SPLs, most certainly a turbulent flow *can* exist, hence here the links are much stronger.

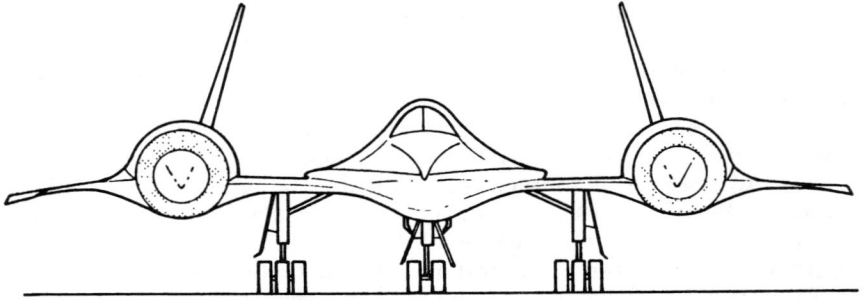

Figure 11.2 Smoothly contoured Lockhead SR71 Blackbird

Much has been made in the past of the relative merits of wood, metal, urethane, glass fibre, and other materials used for the manufacture of horns. Wood was long held by many designers to be superior to metal, due to more benign resonances and a less harsh sound. Recently, however, horns of very complex multiple curvatures have been moulded from synthetic, mineral loaded resins and glass fibre. Much of the mystique of wood still lingers in the recording industry, but wood does not easily lend itself to the manufacturers of complex, ever-changing contours. Again, the cross-correlation of certain shapes being made from certain materials and not from others may have led to the apportioning of certain sonic characteristics to the materials, which should have been attributed to the shapes.

For example, many wooden horns, for manufacturing reasons, terminate in a square section throat and have a sharp, angular lip at the mouth. The square throat must somehow be mated to the round driver, and many very expensive wooden horns have abrupt cross-sectional changes in this region. Does this influence the velocity and direction of the waves, and is it uniform with frequency? Are eddies, or short term rarefaction and compression distortions, present? By the time that the wavefront reaches the mouth, what effects do any of these properties have on the phase correlation, or on any possible cancellations? Have any of the sonic characteristics produced by such geometrical problems been wrongly attributed to properties of wood?

It was in the light of all these questions that the research was undertaken which culminated in the listening tests described in Chapter 12. I had been designing a range of studio monitoring systems, but in four years of intensive searching, I had failed to find an entirely suitable proprietary, purpose-designed, mid-range horn. Eventually, a Malcolm Hill design, modified by ASS, was chosen as being closest to my ideals, but it had subsequently been difficult to assess and quantify the relevant parameters which set this horn apart from the others. It did, however, have the smooth contouring which corresponded well with the aerodynamic parallels. The intention of the listening tests was to attempt to relate the physical properties of horns to their subjective sound qualities, and subsequently to use the correlations to design a new horn. The original criteria for the proposed new horn were:

1 To be a high fidelity, mid-range, studio monitoring horn/compression driver system.
2 To be capable of 125 dB at 1 metre.
3 Design frequency range 800 Hz to 7 kHz.
4 The amplitude response should be smooth and free of any significant peaks for at least one octave either side of the design frequency range – a smoothly falling response would be acceptable.
5 Directivity of approximately 100° horizontal by 40° vertical, held as equal as possible over the design frequency range.
6 The size should be within practical limits for studio purposes, say 30" × 12" maximum front face area, preferably as small as possible but without compromise to performance.
7 Minimum inter-modulation, harmonic and phase distortions.
8 Natural subjective sound quality with minimum colouration; to be assessed by a consensus from listening tests.

11.2 Design approaches

I believed that the only workable approach was firstly to find out what sounded good; secondly, to use scientific means to find out *why* it sounded good; and thirdly, to discover the relevance of this data and apply it in a practical way to future designs. Martin Colloms, in his book *High Performance Loudspeakers*, gives a very great deal of evidence in the chapter on 'Loudspeaker Assessment' to support the lack of correlation between measurable non-linearities and subjective sound quality in loudspeakers, and it was also decided here that the whole project would have to be constantly monitored by listening tests. Only the ear is the ultimate arbiter, as it is coupled to the most advanced computer on earth – the human brain! Research work began with mathematical, numerical, theoretical, practical, and audiological approaches. Careful consideration was given to the inter-relationship of these disciplines, and the myriad of cross-connections between them.

11.3 Practical research

One of the biggest hurdles was still the problem of correlating what is measured with what is perceived. The individual shapes are somewhat akin to the individual species of animals. If an ape has a mutant offspring, the offspring will be a mutant ape. If a lion has a mutant offspring, that offspring will be a mutant lion. Never is the genetic mutation of the fetus of an ape such that it is born a lion, or a giraffe! Likewise with the horns: if a horn had a particular characteristic sound, then even if its physical shape was significantly deformed, it always seemed to sound like a modified version of the original horn.

Just as the alchemists had failed to turn base metals into gold, we had not been able to change any characteristic sound of one horn into that of any other horn. We tried to measure the overall flare rates of individual horns, then mould into the walls of other horns shapes which would modify the flare rate of one horn into the flare rate of another. Would this make one horn sound more like the other? After all, great importance had historically been placed on accurate and controlled rates of flare. To our surprise, we could engineer gross disturbances in the flare rates with pitifully small effect on the perceived sound. Things *were* disturbed, however, by the installation of pillars and dividers, especially when placed on the central axis of the horn (Fig 11.3). Quite astonishingly, in conventional horn thinking this has not been taboo.

Initially, the listening to and measurement of the 'modified' horns had taken place on axis. Nonetheless, it was soon noticed that if an asymmetrical disturbance was introduced into a horn, its presence was noticeable from a listening position at a normal (90°) to the disturbed wall (Fig. 11.4). So, the precise overall flare rate was appearing to be of little consequence, but relatively minor side-wall disturbances were easily noticeable when listening off-axis at the opposing side of the horn. There were two immediate questions which sprang from the above observations. Firstly, they fitted in well with the original trigger for the research, the super-smooth contouring of the Lockheed Blackbird (Fig. 11.2). Secondly, if the ASS horn (Fig. 11.5) was sounding relatively neutral, just what would happen if the 'smoothness' of contour were to be taken to its extreme: a perfectly axisymmetric horn, free of all dividers or other obstructions? From the observation that minor disturbances to the shape at any point at a normal to the listening position were readily noticeable, it was construed that any abrupt, angular junction between the side-walls, top or bottom of a rectangular horn must cause a disturbance in the response at any points at normals to those boundaries.

There had also been other pointers to the desirability of axial symmetry which had a habit of appearing at frequent intervals. Whilst Keith Holland had been working with Tonni Johansen at Trondheim University, where Johansen was doing a mathematical study of horns, involving the numerical modelling of finite wall sections for directivity prediction (both Holland and Johansen were on their way to PhDs for their respective work), several people kept pressing the question, 'Why not axial symmetry?' Furthermore, the Tannoy Dual Concentrics have rarely, if ever, been accused of sounding 'horn-like', yet above around 1 kHz, they are axisymmetric horns.

It had also long been known that, in partially reverberant conditions, any wide discrepancy in polar pattern with varying frequency, as can be experienced when mixing direct radiators and rectangular horns, would cause the reverberant field to have a differing 'frequency response' to the direct, on-axis response. Such a discrepancy, if severe, can cause undesirable characteristics to be attributed to that loudspeaker system when used in reflective/reverberant rooms. This point was strongly made by some readers of my *Studio Sound* articles on the subject. Obviously, the room acoustics will have a great bearing upon the perception of any such effects, and when the room and monitor system are part of one, integrated design, these things can be accounted for at the design stage, putting conventional horns at no disadvantage on this count. However, the use of axisymmetric horns, whose directivity properties more closely match those of any direct radiators with which they may be mated, would greatly alleviate this problem. Many proprietary systems for general use could well benefit also from axial symmetry.

So why have we not seen more of axisymmetric designs? I suppose that the vast majority of all large horns are used in public address, cinema and sound reinforcement applications, where pattern control is of prime

AXIAL PRESSURE AMPLITUDE DISTRIBUTION

Date : 14-6-89   30 Hz per division

Sample type : 8A

Speed of sound : 347.7 m/s

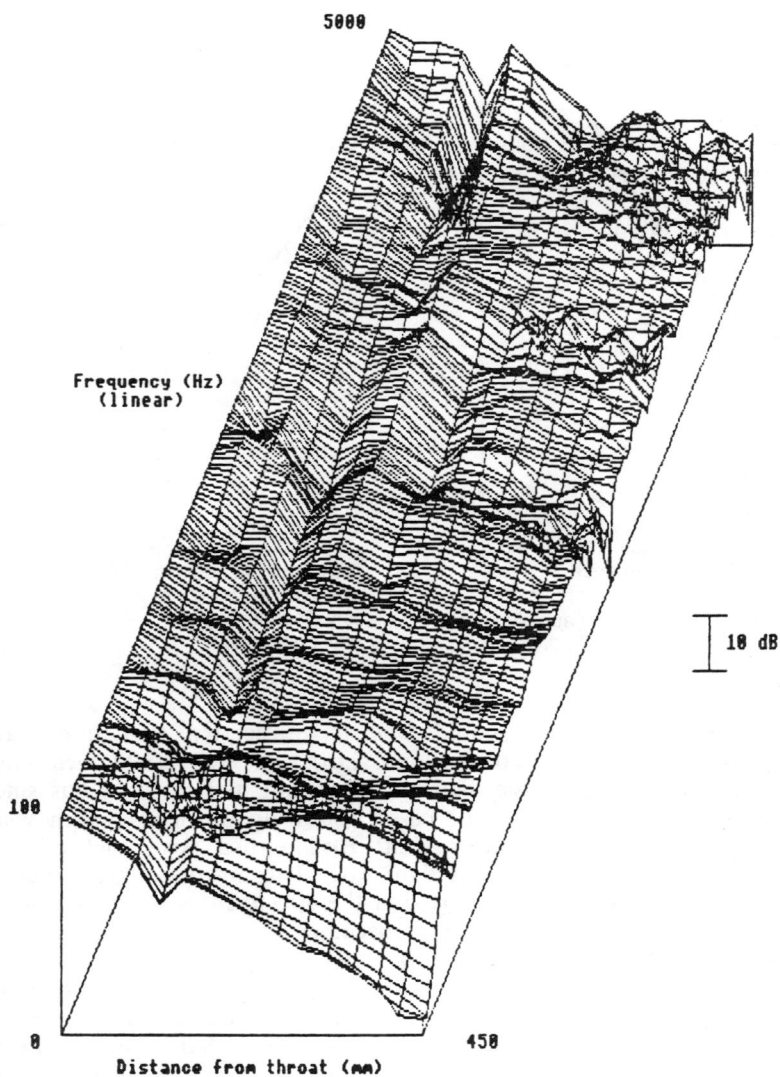

Figure 11.3 Response disturbances in horn with pillars. Response off axis normalised to a corrected on-axis response

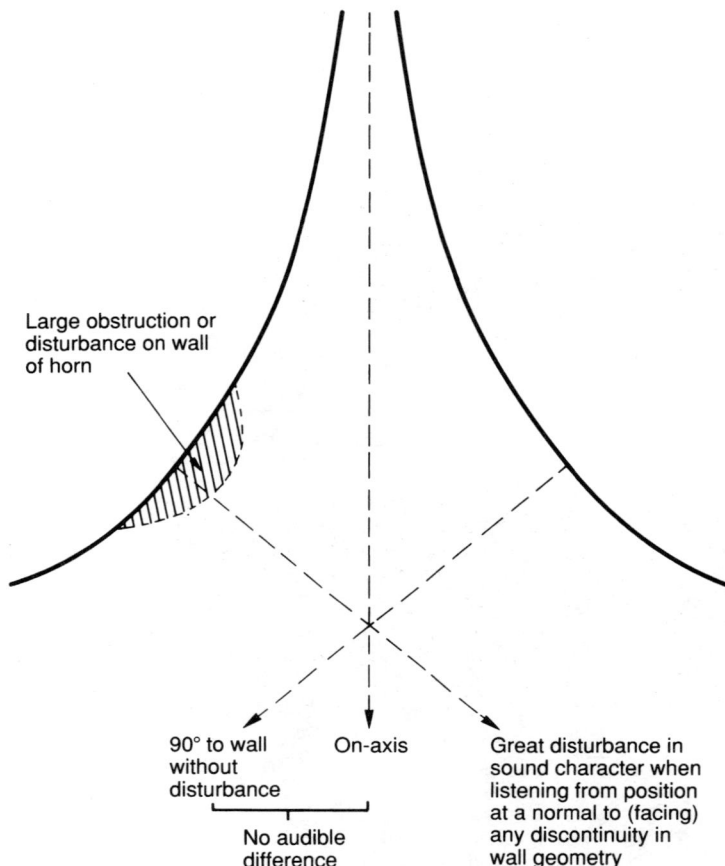

Figure 11.4 Disruption of flare rate

consideration: a 'mind-set' could well have dictated that horns are rectangular. Furthermore, conventional horn concepts could well have dictated an inordinately large diameter of mouth, but research has since shown that this need not be the case. Following the leads which were pulling us towards axial symmetry, Keith Holland had axisymmetric horns made with opposing extremes of flare rate. Throat cut-off frequency is solely a function of the rate of flare, but the effectiveness of the mouth to terminate smoothly to the outside air becomes untenable once the width, height or diameter of the mouth approximates to one wavelength.

Harking back to the intuitive feeling that short horns were desirable, we concluded that a short horn with a relatively wide mouth would probably be appropriate, but this would require a rapid rate of flare. The rapid rate of flare would push up the throat cut-off frequency, and we had already elected to avoid the cut-off frequency by at least an octave in order to avoid the phase and amplitude irregularities associated *with* the cut-off. By these criteria, many of our 'desirable' properties were beginning to appear as though they could not be 'welded' together in one horn.

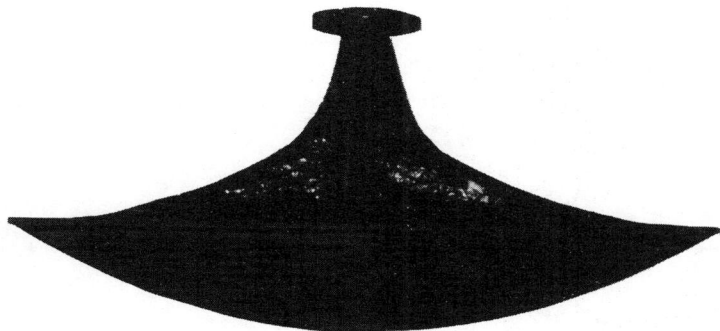

Figure 11.5 Smoothly contoured horn

Undaunted, Keith Holland drew up the axisymmetric designs and ordered them from the ISVR's model makers. The first two were chosen with differing compromise points. The first, the AX1, was a short horn with a flare rate giving a cut-off in the region of 400 Hz. In order to remain short, the low rate of flare dictated a small diameter mouth, which would probably only be rendered usable by the addition of accoutrements such as slant plates or other termination aids. A low rate of flare in conjunction with an adequately sized mouth would inevitably dictate a long horn, with greater possibility of phase dispersion and other unwanted properties. The second model (Fig. 11.6) had a cut-off frequency far too high for our original design criteria, especially if we still intended to maintain a linear response to one octave below the lower limit of our desired operating range. It was a short horn with a rapid flare and very smooth mouth termination, and was labelled AX2.

Mathematical modelling *had* suggested, however, that the latter horn may well be a viable proposition, as the mouth termination seemed to be very close to what was considered to be theoretically optimal. The correlation of the mathematical prediction with the actual, measured device was remarkably accurate. The mathematical modelling had been devised by Holland[1] after an in-depth investigation into the phase plots of waves travelling down horns. Conventional horn theory dictated that the flare must expand in cross-section at a precise mathematical rate as the sound wave travels down the horn.

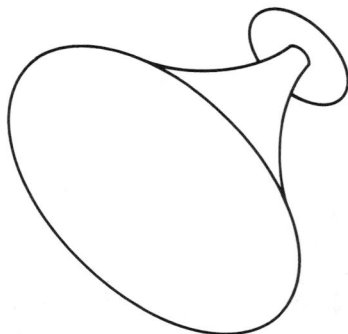

Figure 11.6 Axisymmetric horn geometry

A horn propagates sound, not with the piston action of a direct radiator, but with the breathing action of an expanding and contracting chest or balloon. The shape develops gradually from the virtual plane wave at the throat, to a section of a more or less spherical wave at the mouth. Holland's one-dimensional computer modelling, however, concentrated on the fact that the expansion of the horn must take into account the changing shape of the propagating wave, for in reality it is the *wave* which must expand at the appropriate rate, rather than any particular cross-section of the horn. His previously measured phase contours within the horn gave him the necessary mathematical weaponry to attack the horn shape problem.[2]

The measurements and aural appraisal of the short, fast flaring horn were very revealing. Measurement of throat impedance showed the horn to be almost devoid of mouth reflexion problems. The throat impedance approximates to the 'frequency response' of the horn, as it influences the loading which the horn presents to the driver diaphragm. The unusual smoothness of response was attributed to the flare rate being high enough to allow the horn to blend with a baffle, without any abrupt change of angle at the boundary. This, together with the newly calculated rate of flare, minimised the phase distortions inside the horn, and allowed phase to remain largely coherent as the waves left the horn to enter the listening room.

I was beginning to think of the sound waves leaving the horn in a way not dissimilar to that in which a bubble leaves a child's bubble-blowing kit. To test my 'bubble theory' further, I twisted pieces of wire into shapes roughly approximating to the mouth geometry of various horns. Dipping these wires in bubble-blowing fluid, I was not too surprised to see that the circular shape produced the largest and most reliable bubbles. Many of the other shapes saw the bubbles-to-be tearing themselves apart as they attempted to leave the corners of the wires. It was rapidly becoming apparent that any boundaries formed by the different angles of a conventional horn mouth would make it very difficult for a clean, spherical/spheroid expanding-wave section to transfer itself from the horn to the room.

A surprise bonus from the short, high-flare axisymmetric horn was that it was usable from a sonic standpoint all the way down to its cut-off frequency. The conventional necessity to remain an octave or so above the cut-off was no longer a prerequisite, as the new axisymmetric device was almost free of vices in the cut-off region. It had an almost perfect 12 dB/octave roll-off, which begged the question as to whether or not, in future designs, a horn could be made such that the crossover frequency coincided with the cut-off, and that the acoustical roll-off could be incorporated in the overall roll-off when connected to the electrical crossover of the system.

Initial listening tests showed the horn to be very smooth and neutral, corresponding very closely to the measured characteristics. It was interesting to note that the new horn sounded not too unlike a smoother version of the original ASS horn, which had inspired the early work and the thoughts about acoustical/aerodynamic similarities. Keith Holland had taken many of the basic parameters of our original horn, pushed them towards our most desirable parameters, and the result actually sounded like a smoothed out version of our starting point. Once again, the 'mutant' horn sounded like a close relative of the horn from which its concept had been derived. Despite looking totally different from its predecessor, sonically, it remained of the same species. The ape still had not turned into a lion – it was again merely a different ape!

The widely held belief that a horn of any given flare rate can be any desired shape did not seem to hold true from a sonic point of view. Shape seems to play a very great role in the perceived sonic performance of a horn. Two horns of similar shape, but both having flare rates deviating in different directions from a theoretically desired norm, are much more likely to sound similar than two horns which may adhere precisely to the desired overall flare rate but which possess significantly different shapes.

11.4 Materials and construction

Although extremely difficult to measure by conventional techniques, it had long been suspected at the business end of the industry that the actual material of which a horn may be constructed could have a significant bearing upon the perceived sound character. Before proceeding further, it is worth noting some measurements made in the 19th century[3] on the speed of sound through different woods:

• Beech, along the grain 10 965 ft/sec, cross grain 6028 ft/sec
• Oak, along the grain 12 622 ft/sec, cross grain 5036 ft/sec
• Ash, along the grain, 15 314 ft/sec, cross grain 4567 ft/sec

As mentioned earlier, material of construction has long been held by many to be of sonic significance, though text books have generally suggested that only shape was of importance. A combination of high density and high internal loss would appear to be the most desirable combination for horn construction materials. Metal horns have frequently been considered 'bright' or 'brittle', though I doubt that that could be said of a horn made of lead (Pb). Once again, public address applications have probably precluded the use of lead from the points of view of weight, cost and rigidity in transport.

Wooden horns came into vogue in recording studios in the early 1970s. The hardwoods now seem to have given way to plywoods, and even the plywoods now seem to be gravitating towards their heavier density varieties. Kinoshita's penchant for Japanese apitong ply – as opposed to other apitong plies – almost certainly is a choice based upon density. In the difficult acoustics of a concert hall, delicate subtleties of sound are far less important than lightness and ruggedness, so studios have yet again been the poor relation in the previous history of horn designs. The cost-effectiveness of the research into certain subtle areas of horns has long been non-viable to most horn producers, as they are usually commercial operations and not philanthropic research institutes.

Granted, wooden horns look pretty, and this has no doubt had a part to play in their widespread use in studios. Wood also has less of a tendency to ring than many of the metals used in many horns. High density plywoods do therefore conform with basic tenets of high density and high internal loss. When the high density materials are used in quite thick sections, there is then a lot of mass in the horn itself; it will take a lot of moving. I have known many people involved in the theory of horn design to laugh at such insistence upon the use of particular esoteric materials, but I do feel that there is at least *something* in it.

Sound travelling through the structure of a horn and recombining, even at much lower levels, with the airborne signal at the throat can produce response irregularities and time smearing. The multiple layer-boundaries in plywoods make it much more difficult for a sound wave to propagate than through a solid piece of homogeneous hardwood.

Given modern computer numerically controlled (CNC) woodworking machines, axisymmetric designs should be at no disadvantage if made from wood, except in terms of cost, when compared to other materials such as glass-fibre resins. Such resins, however, especially when heavily mineral loaded, still hold the prospect of offering great possibilities as horn materials.

Another of our basic tenets which still seems to hold good is the short horn philosophy. The shorter the horn, the smaller the amount of time that the sound wave will spend in the horn. In turn, this will reduce the length of time for which the sound will be subjected to any unwanted characteristics which may be imparted by the horn. I do not mean here that all short horns sound better than all long horns, but rather that it is probable that 'shortness' would be one property of our 'ultimate' horn. By short, I am considering a 12″ maximum distance between diaphragm and mouth. Obviously, with these overall restrictions upon horn length, mouth size will limit the flare rate. With the larger mouth size, the coupling to the room will improve at lower frequencies, but the higher flare rate will raise the throat cut-off frequency. With a 12″ diameter mouth giving reasonable coupling down to around 1 kHz, the flare rate will produce a throat cut-off only just below 1 kHz.

Given the workability of the new designs down to frequencies very close to cut-off, the convergence of the mouth termination and the cut-off frequency towards the 1 kHz region was less disastrous than we would have previously expected it to be. Anyhow, anything above 12″ in diameter would probably begin to become somewhat unwieldy, as it would then become difficult to optimally mount the other drivers in the system. Close coupling of the mid and high frequency components of the system would be particularly difficult, unless the mid-range horn was to be modified into an enormous 'bullet' with an HF driver in the tip. In Section 11.1.3, item 6 in the table of our original criteria gave 12″ as the maximum allowable dimension in one plane, exactly for this reason of close coupling to adjacent drivers – less than one wavelength at the crossover frequency.

In order to extend the directivity control of the upper octave of the audible regions, devices such as guides or compound flare rates seem to be required, which would appear to be contrary to the cleaning up of the geometry. If it is that cleaning up process which is so important to sonic purity, then as per our original design criteria, three to four octaves would still appear to be the limit of range for any optimally derived horn system. I do not foresee the physics of horn design allowing six-octave horns of the highest sonic purity. Having said that, however, there are few direct radiators which could be claimed to be performing optimally at the extremes of a five-octave range, so I do not see this restriction as being a limitation which could in any way be construed to leave mid-range horns at a disadvantage. Indeed, the requirement of the physics of any electromagnetic transducer are so opposingly polarised toward the extreme ends of the audible range that undue extension of response would seem to be one of the *last* design priorities in the search for sonic accuracy.

References

1 K. R. Holland, F. J. Fahy, C. L. Morfey and P. R. Newell, 'The Prediction and Measurement of the Throat Impedance of Horns'. *Proceedings of the Institute of Acoustics*, Vol II, Part 7, pp. 247–254 (1989)
2 K. R. Holland, F. J. Fahy and P. R. Newell, 'Axisymmetric Horns for Studio Monitor Systems'. *Proceedings of the Institute of Acoustics*, Vol 12, Part 8, pp. 121–128 (1990)
3 John Tyndall D.C.L., L.C.D., F.R.S., *On Sound*, sixth edition, Longmans Green and Co, page 40 (1895).

Chapter 12

Do all mid-range horn loudspeakers have a recognisable characteristic sound?

12.1 Introduction

What follows is a description of the outcome of the series of listening tests, which were primarily designed to determine whether mid-range horn systems could readily and universally be distinguished from other types of mid-range drive systems. The general techniques could well be applied to other loudspeaker units or systems, and the discrepancies in the results could no doubt show equal levels of overall dissimilarity of perceived sound from such different units.

Loudspeaker systems utilising compression driver/horn combinations for mid-range and high frequency reproduction have been bones of contention for many years in terms of their sonic characteristics *vis-à-vis* direct radiators. Each and every monitor system seems to have both its partisan followers and vociferous critics, so in order to gain some more definitive insight into both measured and perceived performance, Keith Holland undertook, on my behalf, a three year doctorate research programme on horn parameters and their relationship to sonic performance.

Some four years or more after the work began, the findings from the correlation of over 7000 auditioned and measured comparisons yielded results which help to explain not only the reasons for any characteristic differences between 'average' horns versus 'average' direct radiators, but also of other types of loudspeaker units such as electrostatics. They furthermore help to explain how and why certain drive units of either similar or dissimilar genre may or may not sound alike, and put into perspective the relative parts played by pressure amplitude, phase, and the non-linear properties of driver performance.

Do all horns have a recognisable characteristic sound? Are they necessarily always able to be differentiated from other drive systems? What is it about electrostatics which has won the hearts of so many people over so many years, and can their performance be duplicated by other drive systems, even horns? Read on!

It was evident that in order to investigate the full potential for mid-range horns in studio applications, development from first principles would most probably be the only viable path. In the late summer of 1987, Keith Holland began thorough library searches at the commencement of intensive research; searches intended to assimilate as much as possible, from as far back as possible, of just what was known about the sonic aspects of horn geometry or construction techniques. Finite element modelling followed, with single parameter models being evaluated against measurements from the testing of a wide range of actual horns, many of which were relatively common in studio usage.[1] Both linear and non-linear performance data were measured and analysed prior to the first attempts to correlate the subjective and objective measurements.

In the process of this research, a system was devised for both the rapid and accurate measurement of the throat impedance of horns, a paper on which was presented[2] at the Institute of Acoustics (IOA) Reproduced Sound 5 conference in November 1989. The use of this system has continued to produce excellent data on many aspects of the manipulation of horn geometry, giving rapid correlation with effects on throat impedance, and the implied pressure amplitude response when connected to actual drive units. What follows is a description of the listening test concepts and procedures, followed by a discussion of the results, their implications for monitor system similarity or disparity, and indeed what can be done to 'design in' desirable qualities and 'design out' any unpleasantness in future monitor systems. What was 'known' at the outset of these tests was published in *Studio Sound*[3,4,5] during the course of the tests themselves being performed.

12.2 Scientific method

During the latter half of 1989, we set up a series of listening tests in the large anechoic chamber at the Institute of Sound and Vibration Research (ISVR) at Southampton University. The chamber had one-metre foam wedges, and was approximately 11 metres square by 9 metres high. By arranging the flooring grids in the form of catwalks, it was possible to arrange an arc of five test loudspeakers, equidistant from the listener, with a control console to stage right and no flooring grids between the loudspeakers and the listener. Four of the five loudspeakers in the arc were nominated as generic archetypes and labelled A, B, C and D, from the listener's left to right. The centre position in the arc, between archetypes B and C, was reserved for the changeable sample loudspeakers. Sixteen of the changeable samples were employed, whilst the archetypes were, from left to right: A, a Quad Electrostatic; B, a Son Audax 6½″ mid-range driver type PR17 HR100 1AK7; C, a well made and typically representative compression driver/horn combination consisting of a Fostex H351/HA21 horn and Emilar EK175 driver; and D, a Tannoy axisymmetric horn (a 15″ Dual Concentric loudspeaker; the HF driver only being driven).

The choice of the archetypes had been made with considerable care, as to choose a type at any extreme of its genre would not be truly representative of its family. The Quad Electrostatic was chosen as an overall reference, as its generally accepted neutrality had been widely appreciated for many decades. It was also neither the most expensive nor advanced of its type, yet was capable of sonically performing well alongside many other electrostatic devices. The direct radiator chosen as 'B' was of a type in use in monitor systems which, whilst not claiming sonic 'perfection', were well accepted as being highly representative of the performance of a wide range of other loudspeakers. For many people, that is one of the prime considerations in the choice of monitoring systems on which people chose to 'mix for the market'.

The choice of a representative horn was no easy task, as if one of the most neutral sounding horns had been chosen, then, quite probably, most horns in the test would have grouped together very strongly, and if too 'horn-like' a horn were chosen, then the 'better' horns may blur into the other units to a confusing degree. Indeed, 'D', the Tannoy, was itself representative of a horn which was sonically rarely considered to be a horn, so the 'better' end was already well covered. The final choice for 'C' was a well constructed horn of the longer type, which would not be instantly recognisable to any studio engineer, yet would reasonably represent the performance of many of the larger horn loaded monitor systems, mainly of American origin.

The loudspeakers were switched via a comparator unit which, whilst changing over the loudspeakers, also switched pads into the input circuit of the Crown DC 300 power amplifier in order to compensate for differing loudspeaker sensitivities. The levels were adjusted on a pink noise signal, with both measured and aural assessments achieving close correlation. The layout is shown in Fig. 12.1.

A series of nine sound sources, some natural recordings and others computer generated, were sampled and repeated on a digital tape recorder. Each sound in turn was then played twice through the test sample, then twice through A, back to the sample, then B, then sample, then C, then the sample, and finally D: twice through each. Upon completion of the full cycle, the whole process was repeated until the listener could determine to which of the archetypes the sample sounded the closest. If the listener considered the sample to sound similar to more than one archetype, then two or more boxes of the questionnaire could be ticked. If, however, the sample was not considered to sound like any of the archetypes, a further column headed 'None' was made available, which could be ticked if desired. The loudspeakers were hidden from the view of the listeners by an opaque curtain, which was acoustically transparent to around 8 kHz. As the test was primarily on mid-range devices, the sounds were band limited at 24 dB/octave between 1 kHz and 6 kHz.

When one sample had been subjected to all nine sounds, it was replaced by another sample, and the test was repeated until either the listener was tiring, or all sixteen test samples had been exhausted. Samples were chosen in random manner to avoid a 'learning curve' biasing any results towards better discrimination for the later samples.

In other words, if the tests always began with sample one and ended with sample sixteen, the listeners may have been more discriminating as they progressed to the higher numbered samples due to familiarisation with the procedure, or they may even have become more confused due to fatigue. Either way, a random order was desirable.

A large selection of listeners was utilised, but only one listener at a time took part in the tests. Due to the magnitude of the task of listening to nine sounds through each of sixteen sample loudspeakers, each cross-referenced to four fixed units (a minimum of $4 \times 9 \times 16 = 576$ operations), the listeners were given frequent breaks for tea, coffee, lunch or whatever. In some instances, several people were used to complete one questionnaire. The tests continued for a period of four months, until sufficient data had been collected to enable meaningful analysis. After the tests had been concluded, Keith Holland and I both analysed the results – initially, each without reference to the other, only conferring once the results had been tabulated and assessed.

The anechoic chamber had a noise floor of 17 dBA and the tests were carried out at an L_{eq} of 57 dB. The peak signal level of the sounds with the highest crest factors was 84 dB at the listening position, 3 metres from the

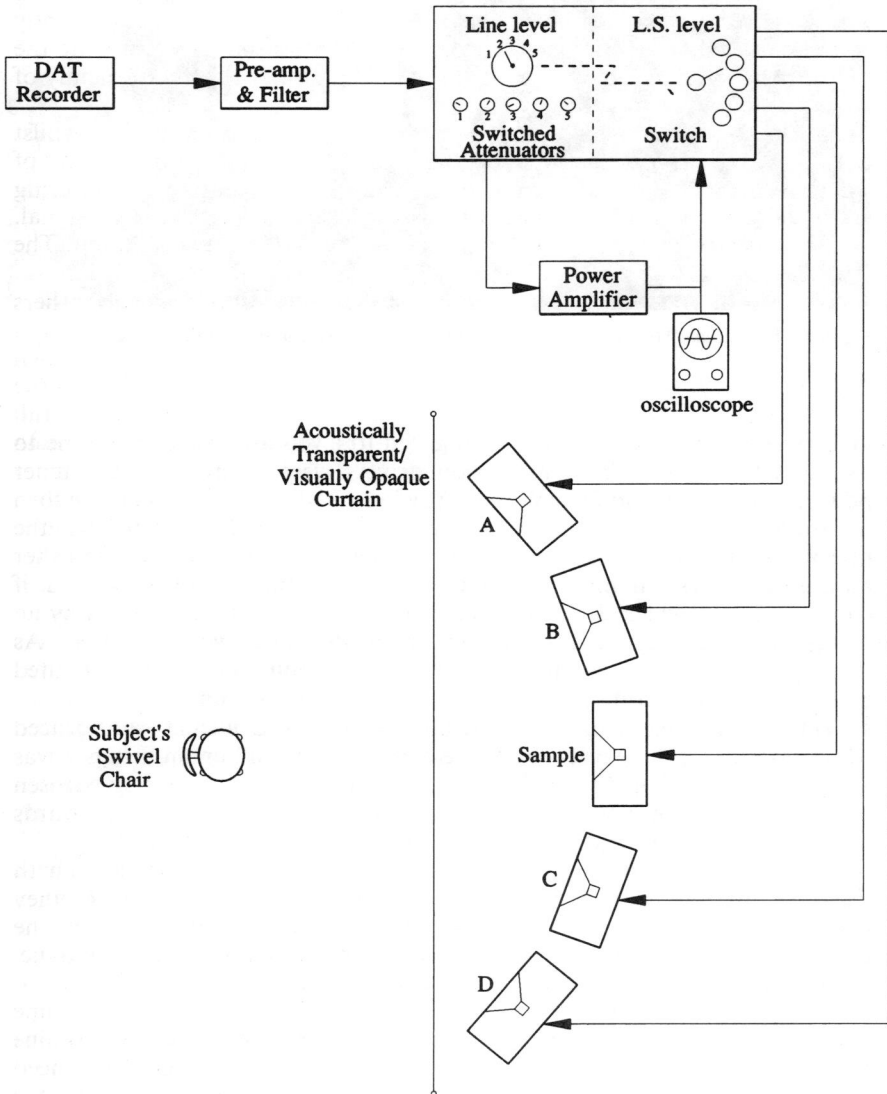

Figure 12.1 (a) Layout of test set-up in anechoic chamber

Figure 12.1 (*continued*) (b) the test set-up in the large anechoic chamber at the ISVR, Southampton

loudspeakers. No unit was being subjected to levels anywhere approaching its capacity; indeed, the compression drivers were receiving a maximum power input of around 20 mW, whilst being rated at 100 watts continuous programme. The least sensitive device, the Quad ESL, was receiving a maximum peak input of around 6 watts, being some 25 dB less sensitive than the compression horns. At no time were the subjects asked to make any comments as to which device they preferred or thought to be most accurate – they were only requested to indicate similarities. They were, however, free to make any notes or comments on their questionnaires, though this was entirely at their own discretion. Incidentally, the maximum SPLs were not set entirely on engineering grounds, but also by the University authorities, who deemed that these tests constituted experiments on human beings.

Some listeners ticked several boxes frequently, others rarely. Some listeners made much use of the 'None' column whilst, once again, others rarely took advantage of its presence. Ticks were sometimes accompanied by 'ish', or were drawn dotted or given a half. Occasionally, comments such as 'clear' or 'natural' were noted for some samples, whilst others were given less flattering descriptions. After ten completed tests, a preliminary assessment of the results was studied. The listeners had been students, lay persons, audio magazine editors, acoustics consultants, academics, record producers and musicians.

The first attempt at assessing the data was to gather together a total list of the unqualified solo ticks – the results where the listeners had shown no doubt as to their opinions of similarity, or otherwise as shown by the 'None' column. The spectra of the nine sounds are shown in Fig. 12.2(a)–(i).

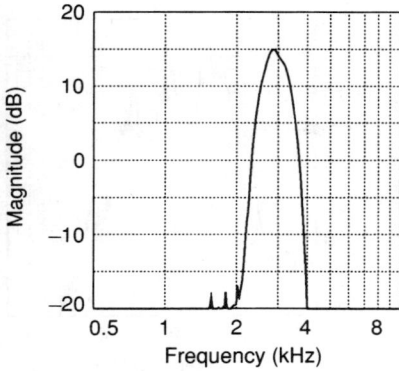

(a)   'Chirp'

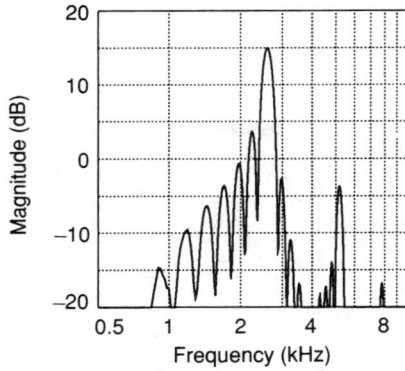

(b)   'Burst'

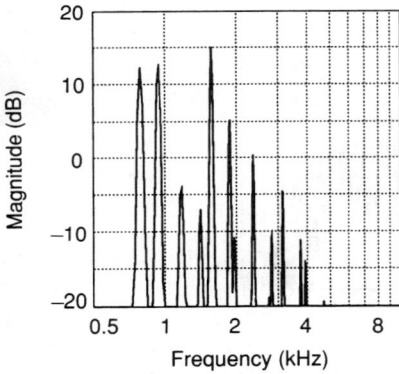

(c)   'Flute notes'

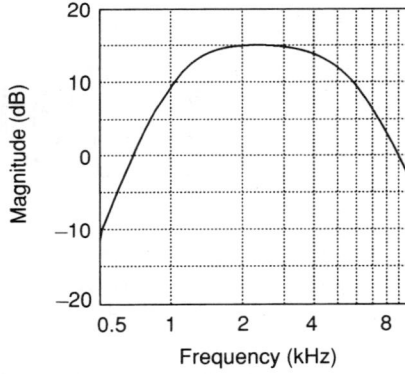

(d)   'White noise'

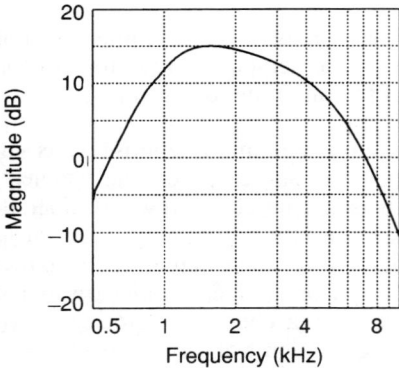

(e)   'Pink noise'

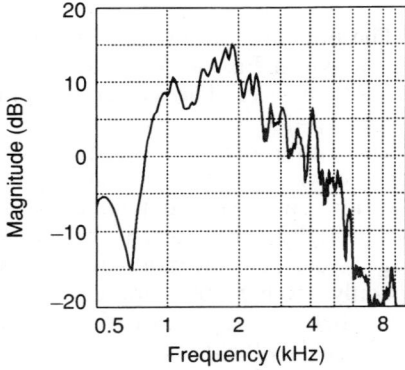

(f)   'Slamming book'

Figure 12.2 Frequency spectra of source signals (through filter)

(g)   'Waterfall'   (h)   'Statue impact'

(i)   'Guitar chord'

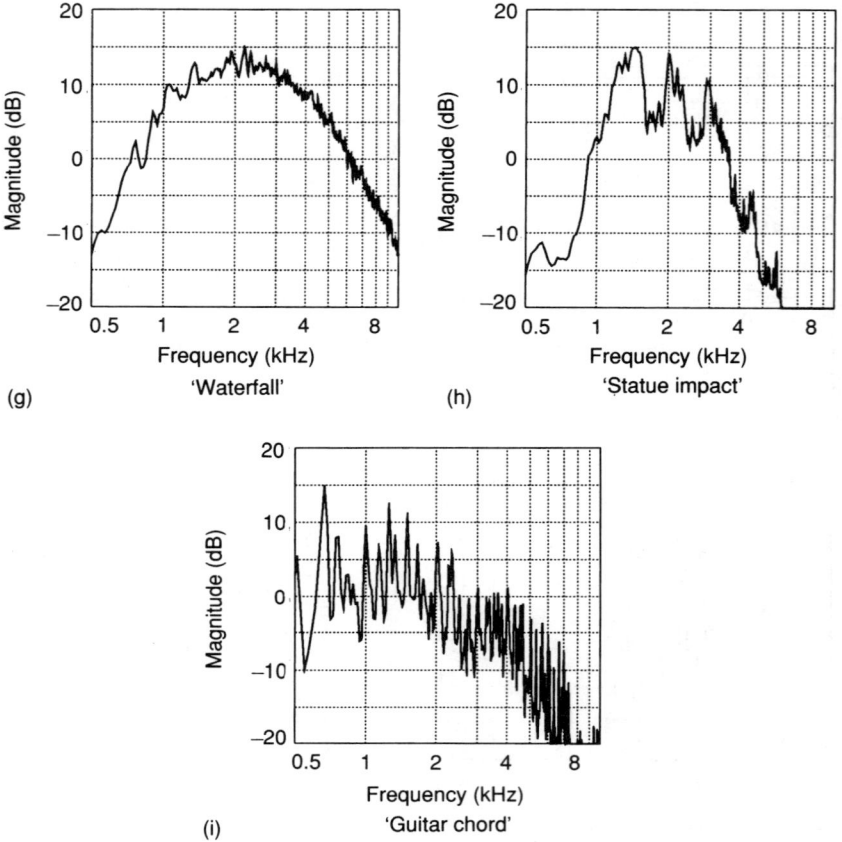

Figure 12.2 (*continued*)

The pressure amplitude response curves on axis at the listening position are shown in Fig. 12.3, (a) to (d) being the archetypes and (1) to (16) the variable samples. The master grouping of the unambiguous results is shown as Fig. 12.4.

These groupings were based on the total number of unambiguous results – in other words, where only one decisive tick was indicated on the listeners' questionnaires, with no qualifications such as 'ish' or 'close'. It was decided to look initially at the more precise answers, rather than to attempt at an early stage to decipher the more arbitrary results. Part of the original question was not only whether one sample device would consistently line up under any one of the archetypes, but also whether similarities to one archetype or another would be dictated by the transient or steady state components of the test signals. The signals themselves were designed to have little or no information content, in order neither to distract the listeners nor to excite any subjective preferences.

Keith Holland subsequently conducted a separate statistical analysis of similarity confidence limits, based on the total number of all ticks after the

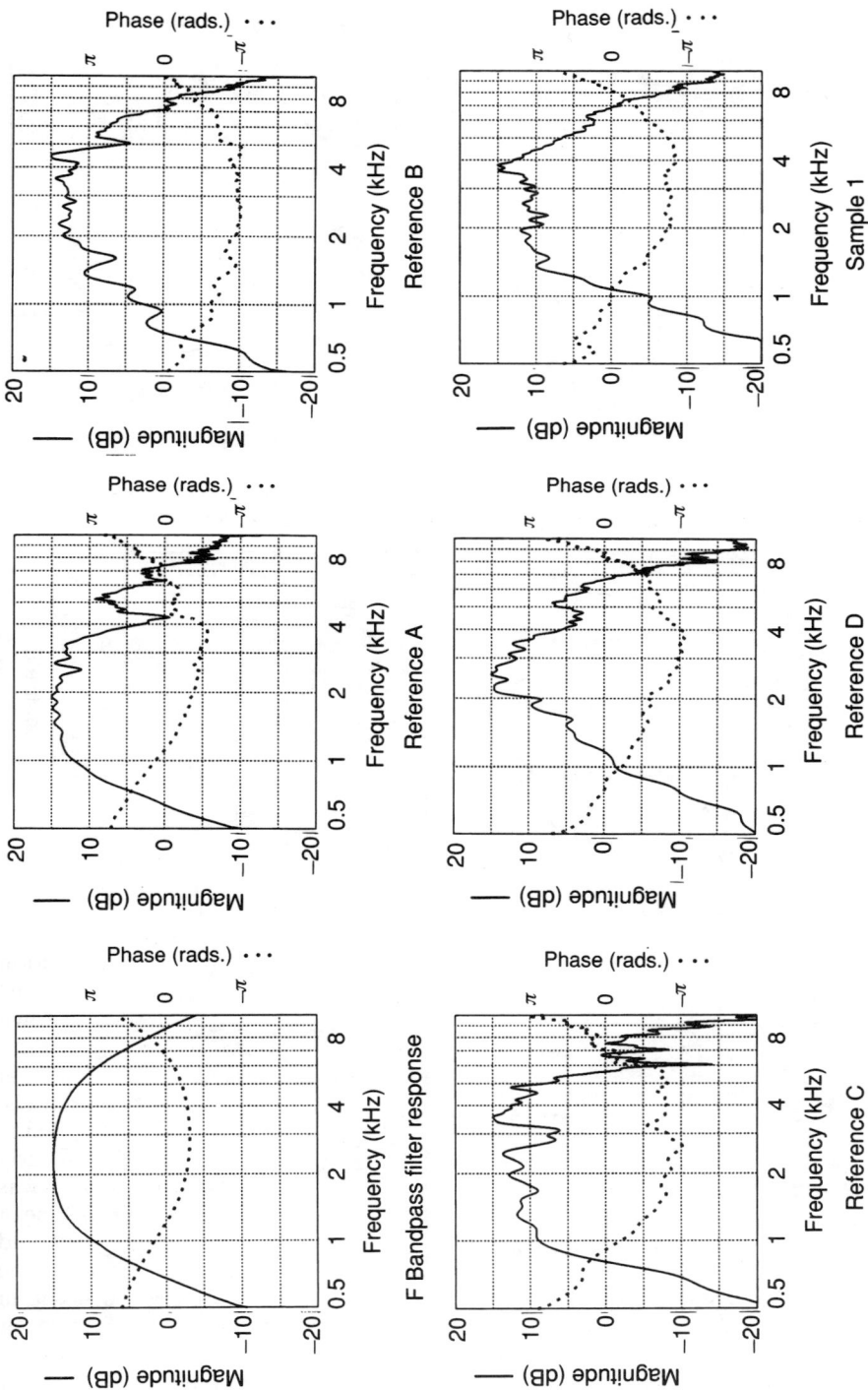

Figure 12.3 Pressure amplitude responses of loudspeaker drive units under test and transfer functions of test loudspeakers

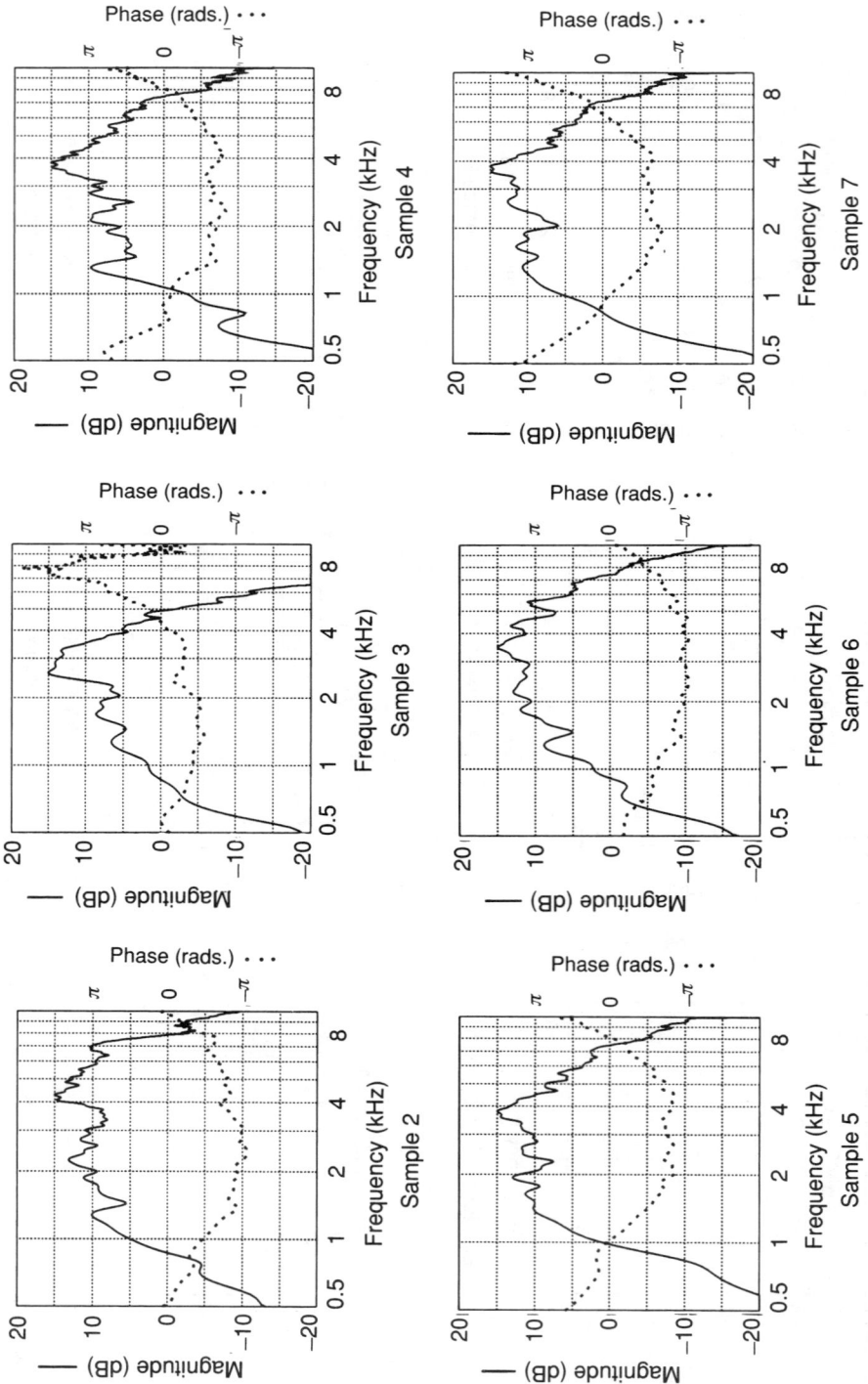

Figure 12.3 (*continued*)

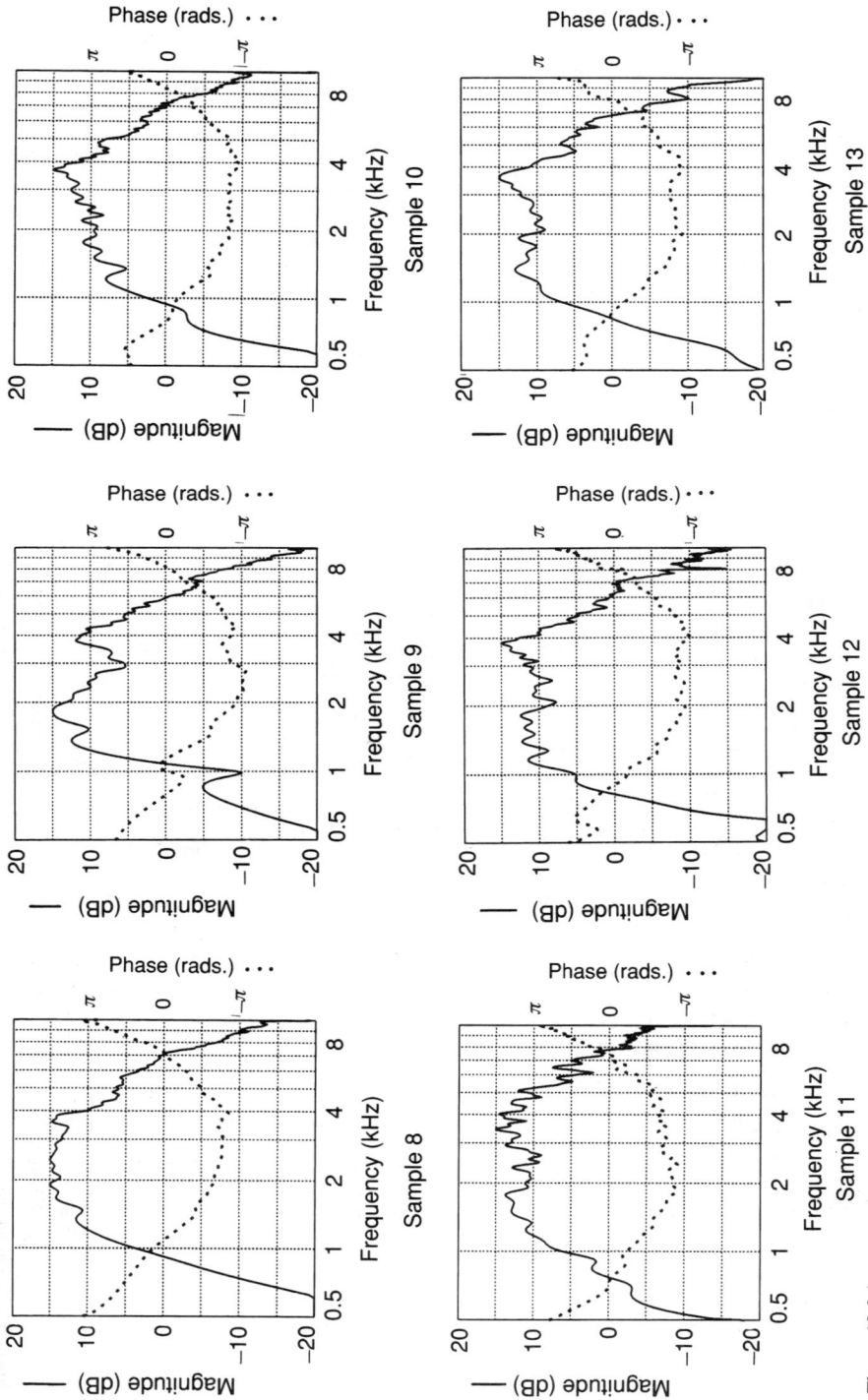

Figure 12.3 (*continued*)

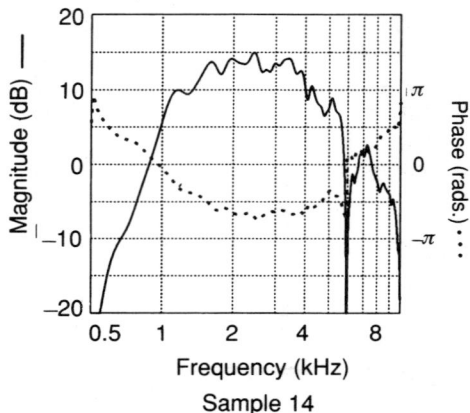

Sample 14

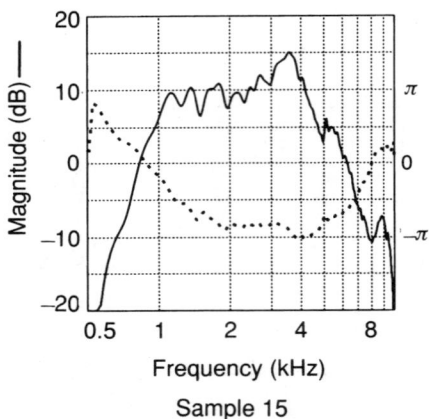

Sample 15

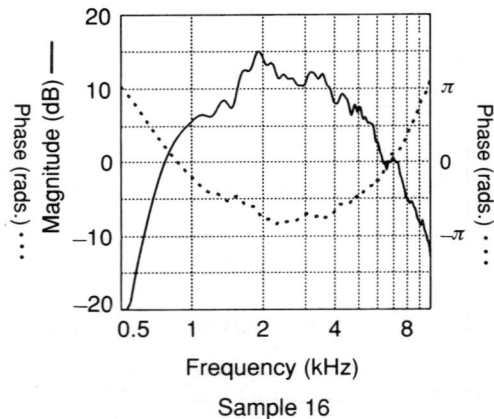

Sample 16

(F)   Electrical response through filter

Fixed sample loudspeakers

- A   Quad Electrostatic – original type
- B   Son Audax PR17/HR100/1AK7 cone mid-range unit
- C   Fostex H351/HA21 horn with Emilar compression driver
- D   High frequency section of Tannoy 15″ Dual Concentric

Variable sample loudspeakers

1   Metal horn with Emilar EK175 compression driver
2   JBL 2105 cone driver
3   JBL 2121 cone driver
4   AX1 axisymmetric horn with EK175 driver
5   Reflexion Arts horn with EK175 driver
6   Son Audax (as 'B')
7   As '5' but with lips sawn off
8   AX2 axisymmetric horn
9   Yamaha horn with EK175 driver
10   Fostex H320 wooden radial horn with EK175 driver
11   JBL 2307/2308 horn/slant-plate combination with EK175 driver
12   Altec sectoral horn with EK175 driver
13   Altec 806C multicellular horn with EK175 driver
14   Starr 'Singing Throat' wooden gramophone horn with EK175 driver
15   Vitavox sectoral horn with EK175 driver
16   JBL 2370 with JBL 2426 driver

Figure 12.3 (*continued*)

gathering of all the test data, the results of which are shown in Fig. 12.5. The two approaches ultimately produced remarkably similar conclusions. Indeed, Figures 12.5(b) and (c) show the statistical confidence results for the 'controls' 3 and 6, and also show that whether the unambiguous ticks alone, the total of all ticks, *or* the ticks weighted for comments were used, the results remained remarkably uniform in their indications.

12.3 The sounds and the drivers

The sounds

The nine sounds were as follows, the spectra of which are shown in Fig. 12.2(a)–(i) respectively:

1 A digitally generated chirp.
2 A digitally generated tone burst consisting of ten cycles at 2.5 kHz, sounding similar to claves being struck.
3 Two notes from an anechoically, digitally recorded flute.
4 White noise.
5 Pink noise.
6 An anechoically recorded heavy book being slammed shut.
7 An outdoor recording of a waterfall, a short section of which was sampled and repeated.
8 A 30 ft high, square section steel tube, open at one end, being struck by a peach stone. The phase dispersion causes the high frequency reflexions to lead the low frequencies, producing a 'laser gun' sound.
9 An anechoic recording of an acoustic guitar chord, sampled and repeated.

Sounds 1, 2 and 3 all had a distinct 'note' content. Sounds 4, 5 and 7 were random noise type signals. Another similar signal which could have been used would have been applause, which clearly would consist of a large number of clapping hands, effectively a series of transients. Similarly, the noise signals, by containing all frequencies, randomly timed, are also theoretically transient-type signals. However, at the 1989 November conference of the Institute of Acoustics in Windermere, Michael Gerzon gave a seminar on the human perception of stereo localisation and the ear's detection of such. In his findings, the mechanisms for transient and steady state localisation were seen to be different, it thus being possible to separate in perceived space the two parts of complex sounds. Under certain conditions, applause can behave not as a transient, or series thereof, as could have been predicted, but as a steady state sound.

Obviously more work needs to be done on the subject of whether sounds 4, 5 and 7 are grouped with the transient or steady state signals, but in the tests they proved most revealing, producing more results in the 'None' column than any of the other sounds. Sound 6, the slamming book, was almost entirely of transient content. Sound 8 was a highly complex signal consisting of an initial impact transient, the resonances of the tube, plus a phase dispersed reflected wave, which was a time/frequency smeared reflexion of the initial transient: it thus contained 'notes' and transients. Sound 9, the acoustic guitar, was again a combination of transients and resonances, with a harmonically related series of notes also clearly audible.

The drivers

Archetype B and Sample 6 were nominally identical units from the same production batch, and were intended to be controls to set the standard for similarity between units. It was thought unlikely that any two dissimilar units in the tests could sound any more alike than the 'twins' B and 6, and therefore the B/6 comparison would set a standard for the minimum audible degree of difference. Results showed that the units were most clearly deemed to be sonically similar (Fig. 12.5). Sample 3 was chosen to define the lower limit of similarity. The unit was a JBL 2121, 10" lower mid-range driver, not specifically suited to the range under test, having a peak around 3 kHz and falling off rapidly above 4 kHz. Less than 25% of ticks on the questionnaires indicated that Sample 3 possessed any specific similarity to any of the archetypes, and those 25% of ticks were spread over all four, showing no clear tendencies whatsoever for this driver to be pigeon-holed with any of the others. From this point of view, the upper and lower similarity control samples, namely 6 and 3, appeared to function well in determining the general audible similarity/difference limits for the whole test.

Remember, the initial aim of the test was to determine whether horns *per se* had any unmistakable sound quality which would render them characteristically recognisable. We wondered if any such properties would be noticeable on all programme material, or possibly only on transient, or maybe steady state programme. Of the sixteen test samples, thirteen were horns, and the remaining three were direct radiators. Two nominally similar compression drivers were used to facilitate rapid changeover of the samples. The compression drivers were chosen for response uniformity and relatively low distortion, and were known to possess a relatively neutral sound. Thus it was hoped that the *horn* differences would predominate, rather than having imposed upon them any common characteristic due to driver irregularities. The one exception to this was Sample 16, which was a complete combination from one manufacturer. The four archetypes included one Fostex horn with an Emilar compression driver, chosen as a representative mean. The combination was neither too outrageously hornlike, nor was it one of the most un-hornlike sounding combinations. D was also a horn, but was the high frequency section from a Tannoy 15" Dual Concentric, where the horn was the cone of the bass driver. In flare, it was not dissimilar to Sample 8, the AX2.

12.4 Preliminary results

Sample 1 – Vitavox Horn

A combination not encountered in studio monitors, this sample received the lowest number of unequivocal, unqualified ticks of all samples tested. Of those unambiguous ticks, never was it deemed to sound like A, only three times was it considered similar to C, each time on Sound 2, the tone burst, and 12 times was it deemed similar to D. Of the 12 ticks for D, none were on Sound 2, the ticks being distributed over the remaining sounds, each receiving between one and three ticks. The most ticks in any column

Signal	Sample 1 A	B	C	D	None	Sample 2 A	B	C	D	None	Sample 3 A	B	C	D	None	Sample 4 A	B	C	D	None
1	–	1	2	4	1	–	2	4	1	–	–	3	1	1	–	–	6	–	2	1
2	–	2	1	3	3	–	4	2	1	2	–	–	1	2	3	–	6	–	–	2
3	–	3	–	4	1	–	7	1	–	1	1	–	1	1	4	–	6	–	2	1
4	–	2	–	1	4	–	3	–	–	5	–	–	–	–	8	–	5	–	–	4
5	–	1	1	2	2	1	4	–	–	4	–	–	1	–	9	–	6	–	–	4
6	–	5	–	3	3	–	4	–	1	2	–	1	1	1	5	–	3	–	–	3
7	–	1	–	2	4	–	3	1	–	2	–	–	–	–	9	–	7	–	–	2
8	–	4	–	5	–	–	5	–	1	–	–	2	–	2	4	–	6	–	1	2
9	–	3	–	3	3	–	9	–	1	–	–	–	–	–	5	–	1	–	2	4

Signal	Sample 5 A	B	C	D	None	Sample 6 A	B	C	D	None	Sample 7 A	B	C	D	None	Sample 8 A	B	C	D	None
1	–	3	2	2	–	–	6	–	–	–	–	5	1	–	–	–	1	–	4	1
2	–	5	1	1	1	–	7	–	–	3	–	5	2	–	2	–	1	–	5	3
3	–	2	1	4	1	–	8	–	–	–	–	3	2	–	1	–	1	1	3	1
4	–	1	–	1	4	–	11	–	–	1	1	2	–	1	5	–	1	1	2	4
5	–	2	1	2	4	–	12	–	–	–	–	1	–	1	4	–	1	–	2	6
6	–	2	1	2	1	–	8	–	–	1	–	3	2	–	1	–	2	1	1	3
7	–	–	–	2	6	–	10	–	–	–	–	–	–	–	4	–	–	–	2	8
8	–	6	–	1	2	–	9	–	–	1	1	1	2	–	–	–	1	1	1	3
9	–	3	–	3	3	–	11	–	–	–	–	4	1	1	3	–	2	–	3	4

Signal	Sample 9 A	B	C	D	None	Sample 10 A	B	C	D	None	Sample 11 A	B	C	D	None	Sample 12 A	B	C	D	None
1	–	3	2	–	–	–	9	–	–	–	–	3	1	–	2	–	3	2	1	–
2	–	3	2	–	2	–	7	–	–	1	1	1	3	–	2	–	1	2	3	2
3	–	–	4	–	–	–	3	2	–	1	–	3	2	–	–	–	3	5	–	–
4	–	–	–	1	8	–	2	–	–	5	–	–	–	–	6	–	–	1	1	5
5	–	1	–	–	9	–	4	–	–	3	–	1	–	–	2	1	–	1	–	6
6	–	2	–	–	3	–	4	1	1	2	–	4	1	–	1	–	2	4	–	3
7	–	–	–	1	5	–	5	–	1	3	–	–	–	–	4	–	–	3	–	4
8	–	2	–	1	3	–	4	–	1	1	–	4	–	1	3	1	2	5	–	1
9	–	–	4	2	2	–	3	–	4	1	1	6	–	–	1	–	–	5	2	1

Signal	Sample 13 A	B	C	D	None	Sample 14 A	B	C	D	None	Sample 15 A	B	C	D	None	Sample 16 A	B	C	D	None
1	–	4	–	1	2	–	2	–	3	1	–	6	1	1	–	2	1	4	–	1
2	1	5	2	–	1	–	3	1	1	1	–	4	–	–	2	–	3	–	–	–
3	1	1	2	–	1	1	1	3	1	1	1	2	3	–	2	1	–	6	–	2
4	–	–	1	–	7	–	–	–	–	5	1	–	–	1	6	2	–	–	–	6
5	–	–	–	–	7	1	–	1	–	5	–	–	1	1	5	–	1	1	–	7
6	3	2	3	–	2	1	1	1	–	5	–	1	1	–	7	–	3	4	–	–
7	2	–	1	–	4	1	1	–	–	5	2	–	–	1	5	–	1	–	–	7
8	3	–	4	–	2	1	–	6	–	1	1	1	3	–	4	2	–	5	–	2
9	2	–	3	1	3	–	–	6	–	3	1	1	2	–	3	–	2	4	–	2

Figure 12.4 Listening test results: unambiguous ticks

were for Archetype B, the Audax cone driver. These 19 ticks were widely distributed for all of the sounds except 7, the waterfall. Only once throughout the entire test was Sample 1 ever deemed to sound similar to any of the archetypes when listening to the waterfall, when one listener considered it to sound like D, the Tannoy. Sample 1 therefore was never considered to sound

Signal	Sample 1 (%) A	B	C	D	None	Sample 2 (%) A	B	C	D	None
1	–	25	25	97	–	–	65	25	–	–
2	–	67	–	25	–	–	–	95	–	–
3	–	87	24	87	–	–	100	–	–	–
4	–	96	–	–	87	–	73	–	–	100
5	–	22	–	–	22	–	95	–	–	95
6	–	91	–	25	–	–	99	–	–	–
7	–	87	–	64	87	–	99	–	–	67
8	–	98	–	90	–	–	100	–	–	–
9	–	25	–	91	–	–	100	–	–	–

Signal	Sample 3 (%) A	B	C	D	None	Sample 4 (%) A	B	C	D	None
1	–	100	–	87	–	–	100	–	–	–
2	–	–	–	25	–	–	100	–	–	–
3	25	–	–	–	67	–	100	–	25	–
4	–	–	–	–	100	–	100	–	–	73
5	–	–	–	–	100	–	100	–	–	93
6	–	28	–	–	93	–	99	–	–	25
7	–	–	–	–	100	–	100	–	–	28
8	–	67	–	67	25	–	100	–	–	–
9	–	–	–	–	100	–	25	–	67	25

Signal	Sample 5 (%) A	B	C	D	None	Sample 6 (%) A	B	C	D	None
1	–	96	24	64	–	–	100	–	58	–
2	–	100	25	–	–	–	100	–	–	–
3	–	87	24	87	–	–	100	–	22	–
4	–	87	–	24	87	–	100	–	–	–
5	–	67	–	–	67	–	100	–	–	–
6	–	85	–	22	–	–	100	–	–	–
7	–	25	–	–	98	–	100	–	–	–
8	–	100	–	–	–	–	100	–	–	–
9	–	66	–	89	66	–	100	–	–	–

Signal	Sample 7 (%) A	B	C	D	None	Sample 8 (%) A	B	C	D	None
1	–	100	24	–	–	–	–	–	100	–
2	–	99	–	–	–	–	–	–	99	30
3	–	60	85	–	–	–	22	–	85	–
4	–	–	–	–	99	–	–	–	–	91
5	–	24	–	–	96	–	–	–	28	100
6	–	99	89	–	–	–	64	24	24	24
7	–	–	–	–	98	–	–	–	–	100
8	–	96	64	–	–	–	–	–	67	67
9	–	91	–	–	–	–	–	–	67	67

Figure 12.5(a) Similarity confidence indices: total number of ticks

like A, only on the toneburst was it ever considered similar to C, and whilst not being strongly representative of any of the archetypes, the greatest similarity was deemed to be to B, the Audax direct radiator. Comparisons between transfer function measurements show an amplitude response most similar to B.

Signal	Sample 9 (%)					Sample 10 (%)				
	A	B	C	D	None	A	B	C	D	None
1	–	96	22	–	–	–	100	–	–	–
2	–	91	25	–	–	–	100	–	–	–
3	–	22	99	–	–	–	96	60	–	–
4	–	–	–	–	100	–	90	–	–	100
5	–	–	–	–	100	–	99	–	–	66
6	–	96	–	–	24	–	91	–	–	–
7	–	25	–	–	97	–	98	–	–	25
8	–	89	–	–	66	–	100	–	–	–
9	–	–	67	–	–	–	99	–	75	–

Signal	Sample 11 (%)					Sample 12 (%)				
	A	B	C	D	None	A	B	C	D	None
1	–	99	–	–	–	–	97	25	66	–
2	–	64	96	–	–	–	–	–	25	–
3	–	100	87	–	–	–	90	100	–	–
4	–	24	–	–	99	–	–	25	–	91
5	–	85	–	–	22	–	–	–	–	91
6	–	100	24	–	–	–	28	96	–	28
7	–	–	–	–	96	–	–	25	–	67
8	–	98	–	–	–	–	–	99	–	–
9	–	100	–	–	–	–	–	99	–	–

Signal	Sample 13 (%)					Sample 14 (%)				
	A	B	C	D	None	A	B	C	D	None
1	–	91	–	–	–	–	66	–	89	–
2	28	93	28	–	–	–	67	67	–	–
3	–	22	84	–	–	–	–	98	–	–
4	–	–	25	–	100	–	–	–	–	99
5	–	–	–	–	100	–	–	25	–	99
6	28	28	28	–	–	–	–	–	–	99
7	–	–	25	–	100	–	–	–	–	95
8	–	–	91	–	–	–	–	100	–	–
9	–	–	67	–	25	–	–	100	–	28

Signal	Sample 15 (%)					Sample 16 (%)				
	A	B	C	D	None	A	B	C	D	None
1	–	100	–	–	–	–	–	99	–	–
2	–	99	–	–	–	–	96	84	–	–
3	–	67	90	–	–	–	–	99	–	–
4	–	–	–	–	100	25	–	–	–	99
5	–	–	24	–	96	–	–	–	–	100
6	–	–	–	–	99	–	89	97	–	–
7	–	–	–	–	99	–	–	–	–	100
8	–	–	67	–	67	–	–	99	–	–
9	–	–	91	–	25	–	–	91	–	–

Figure 12.5(a) (*continued*)

Signal	Total (%)					Weighted (%)					Unambiguous (%)				
	A	B	C	D	None	A	B	C	D	None	A	B	C	D	None
1	–	100	–	58	–	–	100	–	–	–	–	100	–	–	–
2	–	100	–	–	–	–	100	–	–	–	–	100	–	–	30
3	–	100	–	22	–	–	100	–	22	–	–	100	–	–	–
4	–	100	–	–	–	–	100	–	–	–	–	100	–	–	–
5	–	100	–	–	–	–	100	–	–	–	–	100	–	–	–
6	–	100	–	–	–	–	100	–	–	–	–	100	–	–	–
7	–	100	–	–	–	–	100	–	–	–	–	100	–	–	–
8	–	100	–	–	–	–	100	–	–	–	–	100	–	–	–
9	–	100	–	–	–	–	100	–	–	–	–	100	–	–	–

Signal	Total (%)					Weighted (%)					Unambiguous (%)				
	A	B	C	D	None	A	B	C	D	None	A	B	C	D	None
1	–	100	–	87	–	–	100	–	67	–	–	94	–	–	–
2	–	–	–	25	–	–	–	–	30	30	–	–	–	39	90
3	25	–	–	–	67	–	–	–	–	67	–	–	–	–	97
4	–	–	–	–	100	–	–	–	–	100	–	–	–	–	100
5	–	–	–	–	100	–	–	–	–	100	–	–	–	–	100
6	–	28	–	–	93	–	–	–	–	97	–	–	–	–	99
7	–	–	–	–	100	–	–	–	–	100	–	–	–	–	100
8	–	67	–	67	25	–	25	–	67	25	–	–	–	–	78
9	–	–	–	–	100	–	–	–	–	100	–	–	–	–	100

Figure 12.5(b) Confidence in results for 'similar' control (Sample 6 vs Reference B).
(c) Confidence in results for 'non-similar' control (Sample 3). It can be seen from Fig 12.5(b) that the confidence in the 'similar' control, i.e. the result that Sample 6 is similar to Reference B, is 100% for every signal regardless of the way in which the ticks are interpreted. As indicated by a study of the raw test data, the 'similar' control is clearly demonstrated to be effective. The results for the 'non-similar' control (Fig 12.5(c)) show the confidence in the result that Sample 3 is not similar to any of the references to be signal dependent. Good confidence (100%) is shown for signals 4, 5, 7 and 9, fair confidence (67–99%) for signals 3 and 6, and poor confidence (0–78%) for signals 1, 2 and 8. On the whole, the non-similar control worked reasonably well, but is dependent to some degree on the test subjects' interpretation of the word 'similar'.

Sample 2 – JBL 2105 – 5" Cone Loudspeaker

Over 80% of the opinions expressed for this sample were clear-cut, without qualification or ambiguities. Less than 15% were given to A, C and D with about 20% of the results being entered in the 'None' column, most noticeably on the white and pink noise sounds (4 and 5). By far the largest portion of the clear-cut opinions expressed indicated an overwhelming similarity to B, the Audax direct radiating cone mid-range driver. This result was not unexpected, as the drive units were relatively similar both in overall design specification range and physical make-up; however, only around half of the overall number of ticks on the questionnaires were clearly and unequivocally for a likeness to B.

Sample 3 – JBL 2121 – 10" Cone Loudspeaker

This unit was the lower limit 'control' of the similarity tests. Only around 20% of questionnaire ticks were indicating any strong likenesses to A, B, C or D. Even this 20% showed no clear-cut likenesses, being spread over all four archetypes, and on the white noise and acoustic guitar sounds, both very discriminating, never was this unit chosen as being like any of the archetypes. The results were not surprising, and clearly establish the test's lower limit for audible similarity.

Sample 4 – Keith Holland's Axisymmetric AX1 Horn/EK175

No unequivocal decisions were made which showed this combination to be deemed similar to either A or C, and only five such ticks were entered in the column under D. Almost 70% of the unequivocal ticks suggested a similarity to the direct radiator B, with only the acoustic guitar sound showing any question of doubt. Decisions on Sample 4 were generally quite clear-cut. The results strongly suggested that Sample 4 was not being considered to sound similar to the archetypal horn C or the Dual Concentric D. It was being heard as similar to the cone driver B. Sample 4 was originally designed as a 'bad' horn, to make measurements on mouth reflexion problems. Physically, it was in no way similar to B, and also possessed a lumpy throat impedance characteristic.

Sample 5 – Reflexion Arts RA1/EK175

The RA1/EK175 was the combination which brought about the origins of this research. It was used in the Reflexion Arts monitor systems, which were being acclaimed in many circles as not sounding like typical horn loaded monitors. The results showed around 75% of the ticks to be clear-cut unqualified decisions, none of which showed a similarity to A. Only about 3% of the indication showed it to be representative of C, the horn, though around 25% of the ticks were placed in the 'None' column. Of the remaining 40 ticks, the split was 28 to 12 between B and D, strongly favouring similarity to the cone driver, but showing small though distinct commonality with D. The combination was clearly not being deemed 'hornlike' in these tests.

Sample 6 – AUDAX PR17/HR100/1AK7

Taken from the same production batch as B, Sample 6 served as a similarity control. Only 1% of the test results showed any ambiguity whatsoever. Less than 10% of the results were placed in the 'None' column, those being mainly on the transient or noise sounds, whilst 90% of the total number of questionnaire ticks quite unambiguously stated a similarity to its twin, B. The result was encouraging in terms of showing the validity of the test, especially as no listener ever indicated a definite likeness to the other Archetypes A, C or D. While the transfer functions of B and 6 shown in Fig. 12.3 are commendably similar, the cepstra show marked differences, which explain more readily the audible differences on wideband signals. The cepstrum analysis technique is discussed in Section 12.12.

Sample 7 – RA1 with Lips Sawn Off / EK175

Sample 7 was as Sample 5 except that the mouth had been modified by sawing off the lips, in an attempt to define the effects of mutating one horn in such a way that it may begin to sound like another. The results showed more general ambiguity than Sample 5, with fewer clear-cut decisions on similarity, but the total number of 'None' ticks remained very close to that for Sample 5. The numbers of ticks in the C and D columns reversed their trends, whilst a few indications of similarity were made with respect to Archetype A. The general trend was still towards B, but was not so well defined as was the case with Sample 5. As a result of the removal of the lips, the mouth termination was altered. The resulting change in reflexion patterns would cause changes in both amplitude and phase responses when compared to Sample 5. The pattern of such changes can be seen clearly by comparison of the cepstra of 5 and 7, much more clearly than from comparison of the pressure amplitude or phase responses alone in Fig. 12.3.

Sample 8 – Keith Holland's Axi-symmetric AX2 Horn / EK175

The AX2 horn of Sample 8 was based on much numerical modelling of 'ideal' mouth termination, the flare rate being based on a well known horn considered by many to sound relatively neutral. The results showed only 10% of ambiguous ticks; strong, clear indications were the general trend. Of the unqualified ticks, around 50% were placed in the 'None' column, whilst of the remainder, the strongest similarity was shown to the Tannoy, D, the overall shape and size of which was quite similar to the sample. No similarity was shown to A, while only around 3% of total indications pointed towards the archetypal horn C. The results implied that if any similarity to the archetypes did exist, then that similarity would be inclined towards D. Investigations into the causes of the rapid reflexion patterns in the cepstrum showed the problem to be 2" from the diaphragm, indicating a problem at the horn/driver interface, (sudden change in rate of flare). The power cepstrum shows a similarity only with reference D, and is unlike that of any other sample. It is also very free from mouth reflexions, as is D (see Fig 12.13, later in this chapter).

Sample 9 – Yamaha Horn / EK175

The 'None' column overwhelmingly received the largest number of ticks for Sample 9. Only around 25% of the possible overall number were clear-cut indications, and were split more or less evenly between B and C. Correlation with any particular sample was generally poor. The horn actually came from a Yamaha A4 sound reinforcement cabinet.

Sample 10 – Fostex H320 Wooden Radial Horn / EK175

The number of unequivocal ticks on the questionnaires for Sample 10 was only around 65% of the total possible, indicating some smearing of the similarities. However, of the clear-cut decisions, the majority were strongly in the B column. Only around 8% of the total possible unambiguous ticks were entered under the horns C and D. This horn quite clearly was not adjudged to sound like a horn has come to be expected to sound. Interestingly, the throat impedance of Sample 10 is somewhat lumpy, and akin to the badly terminated AX1, Sample 4. This horn received several comments about its 'pleasant' sound.

Sample 11 – JBL 2307 with 2308 Slant Plate/EK175

The results showed a general similarity to those of Sample 7, the RA1 with the 'lips' removed, though with a slightly stronger bias towards B. A low total of unambiguous ticks suggested that the similarities, when present, were not particularly strong. Interestingly, without the slant plate mouth diffuser, this horn was both physically and sonically quite similar to Sample 4.

Sample 12 – Altec Horn/EK175

This horn was widely used in Altec monitoring systems of the 1970s. The horn had been deemed by many recording engineers to *be* 'hornlike' when in studio use. In these tests, despite a considerable number of ticks being placed in the 'None' column, over 50% of the unambiguous votes were entered in column C. Unquestionably, this horn was being chosen as similar to our representative archetypal horn. Only around 15% of the total of ticks were split unequivocally between A, B and D. One of the test subjects stated quite certainly that he was listening to a horn.

Sample 13 – Altec Multicellular Horn/EK175

The pressure amplitude response of this sample was remarkably similar to that of Sample 12, and was very respectable by any standards, suggesting that Altec had done their homework when these horns were designed. Somewhat surprisingly, in the listening tests, the two Altec horns produced very differing sets of results. Whilst the overall number of unqualified ticks were relatively similar, as were the indications in columns B and D, Sample 13 showed a much greater number of unqualified ticks in the A column, with a great migration from the C to the 'None' column when compared to Sample 12. There was a generally broad spread of results, with C showing only a marginal bias over A, B and D. Indeed, the total of unqualified ticks in columns A, B and D outnumber the ticks in column C. It would be difficult to conclude that Sample 13 sounds like an archetypal horn, though it is sonically probably more like C than it is like A, B or D. Examination of the power cepstra show this horn to have fewer and smaller late reflexions than Sample 12, which probably contribute to it sounding less hornlike than 12.

Sample 14 – The Starr, 'Singing Throat'/EK175

'The Singing Throat' was a wooden horn from a pre-war gramophone, terminating in a circular throat of just over 1″ diameter. It was included originally as a pure matter of interest, but showed results which were uncannily like those from Sample 13, though with slightly fewer clear-cut indications. The results speak for themselves. Late reflexions are, however, evident in the power cepstrum, and this horn was actually recognised as such on two occasions, though the overall results do not strongly link it to C, the archetypal horn. The horn had many flat sections, and contour changes were abrupt and angular. Flare rate was low, as this was a 'full range' device, and no attempt had been made to terminate the mouth smoothly.

Sample 15 – Large Vitavox Radial Horn / EK175

Absolutely nothing conclusive can be drawn from the very disparate and ambiguous results for this sample, except possibly that it is not representative of anything. It is not a combination found in monitoring, and was entered more as a horn control, as it was previously known to have a characteristic 'ring'. Surprisingly, whilst not being representative of anything, it was not deemed representative of a horn, as we had originally expected from its 'ring'.

Sample 16 – JBL 2370 with 2426 Driver

An all JBL combination, the only archetype to which this showed little similarity was to D, the Tannoy. Results were most similar to the Altec horns, Samples 12 and 13. Of the Archetypes A, B, C and D, C showed by far the strongest bias, somewhere between the results of Samples 12 and 13. From the results, the sample could certainly be said to sound like the horn C, much more than it could be likened to any other archetypes.

12.5 Comments

The differences between the loudspeakers when reproducing sounds 1, 2 and 3, with their 'note' content, were generally the most difficult to separate, even between drive units of radically differing nature. Although the test devices had been level compensated both by measurement and listening with a pink noise signal, the high 'note' content of the first three signals appeared to cause perceived level differences from one unit to another, even when any tonality difference was hard to detect. This was almost certainly due to the 'notes' in the signals coinciding with peaks or dips in the pressure amplitude responses of the individual units. In general, however, on grounds of tonality alone, these three signals, together with sound number 8, were the least easy with which to separate one device from another. The 'noise' signals, Sounds 4, 5 and 7, were proving to sound 'very' different from unit to unit, even sounding noticeably different with the identical units from the same production line.

Sound 9 sounded so different on many units that, to some people, the actual chord inversion appeared to change. This was probably due to the pressure amplitude response differences between the units, which had caused the changes in apparent 'loudness' on Sounds 1, 2 and 3. As the different harmonics and fundamentals in the chord aligned themselves with any peaks or troughs in the response curves, a different emphasis was placed upon the components of the chord, subjectively changing the predominant notes and harmonic structure. The components were harmonically and 'musically' related, but whilst to some listeners the effect was that the chord remained nominally the same, whilst its position on the fretboard appeared to be changing (most noticeably towards a second inversion), other listeners heard only a change in timbre. Once again, Sound 9 possessed both transient and 'note' content.

In general, I had reservations about the pressure amplitude responses alone being responsible for the differences between the samples, because the two Altec horns, Samples 12 and 13, despite showing very similar throat impedance characteristics and pressure amplitude responses (Figs 12.3.12 and 12.3.13), were not correlating particularly well with respect to test Signal 9. Whilst Sample 12 lined up quite strongly under Archetype C, the long horn, no clear preference was shown in the case of Sample 13. Even when subsequently tested with the selfsame drive unit, they *still* did not sonically match.

Obviously, further tests will be necessary to determine the actual degree of pressure amplitude response differences needed to create the 'inversion change' effect, especially in the light of Sample 5, a horn, strongly being likened to B, the Audax direct radiator, on Sound 9, the guitar chord, whilst the two show distinct dissimilarity in their pressure amplitude responses (Figs 12.3 (b) and (5)). It may, however, be that the position of the peaks and troughs, and the way in which they correspond to the signal spectrum peaks, could well have a part to play, but I doubt that that is the whole story. Analysis has shown a statistical similarity of around 70% in terms of amplitude spectra alone.

I must add here that any comments made in the analysis of the samples do not suggest any criticism of any manufacturers' products outside the scope of these tests. For example, the Yamaha horn, considered to be 'strange' in the tests, has been well received in musical instrument use. The Vitavox horns, also deemed 'strange', had been highly respected in sound reinforcement circles, and I have been fully satisfied with them when using them myself for such purposes. The Fostex wooden horn was, I believe, designed for hi-fi use, and whilst I do not consider it to sound 'accurate', it undoubtedly can sound very pleasant and musical. Remember, these tests were set up to extract deeply hidden performance characteristics, which, in order to filter them out for analysis, had to be highlighted and exaggerated for reference to be possible. All of the units involved were of very high quality, as sub-standard units would have confused the issue, especially considering the fact that the whole project related to studio monitoring.

12.6 Summary of initial conclusions

The statistical analysis conducted by Keith Holland on a basis of similarity confidence limits gave total 100% similarity for Sample 6 when compared to B: 6 was indeed a nominally identical driver to B. The non-similar control, Sample 3, produced by far the worst similarity confidence limits of any of the samples used, producing four 100%s in the 'None' column out of the nine signals available. The 'unambiguous ticks' approach which I used for *my* analysis of the data led to equally strong correlation of the 'similar' and 'non-similar' control reference drivers, 3 and 6. Overall confidence in the test results was greatly enhanced by the excellent results from the 'similar' and 'non-similar' control drivers. Statistical similarity tables are shown in Figures 12.5 and 12.6(a), (b) and (c).

Given that Samples 3, 6 and, to a lesser degree, 15 were inserted as controls, of the remaining 13 samples, the Altec and JBL horns, 12 and 16, were clearly identified as sounding like our archetypal horn C. Samples 13 and 14, also horns, whilst not showing quite the clarity of results as Samples 12 and 16, were still adjudged to sound more like horn C than like A, B or D. Sample 2, the JBL direct radiator, was very clearly chosen as being similar to our archetypal direct radiator B. Horns 9 and 11 were very inconclusive. As Sample 11 was the mid-horn from an earlier range of JBL studio monitors, this unrepresentative, unclear set of results could go some way towards explaining their demise. Horns 1 and 7, whilst showing a tendency towards B, achieved only low numbers of clear, unequivocal ticks in the columns, so can probably be grouped with Samples 9 and 11 as inconclusive and/or unrepresentative. The remaining horn samples, numbers 4, 5, 8 and 10, can be concluded as not sounding like the chosen archetypal horn C, nor do they show any distribution of results which could be deemed similar to those of any of the other horns in the tests. Samples 4, 5 and 10 were clearly chosen as being similar to the direct radiator Archetype B, whilst Sample 8 was deemed similar to the Tannoy, D, both of which were axisymmetric devices of roughly equal overall dimensions.

From the above tests, it would not initially seem to be possible to conclude that horns *per se* have any overall, specific, characteristic sound, nor that mid-range horns cannot be made to sound subjectively similar to comparable direct radiator units. A point worthy of note here is that the typically 'hornlike' horns, Samples 12, 13, 14 and 16, were all long horns. The distance between the driver diaphragm and the horn mouth was in the 13 to 30″ region. Of the 'non-hornlike' horns, Samples 4, 5, 8 and 10, the first three were all short horns, with typically 12″ or less between the diaphragm and the mouth. I had felt for some time that length had a part to play in imparting a hornlike character to a horn, and the results of this test seem to confirm that belief. The approximate diaphragm to mouth distances of all of the horns in the test are shown in Fig. 12.7. So yes, quite definitely, size *is* important!

It is worth noting that Sample 10 was the only exception in the grouping of the 'Horns with similarity to B' being less than $12\frac{1}{4}″$ (310 mm), and 'Horns with similarity to C' being over 16″ (400 mm). Indeed, Sample 10 was one of the two horns which showed the strongest similarity to B, yet it was 18″ (440 mm) long. I remember when I first arrived at the ISVR with the horns designated for the project. They were noticed by Dr Frank Fahy (now Professor) who singled out this horn (the Fostex wooden horn) in particular to ask 'What is that?', to which I replied that it was a horn. Frank replied that according to his knowledge of general acoustics (which could be considered second to only a very few members of the human race) that was no horn, but merely a form of waveguide.

As can be seen from Fig. 12.8, the 'horn' consisted of a very short, exponential 'throat extension', followed by a rather abrupt termination into what was effectively the 'waveguide' section, created by a very large pair of almost semicircular wooden 'lips', giving a good directivity pattern in excess of 120° in the horizontal plane. The horn produced a rather undesirable set of throat impedance plots, implying an uneven pressure amplitude response (frequency response) when connected to a driver. Nonetheless, in

Signal	Sample 1 (%) A	B	C	D	None	Sample 2 (%) A	B	C	D	None	Sample 3 (%) A	B	C	D	None	Sample 4 (%) A	B	C	D	None
1	–	–	–	78	–	–	37	**97**	–	–	–	**94**	–	–	–	–	**100**	–	–	–
2	–	–	–	30	30	–	78	–	–	–	–	–	–	39	90	–	**100**	–	–	–
3	–	29	–	78	–	–	**100**	–	–	–	–	–	–	–	97	–	**100**	–	–	–
4	–	37	–	–	**97**	–	29	–	–	99	–	–	–	–	**100**	–	98	–	–	78
5	–	–	–	39	39	–	78	–	–	78	–	–	–	–	**100**	–	99	–	–	77
6	–	**95**	–	30	30	–	**97**	–	–	37	–	–	–	–	99	–	90	–	–	90
7	–	–	–	37	**97**	–	**90**	–	–	39	–	–	–	–	**100**	–	**100**	–	–	–
8	–	78	–	**98**	–	–	**100**	–	–	–	–	–	–	–	78	–	**100**	–	–	–
9	–	30	–	30	30	–	**100**	–	–	–	–	–	–	–	**100**	–	–	–	37	**97**

Signal	Sample 5 (%) A	B	C	D	None	Sample 6 (%) A	B	C	D	None	Sample 7 (%) A	B	C	D	None	Sample 8 (%) A	B	C	D	None
1	–	85	37	37	–	–	**100**	–	–	–	–	**100**	–	–	–	–	–	–	**98**	–
2	–	**99**	–	–	–	–	**100**	–	–	30	–	**98**	–	–	–	–	–	–	**98**	30
3	–	–	–	78	–	–	**100**	–	–	–	–	**90**	39	–	–	–	–	–	**90**	–
4	–	–	–	–	98	–	**100**	–	–	–	–	–	–	–	**98**	–	–	–	–	78
5	–	–	–	–	78	–	**100**	–	–	–	–	–	–	–	**98**	–	–	–	–	**100**
6	–	39	–	39	–	–	**100**	–	–	–	–	**90**	39	–	–	–	37	–	–	85
7	–	–	–	–	**100**	–	**100**	–	–	–	–	–	–	–	**100**	–	–	–	–	**100**
8	–	**100**	–	–	–	–	**100**	–	–	–	–	–	41	–	–	–	–	–	–	90
9	–	30	–	30	30	–	**100**	–	–	–	–	78	–	–	30	–	–	–	30	78

Signal	Sample 9 (%) A	B	C	D	None	Sample 10 (%) A	B	C	D	None	Sample 11 (%) A	B	C	D	None	Sample 12 (%) A	B	C	D	None
1	–	**94**	41	–	–	–	**100**	–	–	–	–	90	–	–	39	–	90	39	–	–
2	–	**85**	37	–	37	–	**100**	–	–	–	–	–	85	–	37	–	–	–	29	–
3	–	–	**100**	–	–	–	90	39	–	–	–	**94**	41	–	–	–	29	**99**	–	–
4	–	–	–	–	**100**	–	37	–	–	**100**	–	–	–	–	**100**	–	–	–	–	**100**
5	–	–	–	–	**100**	–	**97**	–	–	85	–	–	–	–	38	–	–	–	–	**100**
6	–	41	–	–	**94**	–	78	–	–	–	–	**98**	–	–	–	–	–	78	–	30
7	–	–	–	–	**100**	–	**98**	–	–	30	–	–	–	–	**100**	–	–	85	–	97
8	–	39	–	–	**90**	–	**98**	–	–	–	–	78	–	–	29	–	–	**98**	–	–
9	–	–	78	–	–	–	29	–	78	–	–	**100**	–	–	–	–	–	**99**	–	–

Signal	Sample 13 (%) A	B	C	D	None	Sample 14 (%) A	B	C	D	None	Sample 15 (%) A	B	C	D	None	Sample 16 (%) A	B	C	D	None
1	–	**97**	–	–	37	–	39	–	**90**	–	–	**100**	–	–	–	–	–	78	–	–
2	–	**98**	–	–	–	–	90	–	–	–	–	**98**	–	–	39	–	99	–	–	–
3	–	–	41	–	–	–	–	85	–	–	–	–	29	–	–	–	–	**100**	–	–
4	–	–	–	–	**100**	–	–	–	–	**100**	–	–	–	–	**100**	–	–	–	–	**100**
5	–	–	–	–	**100**	–	–	–	–	**100**	–	–	–	–	**100**	–	–	–	–	**100**
6	30	–	30	–	–	–	–	–	–	99	–	–	–	–	**100**	–	85	**97**	–	–
7	37	–	–	–	**97**	–	–	–	–	**100**	–	–	–	–	99	–	–	–	–	**100**
8	30	–	78	–	–	–	–	**100**	–	–	–	–	30	–	78	–	–	**98**	–	–
9	–	–	30	–	30	–	–	**100**	–	30	–	–	37	–	85	–	–	78	–	–

Figure 12.6(a) Similarity confidence indices: unambiguous ticks

Signal	Sample 1 (%)					Sample 2 (%)					Sample 3 (%)					Sample 4 (%)				
	A	B	C	D	None	A	B	C	D	None	A	B	C	D	None	A	B	C	D	None
1	–	25	25	**97**	–	–	67	25	–	–	–	**100**	–	87	–	–	**100**	–	–	–
2	–	67	–	25	–	–	**95**	–	–	–	–	–	–	25	–	–	**100**	–	–	–
3	–	87	24	87	–	–	**100**	–	–	–	25	–	–	–	67	–	**100**	–	25	–
4	–	**96**	–	–	87	–	73	–	–	**100**	–	–	–	–	**100**	–	**100**	–	–	73
5	–	**22**	–	–	22	–	**95**	–	–	95	–	–	–	–	**100**	–	**100**	–	–	**93**
6	–	**91**	–	25	–	–	**99**	–	–	–	–	28	–	–	**93**	–	**99**	–	–	25
7	–	87	–	64	87	–	**99**	–	–	67	–	–	–	–	**100**	–	**100**	–	–	28
8	–	**98**	–	98	–	–	**100**	–	–	–	–	67	–	67	25	–	**100**	–	–	–
9	–	25	–	**91**	–	–	**100**	–	–	–	–	–	–	–	**100**	–	25	–	67	25

Signal	Sample 5 (%)					Sample 6 (%)					Sample 7 (%)					Sample 8 (%)				
	A	B	C	D	None	A	B	C	D	None	A	B	C	D	None	A	B	C	D	None
1	–	**96**	24	64	–	–	**100**	–	58	–	–	**100**	24	–	–	–	–	–	**100**	–
2	–	**100**	25	–	–	–	**100**	–	–	–	–	**99**	–	–	–	–	–	–	**99**	30
3	–	87	24	87	–	–	**100**	–	22	–	–	60	85	–	–	–	22	–	85	–
4	–	87	–	24	87	–	**100**	–	–	–	–	–	–	–	**99**	–	–	–	–	**91**
5	–	67	–	–	67	–	**100**	–	–	–	–	24	–	–	**96**	–	–	–	28	**100**
6	–	85	–	22	–	–	**100**	–	–	–	–	**99**	89	–	–	–	**64**	24	24	24
7	–	25	–	–	**98**	–	**100**	–	–	–	–	–	–	–	**98**	–	–	–	–	**100**
8	–	**100**	–	–	–	–	**100**	–	–	–	–	**96**	64	–	–	–	–	–	67	67
9	–	66	–	89	66	–	**100**	–	–	–	–	**91**	–	–	–	–	–	–	67	67

Signal	Sample 9 (%)					Sample 10 (%)					Sample 11 (%)					Sample 12 (%)				
	A	B	C	D	None	A	B	C	D	None	A	B	C	D	None	A	B	C	D	None
1	–	**96**	22	–	–	–	**100**	–	–	–	–	**99**	–	–	–	–	**97**	25	66	–
2	–	**91**	25	–	–	–	**100**	–	–	–	–	64	**96**	–	–	–	–	–	25	–
3	–	**22**	**99**	–	–	–	**96**	60	–	–	–	**100**	87	–	–	–	**90**	**100**	–	–
4	–	–	–	–	**100**	–	**90**	–	–	**100**	–	24	–	–	**99**	–	–	25	–	**91**
5	–	–	–	–	**100**	–	**99**	–	–	66	–	85	–	–	22	–	–	–	–	**91**
6	–	**96**	–	–	24	–	**91**	–	–	–	–	**100**	24	–	–	–	28	**98**	–	28
7	–	25	–	–	**97**	–	**98**	–	–	25	–	–	–	–	**96**	–	–	25	–	67
8	–	89	–	–	66	–	**100**	–	–	–	–	**98**	–	–	–	–	–	**99**	–	–
9	–	–	67	–	–	–	**99**	–	75	–	–	**100**	–	–	–	–	–	**99**	–	–

Signal	Sample 13 (%)					Sample 14 (%)					Sample 15 (%)					Sample 16 (%)				
	A	B	C	D	None	A	B	C	D	None	A	B	C	D	None	A	B	C	D	None
1	–	**91**	–	–	–	–	66	–	89	–	–	**100**	–	–	–	–	–	**99**	–	–
2	28	**93**	28	–	–	–	67	67	–	–	–	**99**	–	–	–	–	**96**	84	–	–
3	–	**22**	84	–	–	–	–	**98**	–	–	–	67	**90**	–	–	–	–	**99**	–	–
4	–	–	25	–	**100**	–	–	–	–	**99**	–	–	–	–	**100**	25	–	–	–	**99**
5	–	–	–	–	**100**	–	–	25	–	**99**	–	–	24	–	**96**	–	–	–	–	**100**
6	28	28	28	–	–	–	–	–	–	**99**	–	–	–	–	**99**	–	89	**97**	–	–
7	–	–	25	–	**100**	–	–	–	–	**95**	–	–	–	–	**99**	–	–	–	–	**100**
8	–	–	**91**	–	–	–	–	**100**	–	–	–	–	67	–	67	–	–	**99**	–	–
9	–	–	67	–	25	–	–	**100**	–	28	–	–	**91**	–	25	–	–	**91**	–	–

Figure 12.6(b) Similarity confidence indices: total numbers of ticks

	Sample 1 (%)					Sample 2 (%)					Sample 3 (%)					Sample 4 (%)				
Signal	A	B	C	D	None	A	B	C	D	None	A	B	C	D	None	A	B	C	D	None
1	–	25	25	90	–	–	67	25	–	–	–	100	–	67	–	–	100	–	28	–
2	–	67	–	–	–	–	95	–	–	–	–	–	–	30	30	–	100	–	–	28
3	–	87	24	87	–	–	100	–	–	–	–	–	–	–	67	–	100	–	25	–
4	–	67	–	–	91	–	30	–	–	100	–	–	–	–	100	–	99	–	–	75
5	–	–	–	–	67	–	77	–	–	97	–	–	–	–	100	–	99	–	–	95
6	–	91	–	25	–	–	91	–	–	–	–	–	–	–	97	–	99	–	–	25
7	–	25	–	25	91	–	93	–	–	93	–	–	–	–	100	–	100	–	–	30
8	–	91	–	99	–	–	100	–	–	–	–	25	–	67	25	–	100	–	–	–
9	–	73	–	93	28	–	100	–	–	–	–	–	–	–	100	–	30	–	75	75

	Sample 5 (%)					Sample 6 (%)					Sample 7 (%)					Sample 8 (%)				
Signal	A	B	C	D	None	A	B	C	D	None	A	B	C	D	None	A	B	C	D	None
1	–	97	25	66	–	–	100	–	–	–	–	100	–	–	–	–	–	–	99	–
2	–	100	25	–	–	–	100	–	–	–	–	100	–	–	–	–	–	–	98	30
3	–	67	25	67	–	–	100	–	22	–	–	87	96	–	–	–	64	–	96	–
4	–	25	–	–	91	–	100	–	–	–	–	–	–	–	100	–	–	–	30	95
5	–	30	–	–	95	–	100	–	–	–	–	–	–	–	99	–	–	–	–	100
6	–	90	–	25	–	–	100	–	–	–	–	99	67	–	–	–	67	–	–	25
7	–	–	–	28	100	–	100	–	–	–	–	–	–	–	100	–	–	–	–	100
8	–	100	–	–	–	–	100	–	–	–	–	90	67	–	–	–	–	–	25	67
9	–	25	–	67	67	–	100	–	–	–	–	91	–	–	–	–	–	–	67	67

	Sample 9 (%)					Sample 10 (%)					Sample 11 (%)					Sample 12 (%)				
Signal	A	B	C	D	None	A	B	C	D	None	A	B	C	D	None	A	B	C	D	None
1	–	98	25	–	–	–	100	–	–	–	–	99	–	–	–	–	91	–	25	–
2	–	93	28	–	28	–	100	–	–	–	–	67	67	–	–	–	–	30	30	30
3	–	25	98	–	–	–	99	64	–	–	–	99	89	–	–	–	67	100	–	–
4	–	–	–	–	100	–	25	–	–	100	–	–	–	–	100	–	–	28	–	98
5	–	–	–	–	100	–	98	–	–	67	–	67	–	–	67	–	–	–	–	98
6	–	91	–	–	25	–	98	–	–	28	–	99	25	–	–	–	30	95	–	30
7	–	–	–	–	99	–	91	–	–	25	–	–	–	–	99	–	–	30	–	97
8	–	67	–	–	67	–	100	–	–	–	–	98	–	–	–	–	–	99	–	–
9	–	–	67	–	–	–	78	–	78	–	–	100	–	–	–	–	–	99	–	–

	Sample 13 (%)					Sample 14 (%)					Sample 15 (%)					Sample 16 (%)				
Signal	A	B	C	D	None	A	B	C	D	None	A	B	C	D	None	A	B	C	D	None
1	–	91	–	–	–	–	67	–	67	–	–	100	–	–	–	–	–	91	–	–
2	–	95	30	–	–	–	67	25	–	–	–	98	–	–	–	–	97	60	–	–
3	–	–	85	–	–	–	–	98	–	–	–	67	90	–	–	–	–	99	–	–
4	–	–	–	–	100	–	–	–	–	100	–	–	–	–	100	28	–	–	–	100
5	–	–	–	–	100	–	–	28	–	100	–	–	–	–	98	–	–	–	–	100
6	28	28	28	–	–	–	–	–	–	100	–	–	–	–	99	–	90	90	–	–
7	–	–	28	–	100	–	–	–	–	99	–	–	–	–	100	–	–	–	–	100
8	–	91	–	–	–	–	–	100	–	–	–	–	73	–	93	28	–	98	–	–
9	28	–	73	–	73	–	–	100	–	30	–	–	75	–	75	–	–	97	–	30

Figure 12.6(c) Similarity confidence indices: weighted ticks

Signal	Sample 1 A	B	C	D	None	Sample 2 A	B	C	D	None	Sample 3 A	B	C	D	None	Sample 4 A	B	C	D	None
1	–	4	4	7	1	2	5	4	2	–	–	9	2	6	–	–	8	–	3	2
2	–	5	3	4	3	1	5	2	1	2	–	3	3	4	3	–	8	2	2	3
3	–	6	4	6	1	–	8	2	1	1	4	2	2	2	5	1	8	–	4	1
4	–	7	1	3	6	1	4	–	–	7	2	1	–	–	9	–	7	–	1	4
5	1	5	3	4	5	1	5	–	–	5	1	–	1	–	9	–	7	–	–	5
6	–	6	1	4	3	–	7	2	2	3	1	3	1	2	5	–	8	1	3	4
7	–	6	–	5	6	–	7	1	1	5	2	–	–	–	9	–	9	–	–	3
8	–	7	1	7	–	1	9	1	3	1	–	5	1	5	4	–	8	–	3	2
9	–	7	–	6	3	–	9	–	1	–	2	3	–	2	8	–	4	–	5	4

Signal	Sample 5 A	B	C	D	None	Sample 6 A	B	C	D	None	Sample 7 A	B	C	D	None	Sample 8 A	B	C	D	None
1	1	7	4	5	–	1	13	2	6	–	2	9	4	2	–	–	3	2	8	1
2	–	9	4	2	1	–	10	1	2	3	2	7	2	–	2	–	1	–	6	3
3	–	6	4	6	1	–	13	3	5	–	2	6	7	2	2	1	5	3	7	2
4	–	6	1	4	6	–	13	–	1	1	1	3	–	2	7	–	3	1	3	6
5	–	5	2	3	5	–	14	–	2	–	–	4	3	3	7	–	1	1	3	7
6	–	7	4	5	2	–	13	2	4	1	–	8	6	–	2	–	5	4	4	4
7	–	4	1	3	7	2	14	1	2	–	1	3	1	3	7	–	–	1	1	9
8	–	9	2	3	3	1	13	2	3	1	3	7	5	2	–	–	3	2	5	5
9	–	5	–	6	5	–	14	–	3	–	–	6	2	2	3	–	2	1	5	5

Signal	Sample 9 A	B	C	D	None	Sample 10 A	B	C	D	None	Sample 11 A	B	C	D	None	Sample 12 A	B	C	D	None
1	1	8	5	2	3	–	11	1	1	–	3	8	3	–	2	–	7	4	5	–
2	–	6	4	1	3	–	10	2	–	1	2	5	7	1	2	1	2	3	4	3
3	–	5	9	4	1	–	8	6	4	1	2	9	6	–	–	–	6	8	1	–
4	–	2	1	2	10	–	6	–	1	8	2	4	2	1	8	–	3	4	1	6
5	–	2	–	1	9	–	8	1	2	5	2	7	3	2	5	1	3	2	1	6
6	–	7	3	3	4	–	6	2	3	3	–	9	4	2	2	–	3	6	–	3
7	–	4	2	3	7	–	7	2	2	4	3	3	3	2	8	1	1	4	2	5
8	1	6	2	2	5	–	9	1	3	2	–	7	3	2	3	2	3	7	–	1
9	–	3	5	3	3	–	6	–	4	1	1	8	1	2	1	–	2	7	3	2

Signal	Sample 13 A	B	C	D	None	Sample 14 A	B	C	D	None	Sample 15 A	B	C	D	None	Sample 16 A	B	C	D	None
1	2	6	1	3	2	1	5	2	6	2	1	9	2	3	–	3	2	7	–	2
2	3	5	3	–	1	2	5	5	2	1	2	8	2	2	2	3	8	7	1	1
3	4	5	7	3	1	3	3	7	1	1	1	5	6	1	2	1	2	7	1	2
4	1	–	4	–	9	3	–	3	1	7	2	–	2	1	8	4	1	1	–	7
5	3	1	2	1	9	2	–	4	1	7	3	1	4	2	7	–	2	1	1	7
6	3	3	3	1	2	2	1	2	–	6	–	3	3	1	7	1	6	7	1	1
7	3	–	4	–	8	2	1	2	1	5	3	–	2	1	7	2	2	–	–	9
8	3	1	6	2	2	3	–	9	–	1	3	1	5	–	5	3	–	7	1	2
9	3	–	5	1	4	–	1	8	–	3	1	2	6	–	4	–	3	6	1	3

Figure 12.6(d) Listening test results: overall total numbers of ticks

Signal	Sample 1 A	B	C	D	None	Sample 2 A	B	C	D	None	Sample 3 A	B	C	D	None	Sample 4 A	B	C	D	None
1	–	4	4	6	1	2	5	4	2	–	–	9	1.5	5	–	–	7.5	–	3	2
2	–	5	3	3.5	3	1	5	2	1	2	–	2	2	3.5	3	–	7.5	1.5	1.5	3
3	–	6	4	6	1	–	8	2	1	1	3	1.5	2	2	5	1	8	–	4	1
4	–	5.5	0.5	3	6	0.5	3.5	–	–	7	1	0.5	–	–	9	–	6.5	–	1	4
5	0.5	3	2.5	3.5	5	1	4.5	–	–	5	0.5	–	1	–	9	–	6.5	–	–	5
6	–	6	1	4	3	–	6.5	2	2	3	0.5	2	1	2	5	–	7	0.5	2	4
7	–	4.5	–	4.5	6	–	5.5	1	1	5	1	–	–	–	9	–	8	–	–	3
8	–	6.5	1	7	–	1	8.5	1	3	1	–	4.5	1	5	4	–	8	–	2.5	2
9	–	4	–	5	3	–	9	–	1	–	1	1.5	–	1	8	–	3.5	–	4	4

Signal	Sample 5 A	B	C	D	None	Sample 6 A	B	C	D	None	Sample 7 A	B	C	D	None	Sample 8 A	B	C	D	None
1	0.5	7	4	5	–	1	13	2	4	–	1.5	9	3.5	1.5	–	–	3	2	7	1
2	–	8	4	2	1	–	9.5	1	1.5	3	1.5	7	2	–	2	–	1	–	5.5	3
3	–	5.5	4	5.5	1	–	13	2.5	5	–	1.5	6	7	2	1.5	1	5	3	7	1.5
4	–	4.5	0.5	3.5	6	–	12.5	–	1	1	1	2.5	–	1.5	7	–	2.5	1	3	5.5
5	–	3.5	1.5	2.5	5	–	14	–	1.5	–	–	3	2	2.5	7	–	1	0.5	2.5	7
6	–	6.5	3.5	4	2	–	13	1	3	1	–	7.5	5.5	–	2	–	5	3.5	3.5	4
7	–	2.5	0.5	3	7	1.5	14	1	1.5	–	0.5	2	1	2	7	–	–	0.5	2	9
8	–	8.5	2	2.5	3	0.5	13	1.5	2.5	1	3	6	5	1.5	–	–	2.5	2	4	5
9	–	4.5	–	5	5	–	14	–	2	–	–	6	2	2	3	–	2	1	5	5

Signal	Sample 9 A	B	C	D	None	Sample 10 A	B	C	D	None	Sample 11 A	B	C	D	None	Sample 12 A	B	C	D	None
1	0.5	7	4.5	1.5	3	–	11	1	1	–	2.5	7	3	–	2	–	6.5	3.5	4	–
2	–	5.5	3	1	3	–	9.5	2	–	1	1.5	5	5.5	0.5	2	0.5	1.5	3	3.5	3
3	–	4.5	7.5	3.5	1	–	8	5	3	1	2	8.5	6	–	–	–	5.5	8	1	–
4	–	1	0.5	1.5	10	–	4.5	–	1	8	1.5	2.5	1.5	1	8	–	2	3.5	1	6
5	–	2	–	1	9	–	7	1	2	5	1.5	5.5	2.5	2	5	1	2	2	1	6
6	–	6.5	2	2.5	4	–	6	1.5	2.5	3	–	8	4	2	2	–	3	5.5	–	3
7	–	3.5	1.5	2.5	7	–	6.5	1.5	2	4	2	2.5	2	2	8	0.5	1	3.5	1.5	5
8	0.5	5.5	2	1.5	5	–	8	0.5	2.5	2	–	7	3	2	3	2	3	7	–	1
9	–	2.5	5	3	3	–	4.5	–	4	1	1	7.5	1	1.5	1	–	1.5	6.5	2.5	2

Signal	Sample 13 A	B	C	D	None	Sample 14 A	B	C	D	None	Sample 15 A	B	C	D	None	Sample 16 A	B	C	D	None
1	1.5	6	1	3	2	1	5	1.5	5	2	1	9	2	3	–	3	2	6.5	–	2
2	2.5	5	3	–	1	2	5	4	2	1	2	7.5	2	2	2	3	8	6.5	0.5	1
3	4	4.5	7	3	1	2.5	2.5	6.5	1	1	1	5	6	1	2	1	2	7	1	2
4	1	–	3	–	9	2	–	1.5	0.5	7	1.5	–	1	1	8	3	1	1	–	7
5	1.5	0.5	1.5	0.5	9	2	–	3	0.5	7	2	1	3	2	7	–	1.5	1	0.5	7
6	3	3	3	1	2	1.5	1	1.5	–	6	–	3	3	1	7	1	6	6.5	1	1
7	2.5	–	3	–	7.5	1.5	1	1	0.5	5	2.5	–	1	1	7	1	1.5	–	–	9
8	3	1	6	1.5	2	2.5	–	8.5	–	1	2.5	1	4	–	5	3	–	6.5	1	2
9	3	–	4.5	1	4	–	0.5	7.5	–	3	1	2	4.5	–	4	–	2.5	5.5	0.5	3

Figure 12.6(e) Listening test results: weighted ticks

	STATISTICAL CONFIDENCE				NUMERICAL TICK DISTRIBUTION						TENDENCIES FOR SAMPLES FAVOURING 'NONE'			
Figure Column/Sample No.	13(b) (unam B) b	13(c) (total) c	13(d) (weighted) d	tendency e	13(f) (unamb) f	13(g) total g	13(h) weighted h	Number of possible indications i	Indications from numerical totals j	Statistical k	Numerical l	Absolute consistency of results	Significantly unstable results
1	N	B	D	N/B/D	N	B	D	3	B	B558/C557	B62/C59		✓
2	B	B	B	B	B	B	B	1	B	B	B		
3	N	N	N	N	N	N	N	1	B	–	–	✓	
4	B	B	B	B	B	B	B	1	B	–	–	✓✓✓	
5	B	B	B	B	B	B	B	1	B	–	–	✓✓✓	
6	B	B	B	B	B	B	B	1	B	–	–	✓✓✓	
7	B	B	B	B	B	B	B	1	B	–	–	✓✓✓	✓
8	N	N	N	N	N	N	N	1	D Total of ticks actually favours 'D' over NONE	D	D		
9	N	N	N	N	N	B	N	2	B	B	B	✓	
10	B	B	B	B	B	B	B	1	B	–	–	✓✓	
11	B	B	B	B	B	B	B	2	B	–	–	✓✓	
12	N	C	C	C	C	C	C	2	C	B			
13	N	N	N	N	N	C	C	2	C Total of ticks actually favours 'C' over NONE	B	C		
14	N	C	N	N	N	C	N	2	C	C	C	✓	✓
15	N	N	N	N	N	N	N	1	Total of ticks favours C	B	B		
*16	N	C/N	C/N	C	C	C	C	2	C	–	–		✓

SPREAD

Note: A = Quad Electrostatic   B = Son Audax Direct Radiator
C = Fostex/Emilar, Horn/Driver   D = Tannoy 15″ Dual Concentric
N = None of the above

*Results for Sample 16 change for columns (c) and (d) according to whether statistical confidence indices are calculated for results of 90% or above, or 100% only. All other results are independent of which system is used.

Figure 12.6(f) Comparison of numerical and statistical analysis

Sample no.	Manufacturer/ type	Flare material	Flare-rate	Length (mm)	Mouth size
Horns with similarity to reference B					
Ref. B	Son Audax direct radiator	–	–	–	–
1	Vitavox exponential	aluminium	medium	310	medium
4	AX1 axisymmetric*	glassfibre	low	230	small
5	Reflexion Arts	glassfibre	medium	300	medium
7	Reflexion Arts – no lips	glassfibre	medium	240	medium
10	Fostex sectoral*	wood	high	440	large
11	JBL axisymmetric	aluminium	low	250	small
Horns with similarity to reference C					
Ref. C	Fostex sectoral	aluminium	medium	500	large
12	Altect sectoral*	aluminium	medium	500	large
13	Altect multicellular	aluminium	low/med	600	large
14	Starr gramophone	wood	low	650	medium
15	Vitavox sectoral	aluminium	medium	450	large
16	JBL bi-radial*	composite	medium	400	medium
Others					
8	AX2 axisymmetric	glassfibre	high	230	medium
9	Yamaha sectoral	aluminium	medium	310	medium

Those samples marked * showed particularly strong similarity to the reference.

Figure 12.7 Horn loudspeaker samples grouped according to similarity

auditioning before the tests, the horn was generally 'musical' and pleasant to listen to, and, as already mentioned, received some positive comments during the listening tests themselves!

If Professor Fahy's concept of the bulk of this horn consisting merely of large 'lips' is applied, then the true 'horn' section would only be in the order of 6″ (150 mm), making it the shortest horn of all the samples. Samples 4 and 10, being the two horns showing the strongest similarity to Archetype B, would then also be the two shortest horns in the test. When one further understands that a so-called direct radiator cone is actually an extreme case of a 180° conical horn of zero length, then the correlation of the above results is even stronger. In other words, there is no absolute dividing line between horns and direct radiators.

12.7 Interesting subjective points

Whilst the electrostatic, Archetype A, was deemed similar to the samples on only a relatively small number of occasions, it was occasionally noted that with the waterfall, 1 kHz to 6 kHz band-limited as it was, Archetype A sounded more 'wet' than any other unit. The smaller source area units such as the direct radiators and small horns were hardly ever considered similar to A, whilst the horns with larger mouths showed a considerable shift in similarity towards A.

Comments of clarity were made on occasions relating to Archetype A, and also to Samples 8 and 10, whilst Samples 9, 11, 13 and 15 were frequently considered to be 'strange'. The fact that so few similarities were found between the samples and the Quad Electrostatic (ESL) A is, I suspect, due to the fact that few other loudspeakers could approach the Quad in terms of overall transient accuracy. This point could be proved conclusively one way or the other by conducting tests using transfer function convolutions, which should allow one loudspeaker to mimic another; albeit not in real time. Both Keith Holland and I felt that the Quad was the most inherently accurate and natural of the loudspeakers in the test; and, within its output and frequency range, one of the most accurate of loudspeakers in general. In the anechoic chamber, the dipole properties of the Quad were of no consequence, as the rear radiation was absorbed. The frontal, axial response should have been all that was relevant in these tests. Later in the tests, a Quad ESL63 was brought in for comparison purposes, and whilst its performance specifications were different to the ESL used in the test, the general sound was deemed to be more similar to A than to any other archetype or sample. The step function response waveforms of the ESL and ESL63 were, in fact, very similar. A further reference, post test, to a Sony electrostatic again showed clear similarity to the Quads.

Plots of the pressure amplitude responses of all of the loudspeakers involved in these tests are shown in Fig. 12.3, but the similarities or lack of similarities between them do not appear to lie solely in this domain. The phase responses manifest themselves in a convolution with the pressure amplitude responses to modify the waveform of the signal envelope, and some of the differences almost certainly lie in this area. It is the phase/ pressure amplitude convolution which translates the steady state into the transient response (Fig. 12.9).

12.8 Non-linear distortions

The third area of potential difference is in non-linearities, which give rise to harmonic and inter-modulation distortion. One manufacturer in particular has criticised some of my magazine articles by insisting that I have not been paying sufficient attention to harmonic distortions. It is not my lack of attention to the problem, but more that I have looked hard, yet found no direct tie-ups between audible similarities and harmonic distortion, as long as those distortion products are kept within generally accepted limits.

Figure 12.10 shows a table of measured harmonic distortion products from four of the loudspeakers involved. At an SPL of 75 dB at 3 metres, the Quad Electrostatic, the Son Audax cone driver, and the Emilar compression driver on the Axisymmetric 2 horn, were all producing second and

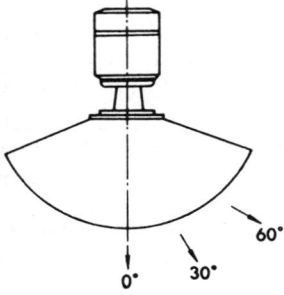
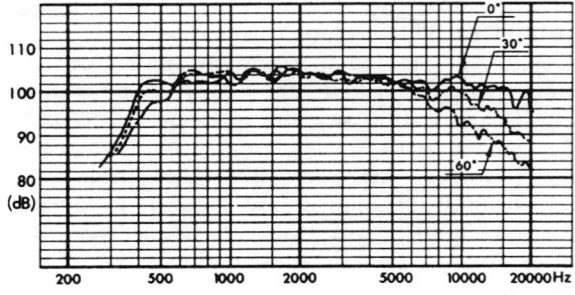

Figure 12.8 Geometry of Fostex horn (Sample 10)

third harmonic distortion products in the region of −60dB, or 0.1%. The JBL 2426/2370 produced around −53 dB of predominantly the 2nd harmonic. All sounded substantially similar with a sine wave. By 95 dB at 3 metres, the Quad had begun to overload, but the Audax, the Emilar and the JBL were still going strong. The distortions had risen to −50 dB for the Audax and −40 dB (1%) for the two compression horns. At 105 dB at 3 metres with a 1 kHz sine wave in the anechoic chamber, the Audax burned out – literally. The horns had begun to produce second harmonics in the −25 to −30 dB region, which may well be due to the onset of air overload distortions in the diaphragm/phase plug cavity. The less pleasant third harmonics were still down in the −30 to −40 dB regions. The overall sound quality was not considered to be unduly harsh or unpleasant.

Whilst harmonic distortions are never a desirable asset, people *can* become carried away with seeking their reduction as an overriding priority. There are several points which I would like to make concerning the relevance of distortion figures. There is little evidence to suggest that typical harmonic distortion is much discernible by the human ear at levels below 0.4%, or about −48 dB. In very many instances, levels of 1% (−40 dB) are only barely on the threshold of audibility. All three of the above drivers cited in the 95 dB measurements certainly *sounded* similar on sine waves.

For those people who consider horn distortion levels unacceptable at the 3% level measured at 105 dB at 3 metres, I must remind them that the typical direct radiator had already given out on a continuous sine wave signal, and that the horns did not sound unduly hard. More importantly, however, for studio monitoring purposes, we are looking for 110, 120, 130 dB levels, not as continuous signal levels, but for transient headroom. I know some crazy people in this business, but I do not know of anybody who makes a habit of listening to steady state signals at 125 dB, so sine wave testing was not continued beyond the above-mentioned 105 dB. The sound character tests of programme up to 130 dB are discussed in the following paragraph. As even loud music is largely a series of transients rather than a continuum, the harmonic distortion effects are less significant than often suggested. There is a masking effect of transients over harmonic distortion such that, at high levels, the extra harmonic distortion would be unlikely to be significantly objectionable. Ear distortions may also mask to some extent the distortions at higher levels, but the ears add distortions even when stood next to a set of loudly struck tympani. The ear canal distortions are hence a part of our high level perception, be it from reproduced *or* original sources.

After concluding our sine wave testing at 105 dB, we began investigations into old but frequently held beliefs that horns tend to 'get harder as you wind them up'. We recorded a range of seven test signals through a selection of horn/compression driver combinations, and also through direct radiators. The recordings were made via a B&K measuring microphone, with a pre-amplifier having attenuation facilities. The recordings were repeated at levels of 70 dB, 80 dB, 90 dB, 100 dB, 110 dB, 115 dB, 120 dB, 125 dB and 130 dB at 3 metres, on axis in an anechoic chamber. Each time that the level was increased by 5 or 10 dB, the pre-amplifier was attenuated by the same amount to provide a constant level on the digital tape.

When the recordings were subsequently played back, either via good

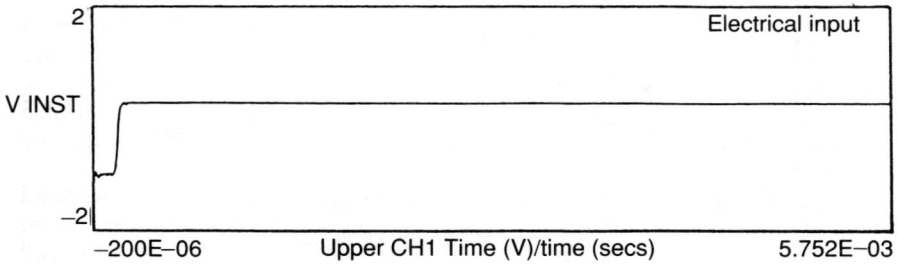

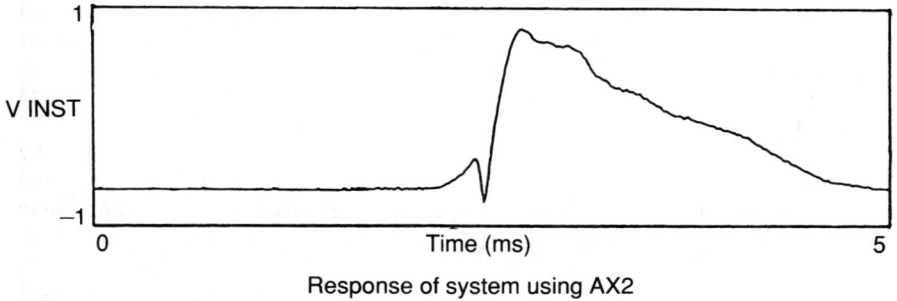

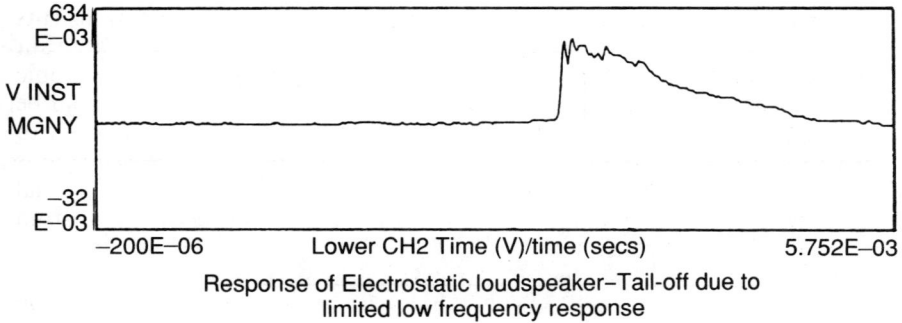

Figure 12.9 Full range step function responses (transient performance)

Response of Tannoy Dual-Concentric

Response of large, widely used, 2-way studio monitoring system

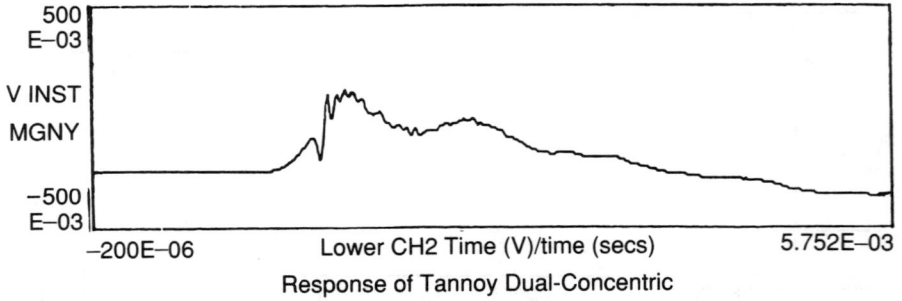

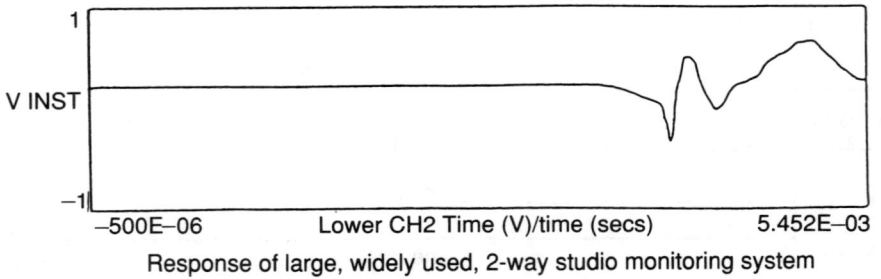

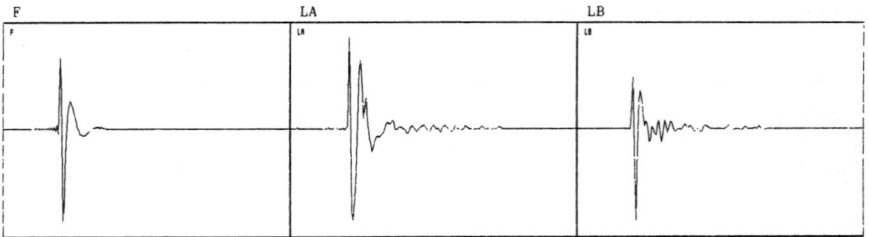

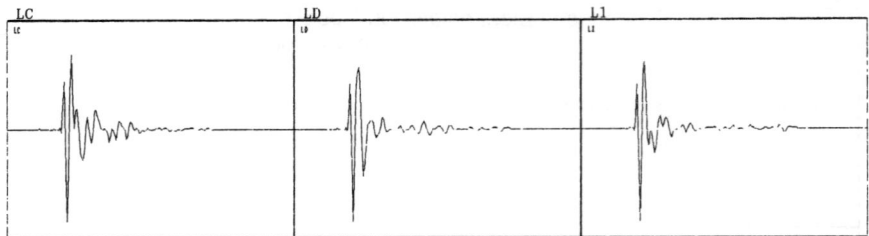

Figure 12.9 (*continued*)

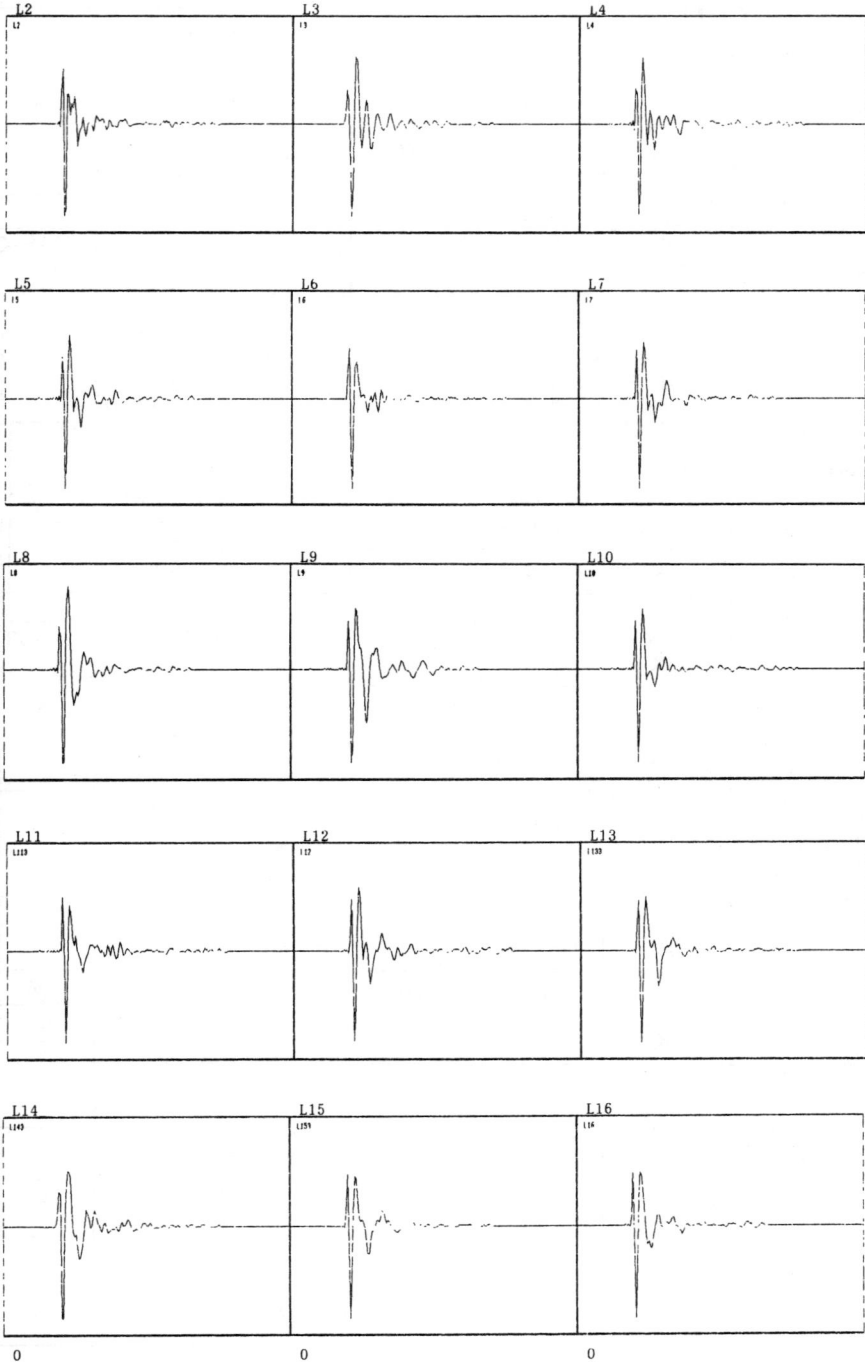

0   0   0

Figure 12.9 (*continued*)

P(3m) = 75 dB SPL

L.S.	f (Hz)	Input (v)	2nd	3rd	4th	5th
QUAD	1K	2.04	− 59	− 60	− 69	− 66
	2.8K	5.6	− 62	− 49	−	−
	5K	3.5	− 66	− 51	−	−
AUDAX	1K	1.01	− 59	− 54	− 6.8	−
	2.8K	0.75	− 61	− 60	−	−
	5K	0.64	− 62	− 59	−	−
JBL	1K	0.2	− 53	− 75	− 77	−
	2.8K	0.18	− 38	− 27	− 46	− 38
	5K	0.19	− 47	− 31	− 59	−
EMILAR	1K	0.24	− 61	− 66	−	−
	2.8K	0.17	− 58	− 73	− 70	−
	5K	0.3	− 47	− 57	−	−

P(3m) = 85 dB SPL

L.S.	f (Hz)	Input (v)	2nd	3rd	4th	5th
QUAD	1K	6.02	− 57	− 57	− 80	−
AUDAX	1K	3.8	− 54	− 49	− 78	− 67
	2.8K	1.7	− 57	− 57	−	−
	5K	2.5	− 47	− 67	−	−
JBL	1K	0.67	− 42	− 44	− 60	− 59
	2.8K	0.44	− 43	− 36	− 49	−
	5K	0.54	− 41	− 37	− 61	−
EMILAR	1K	0.73	− 50	− 55	− 76	−
	2.8K	0.44	− 49	− 65	− 73	−
	5K	0.87	− 38	− 54	− 71	−

P(3m) = 95 dB SPL

L.S.	f (Hz)	Input (v)	2nd	3rd	4th	5th
AUDAX	1K	10.1	− 50	− 44	− 79	− 69
	2.8K	5.7	− 49	− 49	−	−
	5K	7.7	− 37	− 57	−	−
JBL	1K	2	− 41	− 49	− 55	− 53
	2.8K	1.5	− 35	− 53	− 67	− 53
	5K	2.1	− 29	− 55	−	−
EMILAR	1K	2.45	− 40	− 53	− 68	− 59
	2.8K	1.4	− 40	− 55	− 61	− 63
	5K	2.9	− 29	− 48	− 71	−

Figure 12.10 Distortion

P(3m) = 105 dB SPL

L.S.	f (Hz)	Input (v)	2nd	3rd	4th	5th
JBL	1K	7.8	− 26	− 41	− 58	− 51
	2.8K	4.4	− 27	− 43	− 55	− 57
	5K	7.1	− 19	− 37	− 60	−
EMILAR	1K	10.5	− 28	− 32	− 45	− 55
	2.8K	4.4	− 29	− 52	− 57	− 59
	5K	8.8	− 20	− 36	− 51	−

Figure 12.10 (*continued*)

quality loudspeakers or headphones, the timbral differences even between the 70 dB and 115 dB levels were all but insignificant. At a level in the order of 125 dB at 3 metres, the horns rather suddenly produced unpleasant distortions which we are attributing to air overload. At these levels, however, the direct radiators had dropped out of the ratings due to thermal or mechanical failure. Although the horns could not be deemed 'hi-fi' at these levels, it is doubtful that 'hi-fi' *per se* exists at such levels. The main relevant point was that the horn/driver combinations were still not at risk from either thermal or mechanical failure. There was absolutely no evidence whatsoever of mid-range horn loudspeakers showing any tendency towards timbral hardening until well past a point where conventional radiators could no longer sustain such an output. It is therefore true to say that any such criticisms are invalid in terms of comparisons *vis-à-vis* other types of drive unit. Horns are simply more likely to reach the levels where such problems can be noticed.

Ultimately, if frequencies above 7 or 8 kHz are not required in a mid-range driver, dispensation with the phasing plug of the compression driver, or general reductions in compression ratios, can reduce non-linearities to levels generally commensurate with direct radiators. However, even with compression ratios which still afford relatively high sensitivity and good HF response (via the phasing plug), non-linearities do not appear to be controlling factors in any 'hornlike' sound.

12.9 Conclusions on similarity of groupings

12.9.1 Electrostatics

None of the 16 test samples were deemed to sound similar to the electrostatic, A. Only on certain sounds on a few sample drivers by a few people were any suggestions of audible similarity indicated. Initially we began to wonder if there was something wrong or being overlooked, but given the fact that there appeared to be nothing unique about the pressure amplitude responses of the electrostatics, the fact that they were the only units capable of reproducing anything approaching an accurate impulse, square wave or step function indicated that the uniqueness of their sound lay in the time domain. Indeed, the evidence from earlier work in the laboratory was so strong in its implication of time responses playing a large role in subjective assessment that I presented a paper[6] to the same IOA conference to which

Keith Holland had presented the paper on throat impedance measurement.[2] Some comparative step function plots are shown in Fig. 12.9.

12.9.2 Direct radiators

Of the three direct radiators included in the test samples, the two 'serious' contenders were clearly chosen as being similar to the direct radiator reference driver B. As previously discussed, one of the test samples was an identical unit to B, the results giving 100% confirmation of similarity. Even the third direct radiator sample, the 'non-similar' control reference, showed a tendency towards the direct radiator B in what little similarity did exist. The disparity in both size and design of the three direct radiators used in the tests ensured that undue commonality was avoided.

12.9.3 Horns

Of the remaining thirteen samples, all were horns, nine of which were of proprietary manufacture and in general use. Of the other four, one was a wooden horn from a 1920s gramophone, which conveniently terminated in a throat of about 1″ diameter. Another was identical to one of the other nine samples, except that the radial 'lips' had been removed from the mouth; thus maintaining the same throat and flare rate characteristics but disturbing the mouth termination into the room. The other two samples were specially made for these listening tests, one intended to be an example of a 'bad' horn with very irregular throat impedance, the other being an attempt to combine the knowledge of the previous research into a horn with the best throat impedance characteristics which could be achieved.

As already discussed, none of the sample units were deemed to be similar to the electrostatic unit, and only one was deemed similar to the axisymmetric Dual Concentric horn, Sample D. The remainder were spread in their similarity between B, the direct radiator, and C, our 'typical' horn. Essentially, the samples divided between B and C in accordance with their length. Archetype C, the typical horn, had a length of around 24″ (600 mm) from the diaphragm of the driver to the mouth of the horn. Samples with less than 12″ between the diaphragm and the mouth were deemed more or less similar to B, except for Sample 8 which matched well with the Tannoy, D. Those with greater than 12″ between diaphragm and mouth were deemed more or less similar to C.

12.10 Where do the differences lie?

In terms of audible similarity, we could find no evidence of any correlation with non-linear distortions. Indeed, certain samples with non-linearity differences of over 20 dB were deemed to be remarkably similar on a wide range of sounds. Likewise, linear distortions in terms of amplitude (see Fig. 12.3), except where they were grossly different, did not appear to be the controlling factor. In a statistical comparison of the measured pressure amplitude responses of the loudspeakers involved, a confidence correlation of about 70% between pressure amplitude responses and the listening test similarity results could be established. Amongst the horns themselves, neither rate of

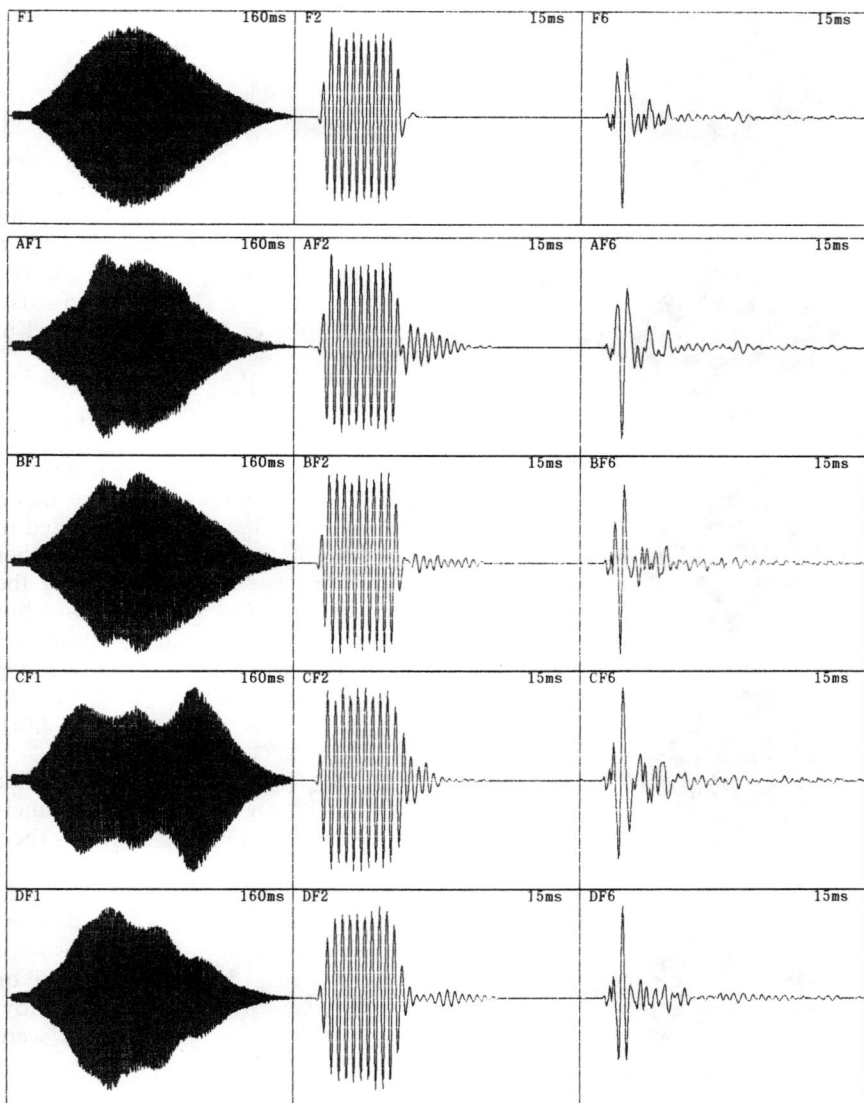

Figure 12.11 Waveform comparisons. Notes on waveforms:
   Number(s) or letter before 'F' denote loudspeaker sample.
   F denotes signal passed through filter (1 kHz to 6 kHz)
   3rd digit denotes sound sample

e.g. 12F1 = Altec horn, through filter, with 'chirp' signal
   DF6   = Tannoy Dual Concentric, through filter, with slamming book

At the head of each group is the electrical input signal waveform for reference

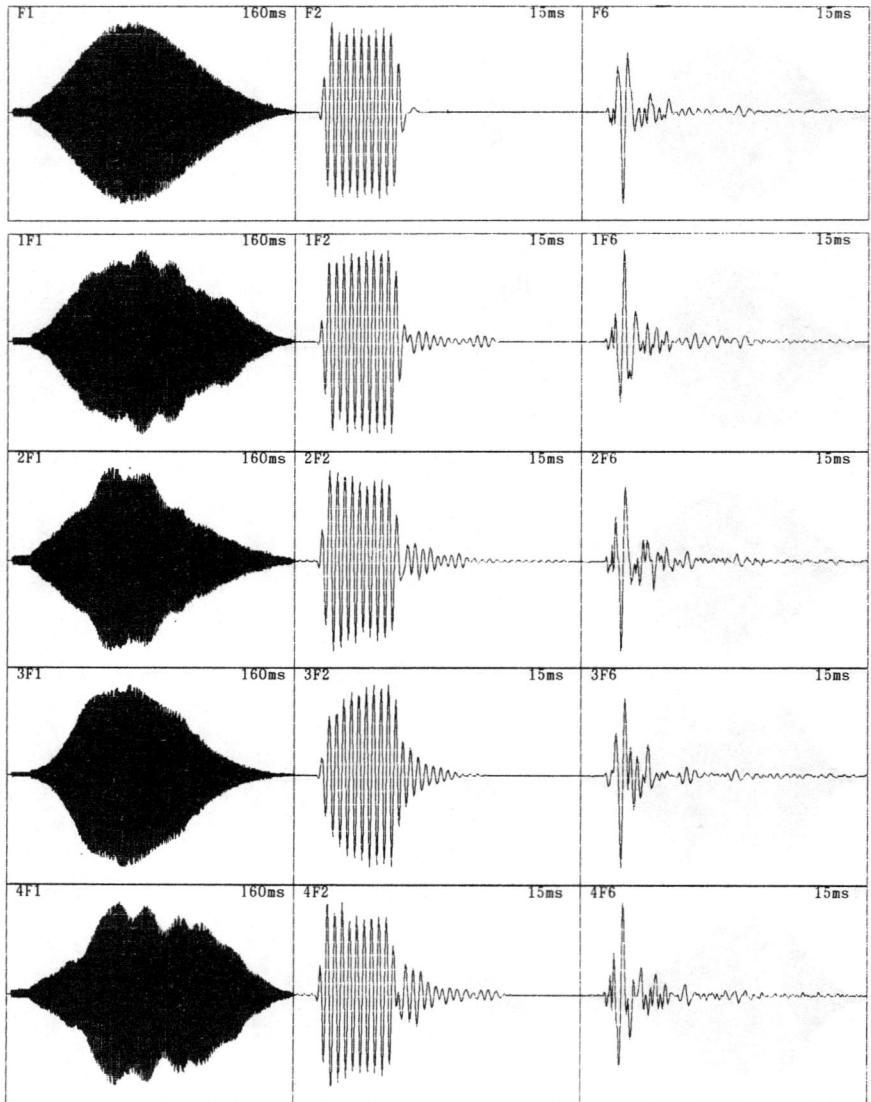

Figure 12.11 (*continued*)

flare/cut-off frequency, nor material of construction, in themselves appeared to have any significant correlation with audible similarity; although it should be pointed out that certain transient signals did excite some timbral anomalies in some of the units auditioned. A metallic characteristic could be detected in some horns on some signals, but only with some of the metal horns. In most cases, improved damping removed the material-related characteristics. A selection of waveform comparisons from the tests is shown in Fig. 12.11.

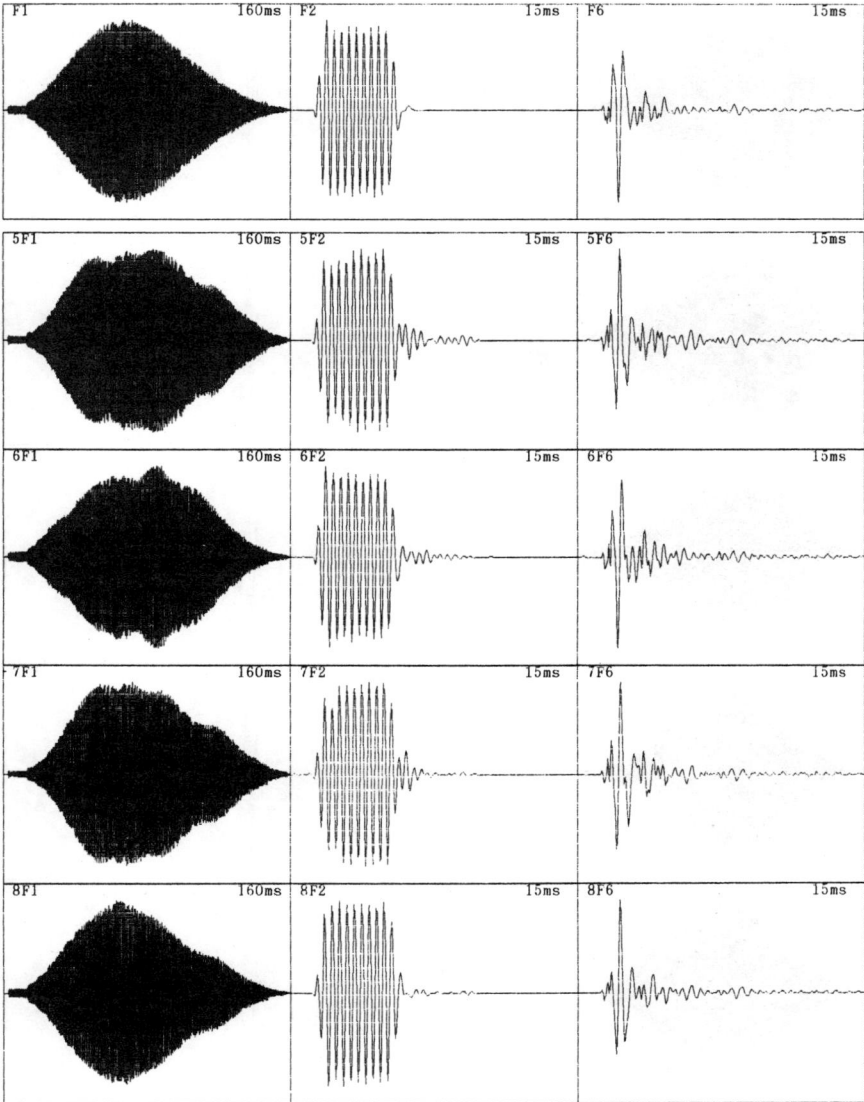

Figure 12.11 (*continued*)

The length of the horn would appear to be important in terms of the distance, and hence time delay, over which the mouth reflexions must travel. Obviously, were the horn long enough, then if a reflexion were present, a distinct echo would be heard. The longer the horn, the larger the mouth would need to be in order to smoothly terminate to the air in the room without any abrupt cross-sectional changes. The 'bad' horn specifically designed for these listening tests had a very abrupt mouth termination

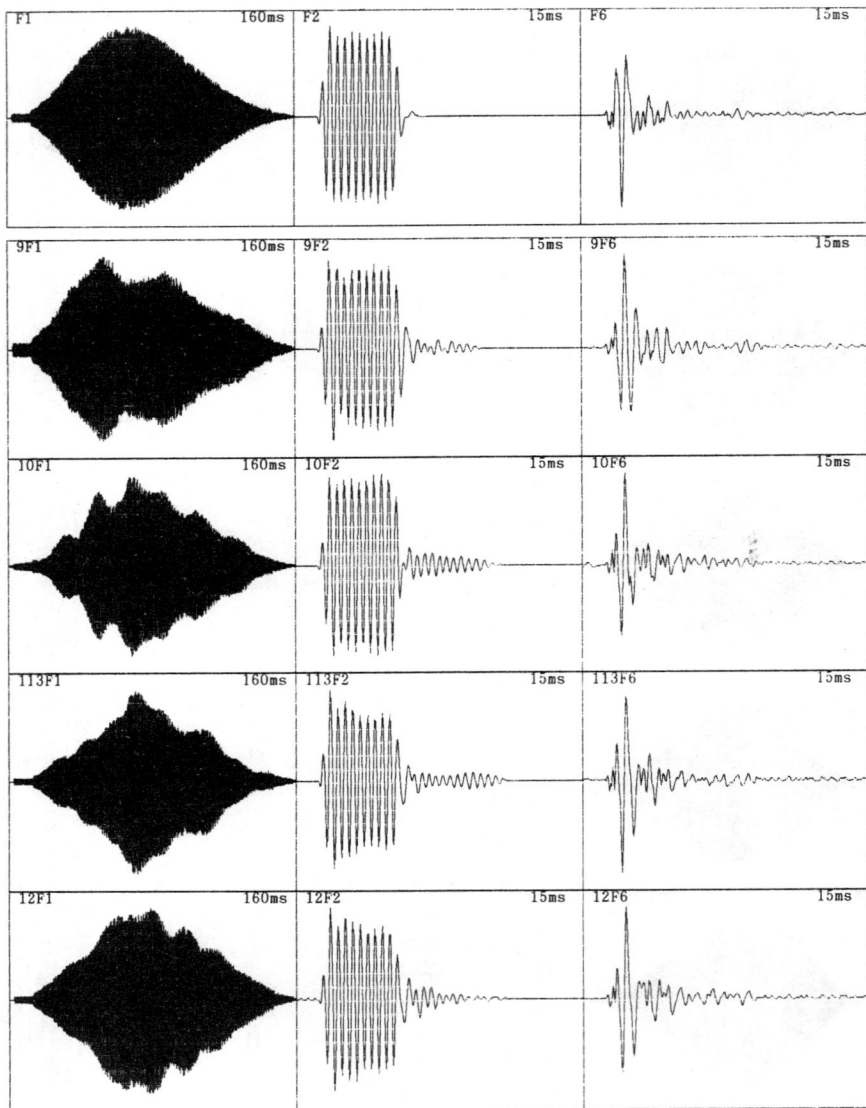

Figure 12.11 (*continued*)

with a diameter of less than 4″ (100 mm), and hence large throat impedance/pressure amplitude (frequency) response irregularities. However, being very short, around 9″, this horn was deemed similar to the direct radiator B with almost 100% similarity confidence. It is also worth noting that it was axisymmetric, and had neither any obstructions in the flare, nor any abrupt angles in its geometry until the sudden mouth termination.

Work being carried out in parallel with these listening tests revealed

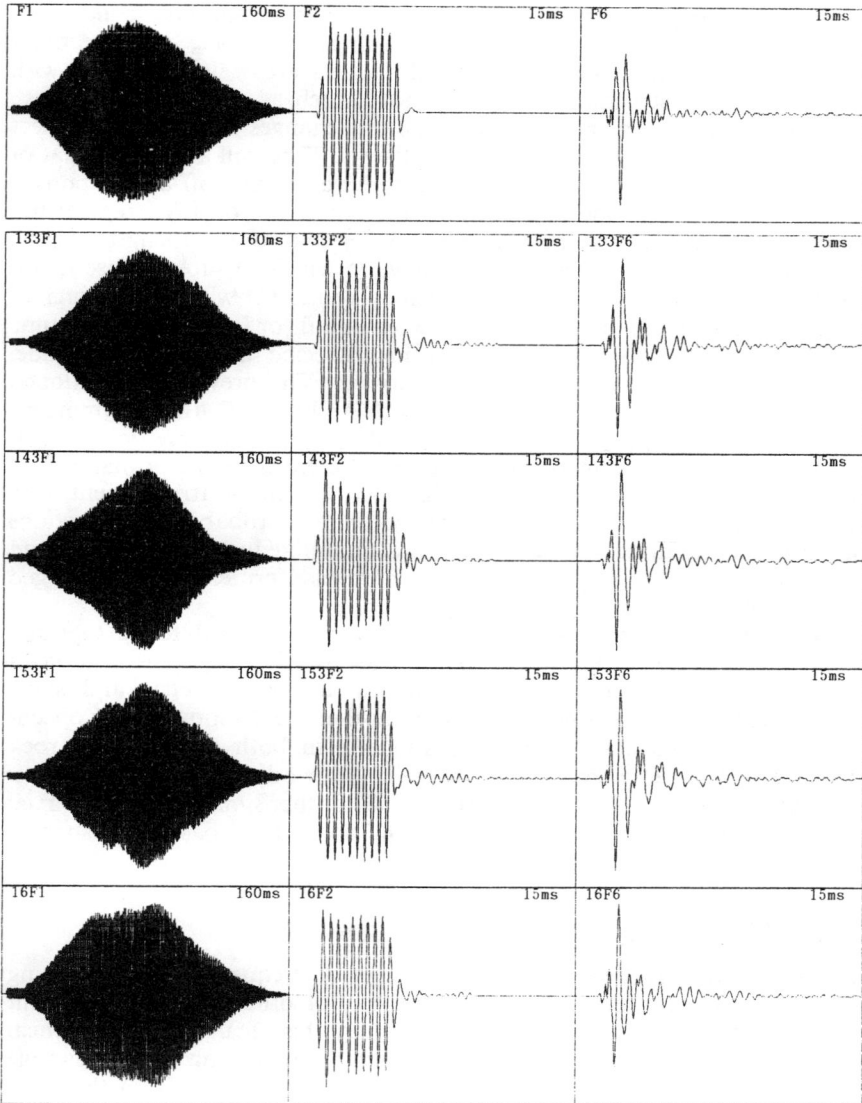

Figure 12.11 (*continued*)

disturbing audible characteristics when listening at a normal to any abrupt irregularity in the horn geometry, such as an irregularity on a side-wall, or even an angle where the side-wall meets the top or bottom of the flare. The implication of this extended to sharp angles at the junction of dividers and waveguides, and also to any pillars and posts which may be present in the throat area for the purpose of resonance damping in the structure of the horn. It is doubtful that all of these irregularities were sonically evident from the perceived tonality in the listening tests in anechoic conditions. But, under reflective or reverberant listening conditions, the directivity properties of a horn, controlled by the shape and size of flare, together with any angles or obstructions within the horn, such as pillars, 'waveguides', slant plates, acoustical resistances, or abrupt changes in shape, will affect the perceived sound quality; especially off-axis. They will also affect that of any reflective and reverberant environment driven by that off-axis response.

It would appear to be not so much the amplitude of any irregularities which contribute to any characteristic horn sound, but the length of time which those undesirable irregularities have in which to superimpose themselves on the desired signal. This is entirely consistent with the findings of any characteristic electrostatic loudspeaker sound (or lack of it) being time domain dependent, rather than being a single specific function of amplitude, phase, or non-linear distortion characteristics. The previously mentioned single sample which strongly correlated with Archetype D, the axisymmetric Dual Concentric horn, was Sample 8, itself an axisymmetric horn of remarkably similar geometry, yet entirely different in terms of material of construction, compression driver design, and mounting arrangement. The specific characteristics of any 'Tannoy' sound are probably also functions of crossover and drive unit design, but it is undoubtedly the overall size and geometry which has largely prevented 'horn-like' criticisms being lodged against them.

None of the short horns were in any respect deemed to be typically 'horn-like'. It is probable that it is for reasons of meeting the criteria of axial symmetry, absence of abrupt angles in the geometry, and short distance from the diaphragm to the mouth, that the Tannoy Dual Concentrics have enjoyed such a long working life in both classical and rock circles, without being sonically 'lumped in' with other horn loudspeakers. Neither have they been generally grouped with other similarly sized co-axial units having discrete horns mounted in the apex of the bass cone, such as the Altec 604s or the Harman-JBL/UREIs.

12.11 Waveform spectral similarity analysis

In order to look at the conventional concept of 'frequency response' being the great arbiter in terms of sound similarity or otherwise, an assessment was made of waveform spectral similarity. By this means, a mathematical assessment was made of the degree to which the waveform of each sample for each sound was similar to each of the Archetypes A to D for the same sound. The results were tabulated, somewhat similar in form to the tables formulated for the 'Similarity Confidence Indices', but with the 'None' column replaced by an 'Absolute' column indicating the degree of similarity of each sample for each sound, when referred to the actual waveform on tape of each signal. Thus, high figures in the 'Absolute' column indicate a close approximation to the original sound, and low figures, a poor reproduction in terms of 'accuracy'.

If pressure amplitude response was to be the major factor in determining audible similarity of loudspeaker units, then the columns with the greatest number of highest marks for each sound should be the ones which correspond to the archetype to which each sample was deemed to be most

	Sample 1					Sample 2					Sample 3					Sample 4				
Signal	A	B	C	D	Abs.	A	B	C	D	Abs.	A	B	C	D	Abs.	A	B	C	D	Abs.
1	56	**125**	36	46	93	82	72	26	77	93	76	54	21	69	62	37	51	32	31	46
2	45	**81**	51	41	95	34	42	31	25	50	22	26	21	40	24	33	39	27	24	40
3	16	24	23	**28**	23	24	36	32	28	42	38	32	29	25	49	19	35	25	30	28
4	24	**44**	37	27	43	17	30	20	18	30	17	16	18	21	18	18	39	28	22	35
5	22	**42**	32	26	37	20	34	24	19	36	21	22	21	26	26	17	36	26	21	29
6	26	**46**	36	25	45	22	39	30	22	41	22	25	24	29	31	18	37	28	22	29
7	24	**47**	38	26	46	18	33	23	18	33	18	19	19	25	21	17	38	29	20	32
8	32	**61**	37	26	54	31	43	31	24	52	21	26	21	31	27	25	41	30	23	36
9	18	34	30	26	29	19	33	23	19	35	24	21	20	22	28	16	34	27	21	27

	Sample 5					Sample 6					Sample 7					Sample 8				
Signal	A	B	C	D	Abs.	A	B	C	D	Abs.	A	B	C	D	Abs.	A	B	C	D	Abs.
1	49	**94**	38	46	78	50	**97**	37	42	69	63	**119**	32	56	105	75	**101**	31	91	159
2	41	67	49	35	86	45	**94**	47	38	104	41	61	40	40	69	50	71	48	55	64
3	16	23	23	25	24	23	67	27	36	40	32	40	40	26	69	18	26	27	**36**	27
4	24	**44**	37	26	51	24	**68**	31	31	54	26	42	34	27	52	37	32	30	39	35
5	22	41	35	24	43	21	66	32	30	47	25	40	38	23	53	28	35	31	32	37
6	25	44	40	24	48	23	71	36	32	49	26	41	41	23	54	34	43	32	32	45
7	23	46	39	24	50	23	72	34	30	53	26	43	36	25	52	37	36	31	37	39
8	33	52	42	23	60	27	66	32	32	50	33	41	39	23	57	38	**56**	36	31	61
9	18	34	34	24	32	20	62	29	30	45	22	37	39	23	44	21	29	32	31	29

	Sample 9					Sample 10					Sample 11					Sample 12				
Signal	A	B	C	D	Abs.	A	B	C	D	Abs.	A	B	C	D	Abs.	A	B	C	D	Abs.
1	44	**54**	42	53	55	62	**100**	29	42	85	48	**67**	28	34	61	56	**81**	31	39	73
2	**46**	45	43	29	52	50	70	38	37	80	**41**	40	37	23	55	**63**	54	43	30	65
3	17	20	24	21	23	26	39	36	30	50	28	37	36	25	58	35	26	**65**	23	55
4	27	27	35	20	31	25	44	36	30	50	25	43	32	24	62	28	33	**39**	24	37
5	24	25	30	19	28	23	44	36	28	50	28	39	35	21	65	33	31	43	21	45
6	26	25	30	18	30	25	47	39	29	53	30	39	39	22	66	36	34	46	21	55
7	27	27	34	20	31	25	46	36	29	51	26	43	35	23	62	30	34	41	23	44
8	32	26	37	17	33	30	48	34	30	56	**41**	40	40	22	72	**47**	34	43	20	61
9	20	24	33	19	26	21	41	37	28	45	27	36	32	21	65	26	29	46	22	38

	Sample 13					Sample 14					Sample 15					Sample 16				
Signal	A	B	C	D	Abs.	A	B	C	D	Abs.	A	B	C	D	Abs.	A	B	C	D	Abs.
1	49	71	32	34	60	79	**118**	32	52	127	44	**68**	32	34	53	73	**138**	33	58	132
2	54	60	49	32	98	55	67	61	37	88	49	59	39	42	65	60	73	45	38	100
3	34	30	53	24	62	22	25	33	28	34	39	30	51	24	54	37	38	35	27	78
4	30	32	39	29	40	**36**	32	36	31	43	26	29	35	27	33	40	44	32	33	62
5	33	32	47	23	46	32	36	39	28	50	29	32	40	23	42	32	42	33	27	61
6	35	35	50	24	55	39	41	41	30	64	30	35	43	24	47	36	39	33	27	59
7	31	34	42	26	44	**38**	36	38	31	51	27	32	36	26	38	39	44	33	31	63
8	43	35	49	22	58	45	**54**	43	28	82	34	37	37	24	49	35	48	32	28	56
9	28	30	49	24	39	26	33	38	28	41	25	30	41	23	36	26	41	32	26	54

Figure 12.12 Waveform spectral similarity (arbitrary units)

similar in the listening test results. Figure 12.12 shows the tables of individual results. For example, for Sample 7, Sound 1, the highest indication of similarity was to Archetype B; for Sound 2, B again, and so forth. The absolute column was not used in this assessment, as we were only looking for the greatest similarity to any one archetype. When the results for Sample 7 were assessed, it could be seen that Archetype B showed the greatest number of 'most similar' highest scores for each individual sound, with C coming second; there were no other strong similarities indicated.

Samples 2, 4, 6 and 10 throughout these tests have consistently shown themselves, by whatever means of analysis used, to line up strongly with the direct radiator B. In terms of waveform similarity, there is no change from this trend: B is clearly their representative archetype. Of the other samples, horns 5 and 7 still show a strong similarity to B, as was also shown from the listening test results. Sample 8 was deemed by listening to be similar to the Tannoy, D, but from its waveform similarity results, by far the strongest indication of similarity is to B, which is clearly at odds with the aural assessments. Likewise, Sample 1 shows a very strong inclination to favour B in terms of waveform similarity, yet in terms of listening alone, the margin of apparent similarity of B over D was only in the region of 52% to 48%, so again we have only poor agreement. Sample 3, the 'non-similar' control, if it had any audible similarity at all, it was to B, yet the waveform results do not favour B on any of the sounds, indicating D or even A to be its closest pairings: clearly again no agreement.

The two other 'ambiguous' short horns, 9 and 11, which showed few clear-cut indications and had only a low percentage of unambiguous ticks in the listening tests, both tended, if anything, towards Archetype B on aural assessment. In terms of waveform similarity, however, Sample 11 goes rather more strongly with B than the listening tests would imply, whilst Sample 9 lines up, if with anything, with C, though the individual similarity figures for each sound are only low. This tendency towards a similarity with C was most definitely not apparent from the listening test results.

What remained were the 'long' horns, Samples 12 to 16. From the waveform similarity results, the all JBL combination, Sample 16, showed a strong and clear resemblance to B; in fact B was most similar to 16 on all nine of the sounds. This result is strongly opposed to the listening test results, almost all assessments of which grouped Sample 16 with the long horn archetype C. Samples 12, 14 and 15 showed reasonable correlation with the listening test results, though 15 possibly shows a more marked resemblance to B than was actually heard. Sample 13, however, the Altec Multicellular horn, clearly reversed its position in comparison with Sample 12, showing a greater tendency towards C than did 12, when the listening tests had clearly shown it to be much less similar to C than was 12.

In general, the waveform spectral similarity assessment showed something in the order of a 70% agreement with the listening test results. The waveform analysis was useful in that it gave some indications, where spectral similarity was high, in terms of to what degree 'frequency response' could be used to explain similarity on certain sounds, but was less revealing on other sounds. Of the remaining 30% or so of disagreement, obviously, with the phase/amplitude convolution defining the whole response, apart

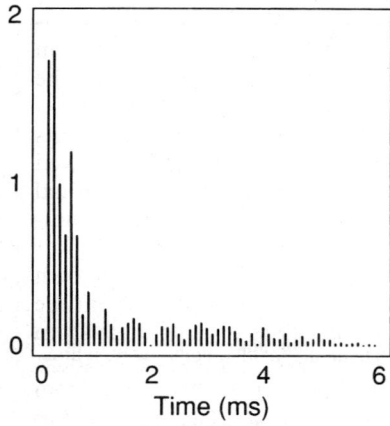

Reference A

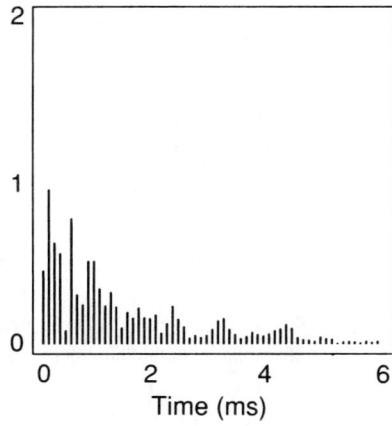

Reference B

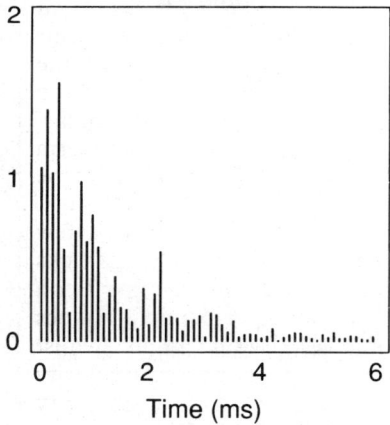

Reference C

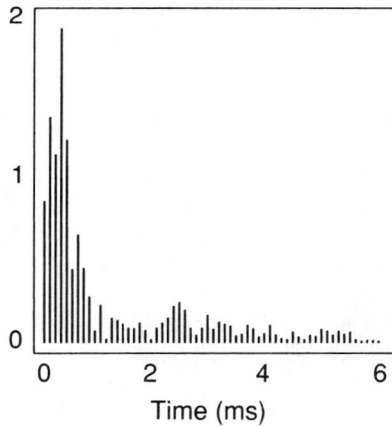

Reference D

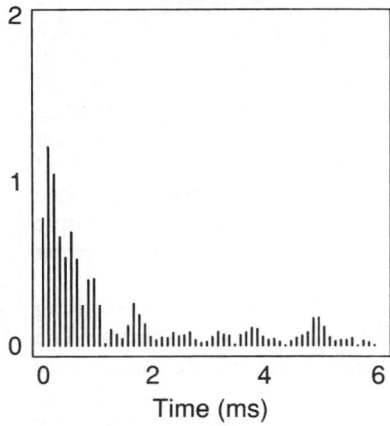

Sample 1

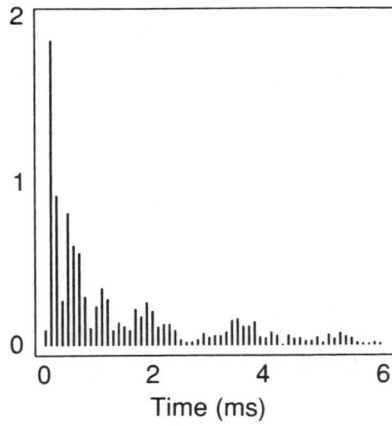

Sample 2

Figure 12.13(a) Power cepstra of test loudspeakers

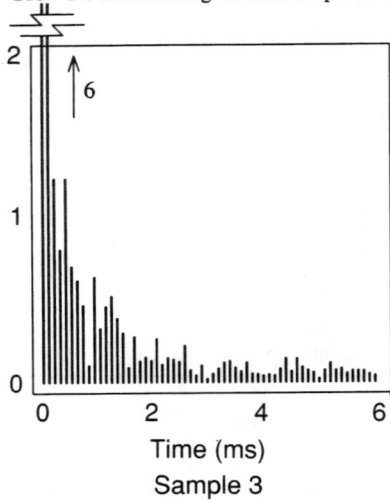

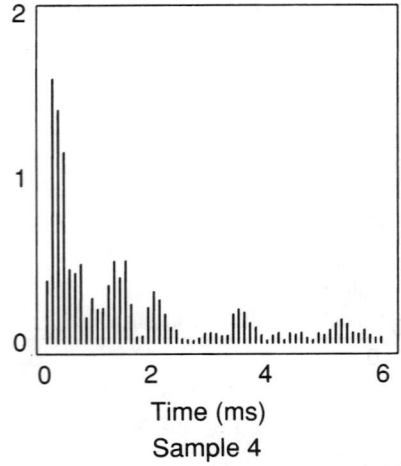

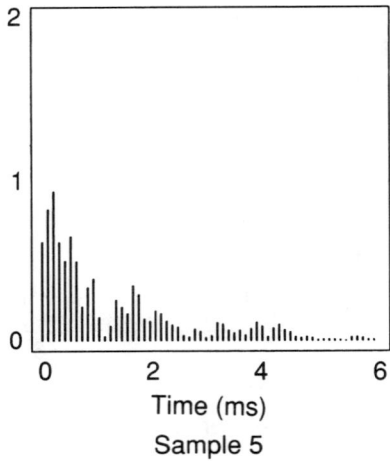

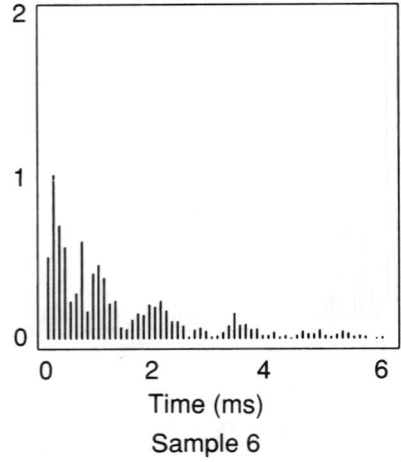

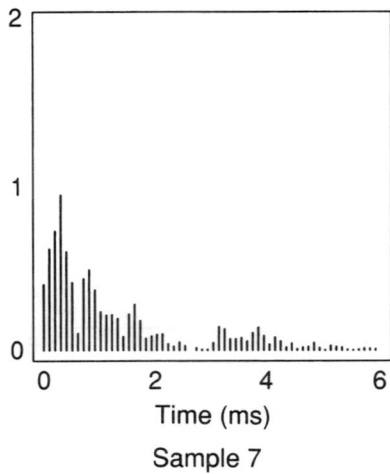

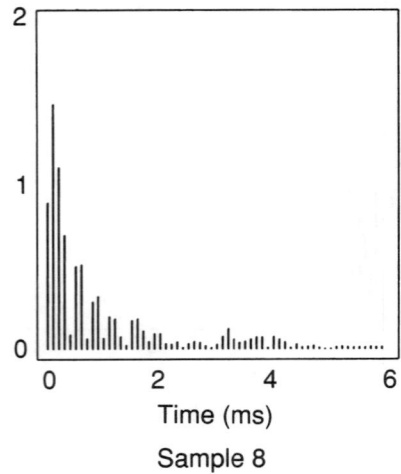

Figure 12.13(a) (continued)

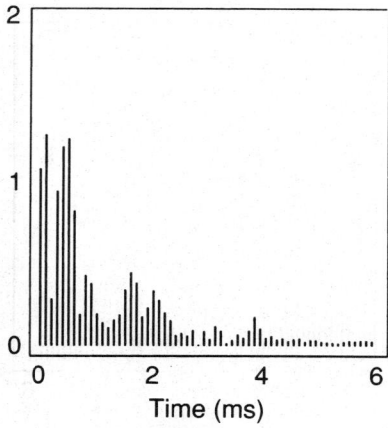

Sample 9

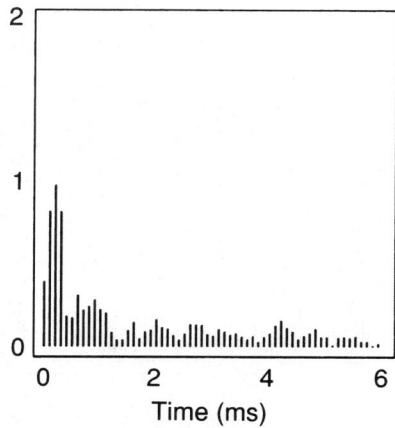

Sample 10

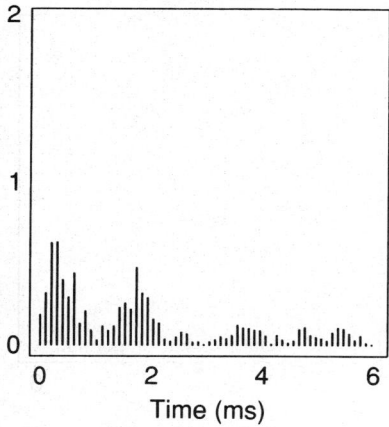

Sample 11

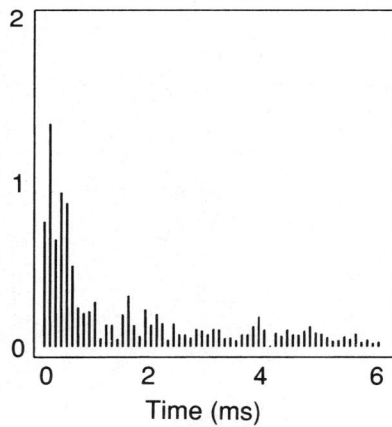

Sample 12

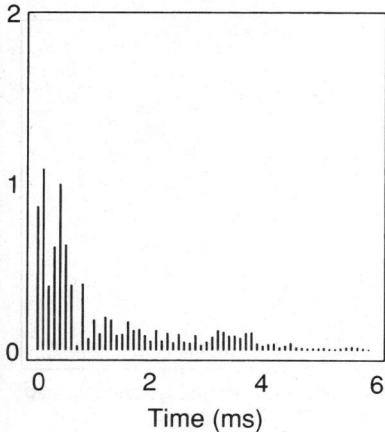

Sample 13

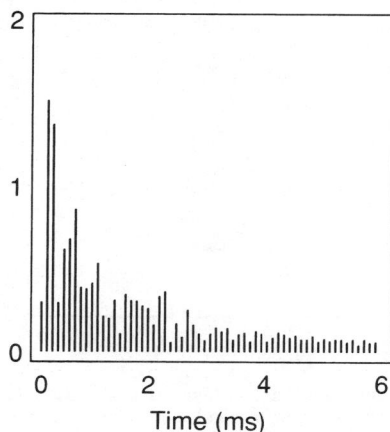

Sample 14

Figure 12.13(a) (*continued*)

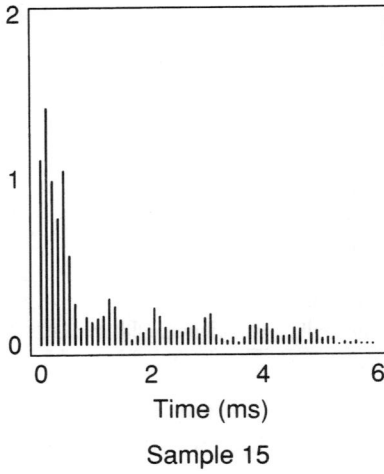

Sample 15

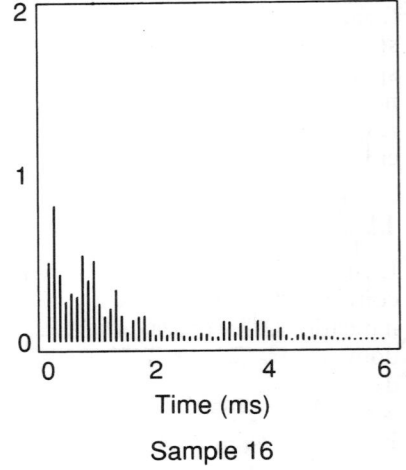

Sample 16

(a)

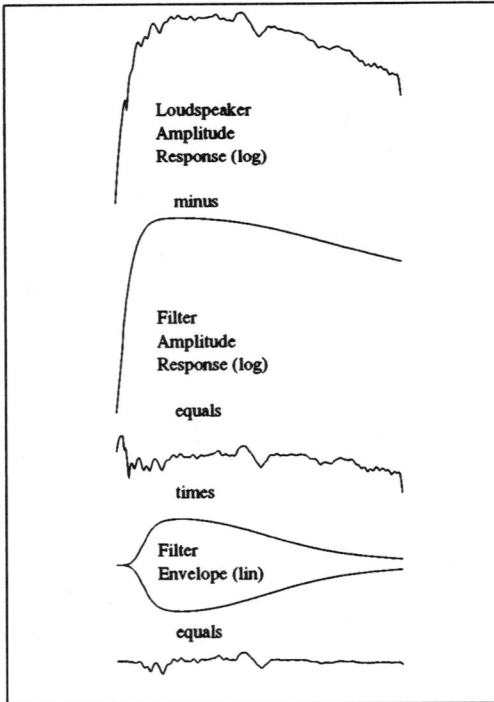

(b)

Figure 12.13 (a) (*continued*). (b) Diagram: representation of processing of loudspeaker transfer function prior to calculation of power cepstrum.

from the already discounted non-linearities, the differences must reflect somewhere in the phase/pressure amplitude responses. However, the above spectral similarity analysis suggests that the causes of the differences are not decipherable from conventional analysis of the pressure amplitude responses, so the answers must be sought elsewhere. Comparisons of the envelopes of some of the actual waveforms are shown in Fig. 12.11.

12.12 The proof

Until the cepstrum analysis was initiated, from the listening tests and conventional measurements alone, all that we had achieved thus far were indications that many of the similarities or differences lay in the time domain. Power cepstra, for those readers unfamiliar with them, were developed in the early 1960s[7] for the enhancement of the detection of echoes in seismic signals. A power cepstrum is effectively the inverse Fourier transform of a logarithmic power spectrum. Cepstrum is an anagram of spectrum, and the associated relative terms, quefrency, lifter, rahmonic, saphe, and gamnitude are also anagrams of the more usual terms to which they quasi-relate: frequency, filter, harmonic, phase, and magnitude. In this strange pseudo-dimensional domain, high pass filters become 'long pass lifters', and so forth, scaling in non-dimensional dB's. Figure 12.13 shows the power cepstra for the twenty test loudspeakers, in which the patterns of reflexions are more evident than in more usual representations of signal parameters.

In the power cepstra plots can be seen distinct echo patterns relating to individual characteristics of each driver or combination. From these plots, the groupings of the timings of the inherent reflexion patterns of each driver or combination can be seen to match well with the groupings of the listening test results. Effectively, the reflexions from the mouths of short horns, being in the 1 ms to 2 ms region, are somewhat similar in timing and nature to the reflexions produced within many direct radiating loudspeakers, and the two had been very difficult to differentiate in the listening tests. The reflexions from the mouths of the longer horns are in the 2 ms to 4 ms region, and are quite distinctly recognisable from the reflexion patterns of the direct radiators. The electrostatic produced few reflexions which could be considered significant above the 1 ms region, and this lack of later reflexions is almost certainly a greatly contributing factor in the sonic transparency shown by such loudspeakers.

Whilst the reflexions in the long horns are generally less severe than those from their shorter counterparts, their additional time separation from the initial signal appears to render them more noticeable than level alone would suggest. As can clearly be seen from the pressure amplitude response graphs and the throat impedance plots, some of the longer horns would appear to be performing very well indeed, yet they unmistakably sounded like horns.

The mouth reflexions in the long horns are generally less severe than those of the short horns, as the mouth is larger, and the termination correspondingly better. This can be seen from the generally smoother transfer functions for the longer horns. However, from the various comments made by the subjects, both orally during the test, and on the

Figure 12.14 Axisymmetric horn geometry

questionnaires, it is clear that the longer horns can be more readily identified, again strongly suggesting that it is the greater separation in time which renders the reflexions more noticeable. Given the correlation between the shorter horns and the direct radiators in terms of the sound character due to delayed responses, it becomes more clear why the Quad Electrostatic was unrepresentative of any of the other loudspeakers in these tests. The response of the Electrostatic is almost entirely free of any resonant hangover or internal reflexions, a fact clearly shown by its characteristic step function response being significantly more accurate than any other driver tested in our experimentation (see Fig. 12.9). There is no suggestion here that any horns leaning towards the direct radiator characteristics are necessarily more desirable than any which may ultimately tend towards the Electrostatic, but merely that at the present time, they may be more representative of the majority of high quality loudspeakers, as opposed to being more representative of 'reality'.

The above point is further borne out by experience gained in studios since these listening tests were carried out. Sample 8 was an experimental horn produced at the ISVR for inclusion in the tests. The major design parameters were based on research work in the two years immediately preceding these tests.[8] The horn had an exceptionally smooth cut-off, tapering off at almost precisely 12 dB/octave below 1 kHz. During further tests on its directivity, especially at higher frequencies, it was mated with a TAD 2001, beryllium diaphragmed compression driver, with a response extending to 22 kHz. The performance results were so outstandingly smooth that the unit has since gone into production for studio monitor systems of very high quality. Further details of the AX2/TAD/Emilar combination responses are discussed in Chapter 2, and also shown in Figs 2.1 to 2.4.

The amplitude and phase responses for the systems in which these horns are fitted bears a remarkable resemblance to those of the Quad Electrostatics, a fact clearly noticeable from the similarity in their step function responses (see Fig. 12.9). The physical shape of the horn is quite similar to D, the Tannoy Dual Concentric with which it was deemed to be most similar in the listening tests. It was in fact the only sample in the tests which showed even a reasonably strong similarity to D. Figure 12.14 shows the general geometry of Sample 8.

The 'outcasts' in these listening tests were Archetypes A and D, yet ironically they were both products of 1950s design philosophies, and were held in high regard in their day. Notwithstanding power output limitations, they are probably the only two British products of that era which could still give a well respected account of themselves alongside modern monitoring systems, and are in fact still to be found performing quality control functions. Both of these loudspeakers fail to meet current mainstream monitoring demands on largely the same points – overall sound pressure level, especially in the production of loud, tight, low frequencies, and a slightly idiosyncratic high top. They have their partisan followers and their critics, but almost without doubt it has been the clarity and transparency of their mid-range performance which has ensured that the general respect for them has now covered four decades. Clearly they must have got *something* right.

Given its superb step function performance, despite the lack of low frequencies, the Quad Electrostatic must be considered to be something of a reference in terms of the envelope of its power cepstra in Fig. 12.13, Sample 14. The only other two loudspeakers in the tests with reasonably similarly smooth plots, and also having a clear absence of obvious reflexions in the

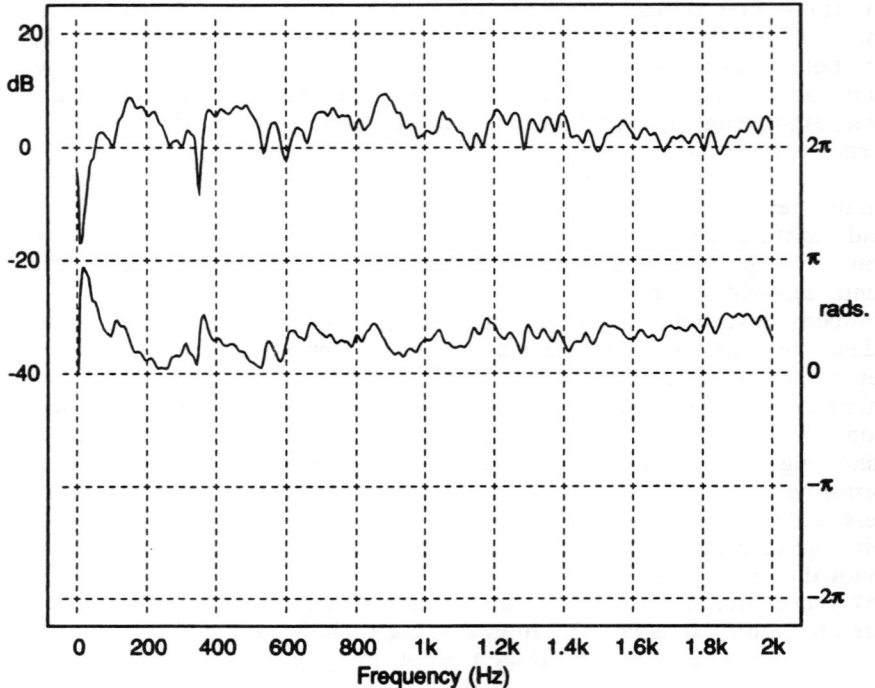

Figure 12.15 Amplitude and phase plots of complete system through cut-off and crossover region

first milliseconds of their responses, are the Tannoy (Archetype D), and the AX2 (Sample 8). The echoes apparent on Sample 8 are not evident when used with the TAD driver, as the taper of the throat of the drive unit almost perfectly blends into the throat of the horn, producing an extremely smooth match.

The sensitivity to disturbances of flare rate in the throat area seemingly must have implications with regard to the Tannoy philosophy of modulating the HF horn (the bass cone) with the low frequency component of the music. In the tests here described, the bass cone (the HF horn) of Archetype D was never driven, but given the discontinuity at the throat, which must exist in order to allow for the low frequency voice coil gap, I do not see how the horn performance can be uniform or unmodulated under the influence of full inward and full outward excursions of the bass cone. This point has been raised by many in the past, but the cepstral irregularities caused by the relatively minor mismatch of flare rates between the Emilar/AX2 throat interfaces must surely be small when compared to the full deflection mismatches in the Tannoys. Definitely more work is needed on this subject.

Sample 3, the 'non-similar' reference, also shows a good cepstrum response, and is known to be used, within its design frequency range, in systems reputed for their open and natural reproduction. As a result of research in the first year of this horn programme, a paper was presented to the Institute of Acoustics in 1989 entitled 'Impulse Testing of Monitor

Systems (The only way to achieving a better understanding of the path to the perfect monitoring environment)'.[6] The presentation of this paper was prompted by the early results of this research programme, showing clear indications of audible grouping of loudspeakers in terms of time characteristics, as opposed to the conventional emphasis on pressure amplitude responses.

The production monitor systems with which Sample 8 is now used has sandwiched the horn between two 15″ low frequency drive units, one above and one below, all three units in the same vertical plane. The 'phantom' source of the two low frequency units is thus located directly over the centre of the horn, so for horizontal movement at least, a largely coincident source is produced. The step function of the combination is shown in Fig. 12.9(a) with the amplitude and phase responses through the 1 kHz crossover region shown in Fig. 12.15. The cut-off characteristics of the horn are utilised as part of an electro-acoustic, composite crossover. The step function is more akin to the Quad Electrostatic than almost any other available electromagnetic monitor system; see again Figs 12.9(a) and (b). Perceived performance is of exceptional clarity, with first class stereo imaging, openness, and many of the other attributes attached to the Electrostatic, together with the very high SPLs available from a horn loaded mid/high frequency loudspeaker, and the deep, tight and powerful bass available from the two 15″ loudspeakers in their 20 cubic foot cabinet. The mid-range clarity is on par with, if not exceeding, that of the Electrostatic.

12.13 Conclusions

The implication gleaned from the results of these listening tests, together with the practical performance of units derived from the research, is that any perceived horn-like nature of a horn is predominantly a function of the temporal separation from the initial event of any characteristic reflexions or disturbances in the response. The majority of these disturbances are caused by the truncation of a horn. In general, the longer the horn, the less the disturbances – if well designed – but the greater will be the temporal separation of any disturbances, which seems to lead to the characteristic 'horn' sound. If the horn was infinite in length and free from internal discontinuities, the disturbances caused by reflexions would not exist, as they would be infinitely separated in time and infinitesimal in amplitude. If the horn was of zero length, we would have a direct radiator. If a horn is well designed and short in length, the reflexion-related disturbances in the response can be similar in both amplitude and time to the irregularities inherent in many direct radiators. Where the response irregularities of direct radiators and horns are generally of the same order and value, audible discrimination between direct radiators and horns is not to be expected.

When the reflexions are occurring very shortly after the initial sound, say less than 0.8 ms, similar mechanisms to those which mask harmonic distortion on transient signals would appear to be similarly masking reflexions. As the time separation of the reflexions from the initial event becomes greater, even though they become lower in amplitude, the ear's ability to detect them would seem to increase. Evidence from the listening tests would appear to indicate that when pressure amplitude responses are reasonably similar, the time related reflexions patterns dominate in the perception of any audible differences between units; and where time behaviour is similar in terms of reflexion patterns, then the pressure amplitude responses would be the main differentiators in sonic performance. The degree to which each must be 'similar' before the other dominates is a function of the transient, steady state, or harmonic nature of the signal. Phase is, of course, the common link between time and pressure amplitude.

As can be seen from the analysis of the results, where no other form of obvious difference seemed to exist between two drive units which had been adjudged sonically quite different in the listening tests, but exhibited reasonably similar amplitude responses, there was usually something amiss between the phase responses of the two units. Conversely, where two units were adjudged sonically similar when general measurements were proving different, a similarity in the phase responses was often present.

12.14 Implication for practical horn design parameters

By the very careful choice of design parameters and construction, horns can be produced for use above 1 kHz or thereabouts, which approach the performance of electrostatic loudspeakers, with their very low levels of deviation from their intended amplitude and phase (and hence time) responses. It would appear, however, that there are finite practical limits to the performance ranges over which horns can produce near optimum results:

1 The cut-off frequency of a horn is a function of its rate of flare. A low cut-off frequency demands a slow rate of flare.
2 If 'horn-like' sound characteristics are to be avoided in a practical horn, the length should not exceed 12″ (300 mm) or thereabouts.
3 Taking 1 and 2 together, if a horn has a low flare rate, and cannot exceed 12″ in length, then given a 1″ throat diameter, and, say, a 250 Hz cut-off frequency, the horn will inevitably have a small mouth area. There will consequently be an abrupt change in cross-sectional area when it meets the outside air. Samples 4 and 11 highlighted this point, showing poor throat impedance linearity, or smoothness, especially near cut-off. Subjectively, although almost always being grouped with the direct radiators in the listening tests, as musical reproduction devices they were not considered smooth, flat, or natural.
4 In order to achieve a smooth and trouble-free mouth termination, from a 1″ throat of a horn not exceeding 12″ in length (diaphragm to mouth), a mouth diameter of around 12″ would seem to be the smallest practical size. This dictates a flare rate which results in a cut-off frequency in the order of 1 kHz, but can yield exceptionally smooth performance through cut-off if carefully designed; even allowing use *through* cut-off, and utilising the acoustic roll-off as part of the electroacoustic crossover (see Fig. 12.15). (The first test system described in the technical press[9], and Chapter 17, did not take advantage of this facility. Compare Fig. 17.7(a) with Fig. 12.15, noting the responses through the 1 kHz crossover point.)
5 To minimise internal disturbances which can cause disruption to both the

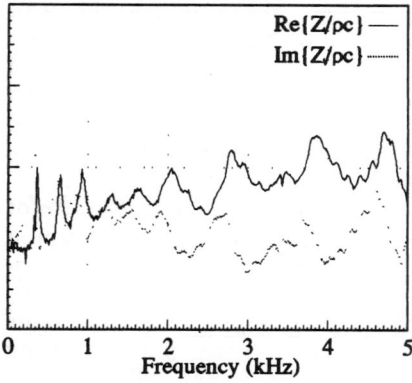

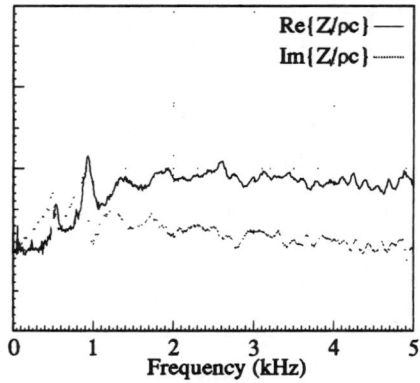

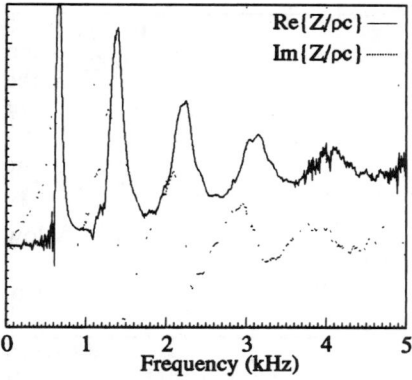

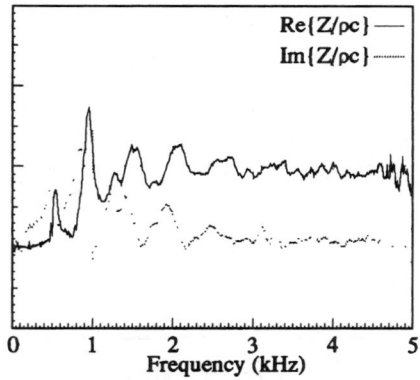

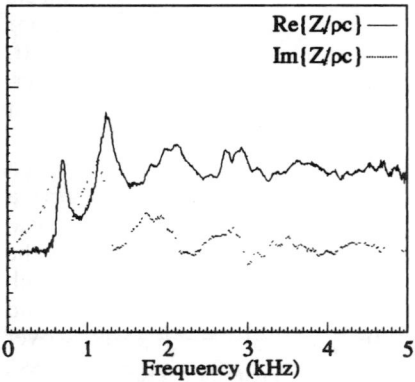

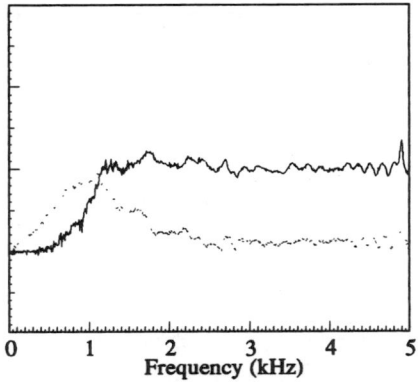

Figure 12.16 Throat impedance characteristics of the horns used in the tests: plotted on a linear scale

Sample No. : 9

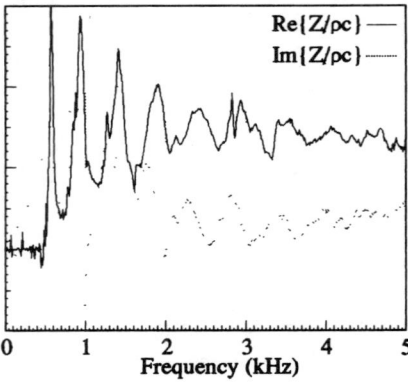

Sample No. : 10

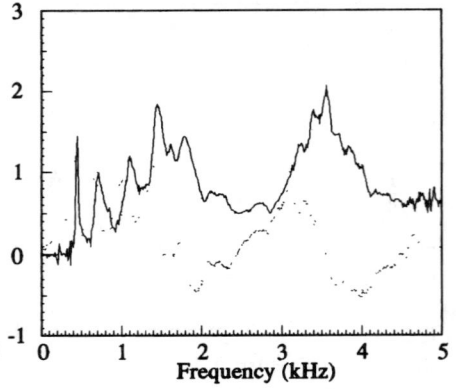

Sample No. : 11

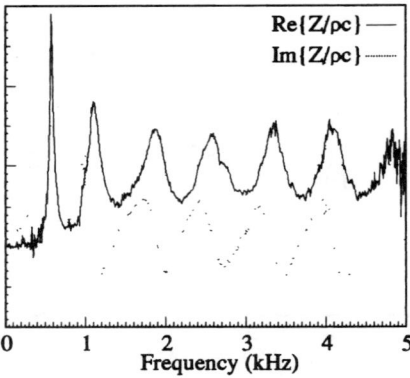

Sample No. : 12

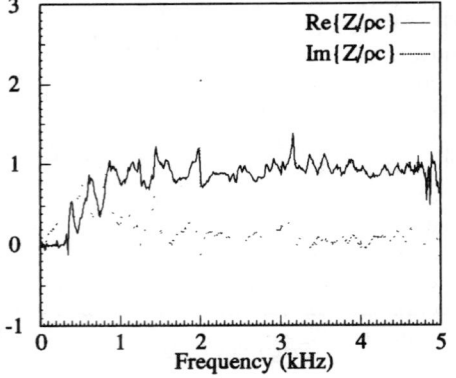

Sample No. : 13

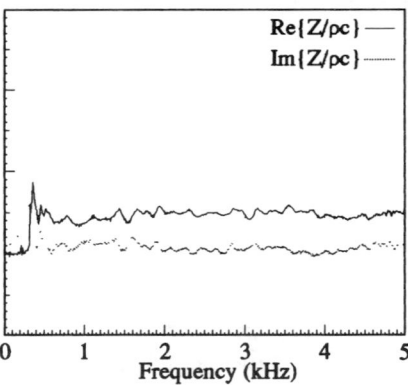

Sample No. : 15

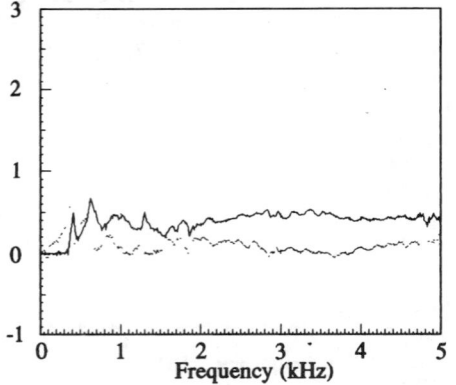

Figure 12.16 (*continued*)

on- and off-axis responses, all corners, angles and obstructions should be removed, rendering axial symmetry and smoothly contouring surfaces. Figures 12.16 and 12.17 show the respective linear and logarithmic throat impedance plots of all but one of the horns used in the tests. The vastly superior characteristics of the AX2 (Sample 8) can be readily seen.

6 'Squashing' the axisymmetric shape into an ellipse would perhaps allow some change in directivity pattern, without undue disturbance of the time response. This may be an option for future designs.

Sample 8, mated with the TAD 2001 drive unit, produced what would seem to be a near optimum response in terms of both phase and amplitude (hence time), smooth directivity, a very smooth overall performance from 1 kHz to beyond 20 kHz, and was deemed to be very musical, natural, transparent, and definitely not hornlike. It is interesting that in physical dimensions, though not in its drive system, it strongly resembled the Tannoy Dual Concentrics from around 40 years before. It would appear that the Tannoy, all those years ago, defined the physical limits for accurate performance, beyond which horns will begin to run into trouble.

Whether Tannoy knew all of this at the time when the Dual Concentrics were first designed, or whether some of their clever and intuitive engineers merely 'saw the logic' of using a duly contoured bass cone for the horn of a co-axial system, and heard that it sounded good, maybe we shall never truly know. In general, however, there is now no doubt that carefully designed horns and drivers can produce both sonic and measured perform-ances as good if not better than the finest dynamic direct radiators, without any hint of a 'horn-like' sound. Indeed, to run a truly seamless 1 kHz to 20 kHz within tight limits, whilst producing a very smoothly controlled directivity pattern (courtesy of a horn emanating a section of a spherical expanding wave, unlike a piston), a well designed horn and driver combina-tion can be superior to the vast majority of direct radiators. When and if very high sound pressure levels are added to the list of requirements, there are few alternatives.

The fact that Archetypes A and D did not correlate well with the samples goes some way to explaining why, whilst A and D are well liked by many people, and quite widely used as quality control monitors, their lifelike performances in some ways may actually have been be a handicap. It may have limited their use as monitors, as they are not as widely representative of most loudspeakers on which commercial recordings are expected to be played. The Son Audax, B, can thus be considered to be more representative in terms of sonic similarity to the domestic systems on which the greatest proportion of recorded music is likely to be heard.

The two horns having minimal mouth reflexions, one long and one short, were not identified as horns, and did not sound similar to the direct radiating reference. Their greater similarity to A and D respectively may imply a more accurate performance than the typical direct radiators, especi-ally in the light of the points relating to A and D in the previous paragraph. The direct radiators in this test should only be considered to be representative of the most common type of loudspeaker, and should not be considered to be any more accurate, in absolute terms, than any other type of drive unit.

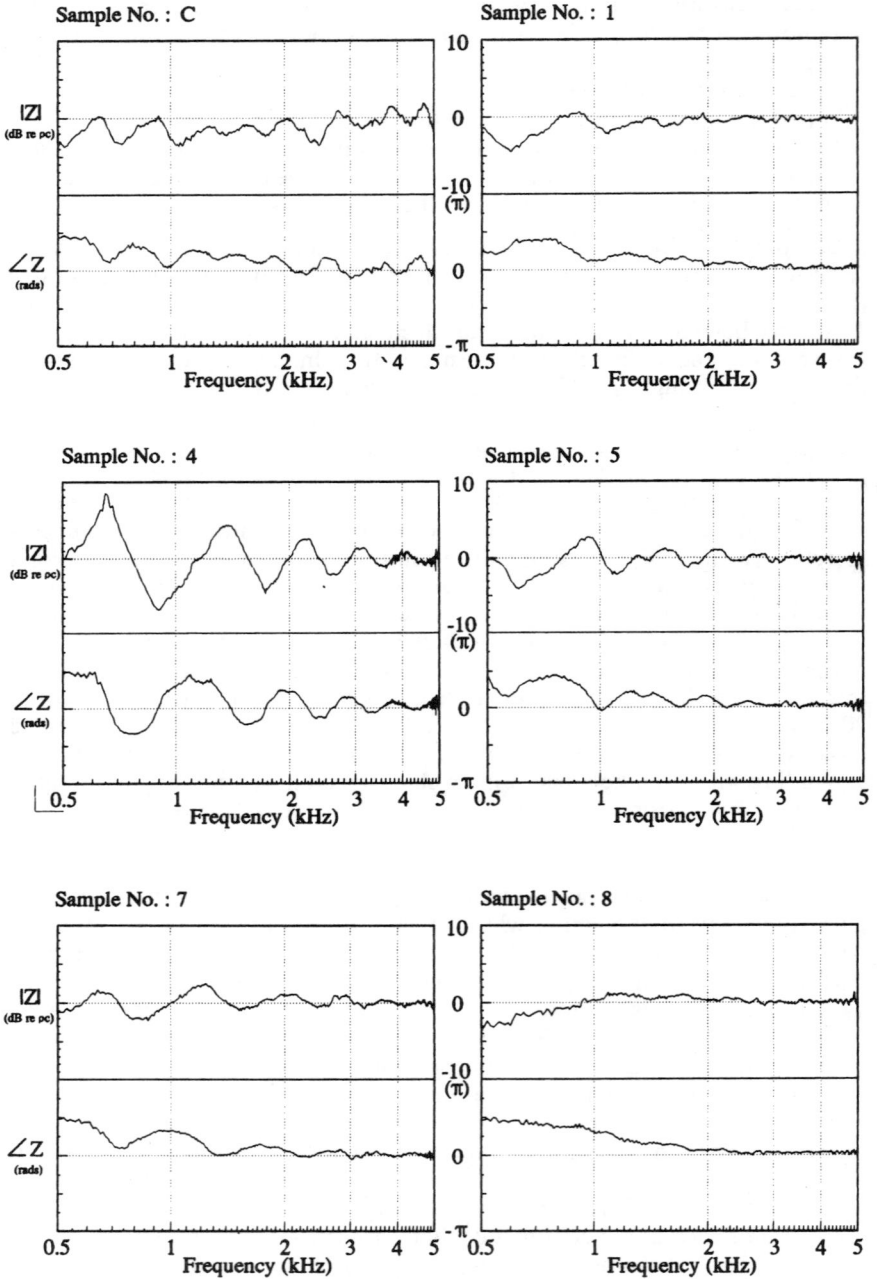

Figure 12.17 Throat impedance characteristics of the horns used in the tests: plotted on a logarithmic scale

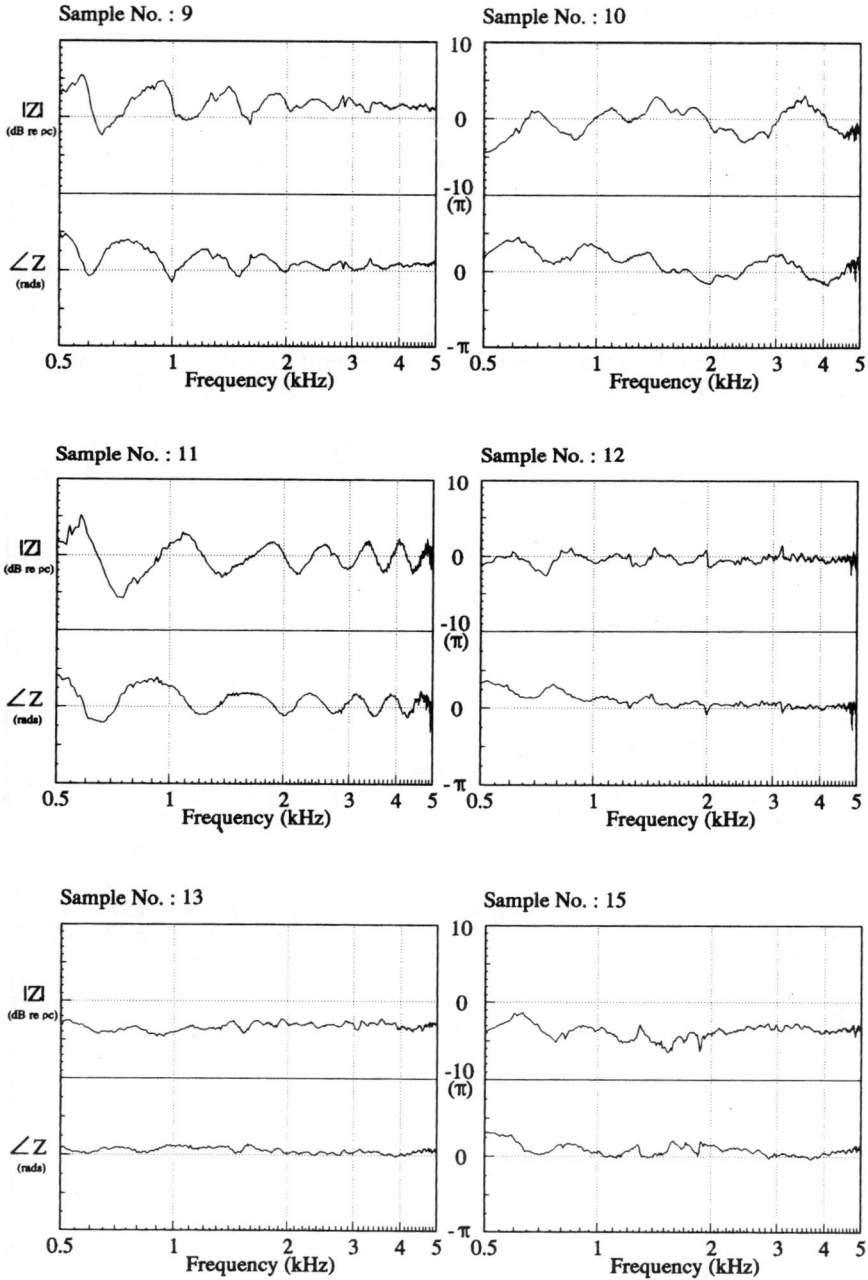

Figure 12.17 (*continued*)

12.15 Summary of results

The above tests are fully documented, eminently repeatable, and open to inspection. The initial findings suggest that short horns can be produced having high efficiency, wide frequency range and benign distortion levels, which are not sonically horn-like, and can be grouped as audibly similar to typical direct radiators. Much of the audible similarity of drive units would appear to be in their time histories, and where a mouth reflexion effect of a horn is in the same order as any inherent reflexions in direct radiator units, then general audible similarity is to be expected. Long horns produce longer reflexion delays, and do group together, whilst electrostatics group due to their rapid and accurate transient (impulse/step/square wave) response. In fact, much of the general audible similarity of all loudspeaker drive units of similar frequency range and general overall quality, irrespective of generic type, lies not in the non-linearities, nor solely in the pressure amplitude response, but in the time domain response as specified by the linear distortions of the convolution of the amplitude and phase responses. This would seem to be especially true when such disturbances have their origins in the patterns of inherent reflexions.

References

1 K. R. Holland, PhD Thesis, Institute of Sound and Vibration Research, Southampton University, UK (1992)
2 K. R. Holland, F. J. Fahy, C. L. Morfey and P. R. Newell, 'The Prediction and Measurement of Throat Impedance of Horns'. *Proceedings of the Institute of Acoustics*, Vol 11, part 7, 247–254 (1989)
3 P. R. Newell, 'Mid-Range Horns', *Studio Sound*, Vol 31, No 9, 62–72 (Sept 1989)
4 P. R. Newell, 'A Background to the Search for a High Definition Mid-Range Horn', Part 1, *Studio Sound*, Vol 31, No 10, 104–109 (Oct 1989)
5 P. R. Newell, 'A Background to the Search for a High Definition Mid-Range Horn', Part 2, *Studio Sound*, Vol 31, No 11, 68–72 (Nov 1989)
6 P. R. Newell and K. R. Holland, 'Impulse Testing of Monitor Systems'. *Proceedings of the Institute of Acoustics*, Vol 11, part 7, 269–275 (1989)
7 B. P. Bogert, M. J. R. Healy and J. W. Tukey, 'The Quefrency Analysis of Time Series for Echoes: Cepstrum, Pseudo-Autocovariance, Cross-cepstrum and Saphe Cracking'. In M. Rosenblatt (ed.), *Proceedings of the Symposium on Time Series Analysis*, Wiley, New York. pp. 209–243 (1963)
8 K. R. Holland, F. J. Fahy and P. R. Newell, 'Axisymmetric Horns for Studio Monitor Systems'. *Proceedings of the Institute of Acoustics*, Vol 12, part 8, 121–128 (1990)
9 P. R. Newell, 'The Non-Environment Control Room', *Studio Sound*, Vol 33, No. 11, 22–29 (Nov 1991).

Chapter 13

Monitor systems – a look at the overlooked

13.1 Output SPL requirements

Since my introduction into the recording industry in the mid-1960s, the importance of adequate system headroom has always played a great part in my thinking. Looking back, this has been an awareness of the necessity to maintain the integrity of transients, which holds true for both the electrical and electro-mechanical links in the recording chain. With the seemingly ever increasing size of control rooms, the demand on the output of the monitor systems has increased alarmingly. This is the result of two major influences. Firstly, digital recording has realistically extended the lower recording limit by a full octave, with the improved transient performance of the recording medium making unprecedented demands on the transient response of the monitoring systems. Secondly, the so called 'double distance rule' states that the sound pressure is subject to the inverse square of the distance. In other words, if you wish to maintain the SPL at your ears, but double your distance from a sound source, the power output from that source must be increased by the square of the distance change: 2 squared, or 4 times, which is of course a 6 dB increase. The 'double distance rule' is a free-field condition, and does not strictly hold for reverberant rooms, but it does hold for the *direct* wave in reverberant rooms.

With an extra 6 dB from the increased control room sizes and, say, 12 dB additional headroom to allow for the higher peaks and lower lows of digital recording technology, this translates into a 20 dB increase in the output capability of a monitor system, compared to the requirements for a conventional, analogue studio of not so long ago. On the subject of power, I am still somewhat non-plussed by the number of engineers who ask me of monitoring systems: 'How much power will they take?' They are trying to ask about the relative, perceived volume level which the loudspeakers will produce, but this is a function of acoustic watts delivered into the room, not electrical watts into the loudspeakers. The relationship is entirely a function of the conversion efficiency of the loudspeakers in their translation of electrical watts in, to acoustic watts out. One hundred watts into a driver of 20% efficiency will produce twice as much sound in the room compared to four hundred watts into a driver of 2.5% efficiency. As a practical example of this, 1 watt into a UREI 815 loudspeaker system, with a 103 dB sensitivity for 1 watt input at 1 metre will be equal in acoustic output to about 200 watts into an ATC SCM10 monitor loudspeaker with an 81 dB/ w/m sensitivity.

The problem of making monitor systems significantly louder is particularly difficult in the mid-range, where design considerations should always be made with due deference to the wavelengths involved. At 5 kHz, the wavelength is just over $2\frac{1}{2}''$, so a prime requirement is to keep the source area small in order to reduce the potential for phase cancellation. Such cancellation is due to the different possible wave paths from a large transducer to the ear. If a 5 kHz signal were generated by a 12″ loudspeaker, the listener would only have to be 3 or 4″ off-axis to find that the distance from the ear to one side of the cone, compared to the distance from the ear to the other side of the cone, would show a differential of around $1\frac{1}{4}''$. At 5 kHz, a $1\frac{1}{4}''$ path-length difference is 180 degrees out of phase, so total cancellation would result. This is the source of the lobing of the polar patterns of direct radiating piston-action loudspeakers.

So the source of the mid-range sound needs to be small. To be practical, it also needs to be light, in order to be capable of accelerating, decelerating, stopping, accelerating, decelerating, and stopping again, up to ten thousand times a second in a power efficient manner. These requirements may well be met by a cone or dome of 4 or 5″ diameter, but this type of driver has a typical conversion efficiency in the order of only one or two per cent. An input of 1 watt of electrical power would typically produce an acoustic output in the order of 96 dB at a distance of one metre. In certain examples, however, this may be in the order of 100 dB, but, conversely, may be 80 dB or less. Soft dome mid-range units have enjoyed a period 'in-vogue', but in the larger systems, they are often on the absolute limit of their power output capability. This has prompted the addition of limiters on some systems, an addition which I, personally, believe is undesirable on monitor systems.

If you are going to hear a 'limited' sound, albeit at high levels, then you are not 'monitoring' the input signal. It is obviously a workable system, as many people do indeed work with such systems, but I feel that it is a last resort measure to drag the last ounce of output from a given philosophy of system. It is not possible to endlessly continue increasing the power handling capacity of direct radiating cones or domes, as a larger coil adds weight, which is contrary to our requirement of light weight for efficient, fast, transient performance. Even without such limitations, higher power from a unit of any given size means more heat to be dissipated. Each 10 dB increase in acoustical output would call for a 10 times increase in input power. A 20 dB increase in acoustical output power would therefore call for a 100 times increase in input power. If we sought a 20 dB output increment from a conventional monitor system, then in direct radiator terms, we could be looking for something in the order of a 10 000 watt, 4″ or 5″ driver. Given 1% efficiency, we would be seeking to get rid of 9900 watts of heat at maximum output power. The dissipation of that amount of power in a small space is not a practical proposition, and effectively limits the maximum SPLs that can be expected from direct radiators suitable for mid-range use.

The other path to follow in order to gain the previously discussed, desirable 20 dB increase in output power, is to seek to increase sensitivity and power efficiency to a similar degree. This involves creating a better match from the driver to the outside world – the fitting of an acoustical transformer – in other words, a horn! Typical sensitivity of mid-range horn/ driver combinations range from 105 to 111 dB for a 1 watt input at a distance of one metre. This in itself represents a 20 dB increase over the 91 dB/w/m of a typical direct radiating mid-range driver which is the increase for which we were originally looking. Despite the fact that some direct radiators are now available with sensitivities in the 100 dB/w/m region, that 10 dB difference still represents a 10:1 power ratio.

13.2 Hidden problems

In late 1985 I installed a set of Reflexion Arts 235, 4-way monitors in Firehouse Studios, North London. There had been a delay in the supply of the prescribed JBL 10″ low-mid drivers, which spanned the range from 300 Hz to 1200 Hz. As a temporary solution, ElectroVoice EVM10Ms were substituted until the JBLs finally arrived. The EVs were fine loudspeakers in their own right, and the system exhibited an acceptably uniform pressure amplitude response, but the monitors were considered somewhat 'brittle' at the extreme top. Reducing the level of the slot tweeter seemed to help, and toilet paper over the slots was used by the engineers as a temporary fix. The slot tweeters operated from 6 kHz upwards, and at the time, there was much conjecture as to the source of the problem. Before the problem could be resolved, the JBL 2122s finally arrived, and upon installation the problem immediately disappeared, to the relative consternation of all concerned. The tissue paper was removed from the slots, which were operating $2\frac{1}{2}$ octaves above the uppermost frequency of the loudspeakers which were changed. Investigations showed that the 'frequency responses' of the EV and JBL over the relevant range were, to all intents and purposes, identical to within very close tolerances. The problem, which was almost certainly due to harmonic distortion difference between the two low-mid drivers, proved to be neither where, nor what, it originally appeared to be – even to experienced ears! In a similar way, so many 'loudspeaker' problems turn out to be in the amplifiers, crossovers, or mixing consoles. One should be very cautious about jumping to 'obvious' conclusions.

In the above problem, the pressure amplitude response was initially deemed to be the fix, but the pressure amplitude response was not the source of the problem. This is at the root of one of my main objections to the use of one-third octave equalisers on monitor systems. They are often seen as the potentially universal panacea for *all* monitor ailments. Endless twiddling of the knobs is undertaken in a futile search for that elusive 'correct' setting. There are so many sources of potential response problems which find themselves incorrectly diagnosed as 'frequency response' problems, that very great care should be taken before any equalisation is undertaken as a 'fix'. In all too many cases, the 'fix' only succeeds in moving the problem elsewhere, which, due to the relief felt at having apparently solved the immediate problem, does not always become noticed until some time later.

13.3 Phase

In 1986, at the ISVR, Professor Taylor performed a demonstration of the sounds of various instruments with their leading edges removed. Without this leading edge transient information, the various instruments involved, varying from violins to oboes, were almost indistinguishable from each other. A fundamental part of the information which makes one instrument readily discernible from another is in the attack of the note, a fact which has long been known to the designers of synthesisers. A poor phase response automatically means an inaccurate transient response. Hence, whilst phase strictly speaking relates only to a symmetrical, steady state signal, the root causes of its effects cannot in practice be separated from transient performance; which is probably why the terminology has become somewhat confused. As music is largely composed of transients and asymmetrical signals, the text book statements about the relatively small audible effects of phase are somewhat misleading. Certain phase discrepancies on steady state signals may well not be readily discernible; but steady state, symmetrical signals have little to do with music, save for the old chestnut of the 'lightly blown flute'.

There are so many old text books, and indeed very worthy text books, which have stated in the past the relative imperceptibility of phase at frequencies above those which cause seriously noticeable cancellations. One must remember, however, that many of these books were written before stereo became an important feature of most music reproduction systems, and especially before any form of essentially phase accurate recording was available. They were therefore not concerned with problems of the disturbance of stereo imaging, nor could they find phase accurate recordings to assess, such as can now be provided by digital sources. Their work was valid in their time and circumstances, but things have now progressed a long way. Remember also that many tests were carried out using sine waves or other typical 'test' signals, and most tests were carried out in significantly reverberant rooms. The Schroeder 'Phase Organ' experiment mentioned in Chapter 6 should now be taken as the definitive experiment on the audibility of phase. It relates very well to listening to music under modern monitoring conditions, with relatively dead control rooms, more phase accurate stereo monitors, and digital recording systems.

13.4 Absolute phase

There are, to the best of my knowledge, three loudspeaker manufactures still adhering to the principle of 'absolute phase'. The principle implies that a vocalist, in a studio and facing the control room and engineer, when making a plosive 'p' or 'b' sound, would cause the diaphragm of the microphone to move inwards in response to the positive pressure wave emanating from the mouth. It follows from this that the loudspeakers in the control room should move outwards, in order to create a positive pressure in the control room. To achieve this, and assuming (though not always the case) that the electronic components of the system are 'in phase', the loudspeaker cone should always move outwards in response to a microphone moving inwards. The three loudspeaker companies who I believe have adhered to this concept are JBL, Tannoy and Quad.

Although Tannoys and Quads only rarely appear to be intermixed with drive units of different manufactures, the same cannot be said of JBL. Drive units of JBL origin appear in the monitor systems of many specialist manufactures, such as Eastlake, Westlake, Urei, Reflexion Arts, and many others. These manufacturers use JBL units in conjunction with drive units of other manufacturers such as Altec, Emilar, Gauss, TAD, and Electro-Voice. I have come across many instances in the past where studio engineers have unwittingly attempted to substitute, say, a TAD compression driver instead of a JBL, or a JBL bass driver instead of a Gauss. In so many cases it has not been realised that such a substitution must be accompanied by a polarity reversal of the loudspeaker connections, to avoid disturbance of the system design around the crossover frequency. A positive voltage on the red terminal of a JBL causes the cone to move inwards, or the diaphragm away from the phasing plug. This is in the opposite sense to Gauss, ElectroVoice, and most other manufacturers, where a positive voltage to the red terminal causes a positive pressure wave, or in other words, the cone moves outwards.

Is the consequence of such unwitting phase reversed substitutions disastrous? Somewhat surprisingly, not necessarily! Although the loudspeaker system would not be functioning as the manufacturer intended, that could well be said to be the case merely by *any* substitution of components – in or out of phase. The answer to the conundrum tends to lie in the type of crossover used, but before looking at the crossover problem, let us recap on the roots of the absolute phase concept.

The human ear responds mainly to the rarefaction half-cycle of a sound wave. It is an extremely non-linear organ. Strong evidence has existed for some time that a drum beat sounds different, depending on whether the initial transient pressure wave is positive or negative going – in other words, whether the initial attack causes the loudspeaker cones to move inwards or outwards. This is the principle of 'absolute phase' to which Tannoy, JBL and Quad have adhered. If you are ever dealing with a non-factory re-cone, be careful! Always double check that the coil has been wound in the correct sense. The adherents to this principle believe that the phase/transient polarity should be maintained through the entire record/reproduce system, from recorded instrument to listener's ears.

Returning to the subject of the potential sonic disaster factor of such interchanges, this can be dependent on the crossover. It depends on the slope, and on the design. Figure 13.1 shows the step response of the low frequency output of a JBL 5234, 12 dB/octave, electronic crossover. The top scale is the step input, the lower scale is the output. The 5234 is a superbly engineered unit which, being of excellent quality and typical of its genre, will very clearly show the normal characteristics. Figure 13.2 shows the high frequency output of the same crossover, responding to the same step input. Figure 13.3 shows the summation of the two outputs – a respectably good approximation to the input signal. Figure 13.4 shows the pressure amplitude response or 'frequency response' of the combined outputs, along with a representation of the phase response. The 'frequency response' shows a distinct dip at the crossover point, which can clearly be seen to relate to the sudden break in the phase response as the crossover frequency is approached. With all conventional 12 dB/octave crossovers,

(a)

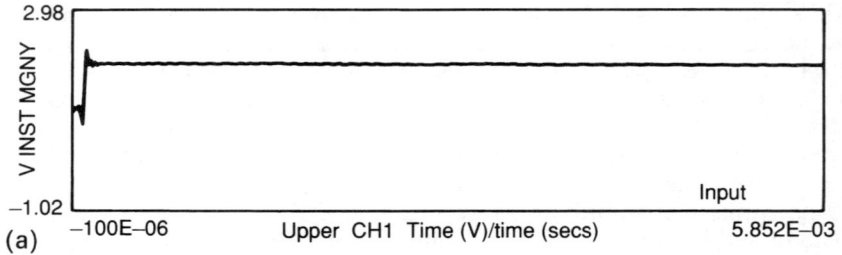

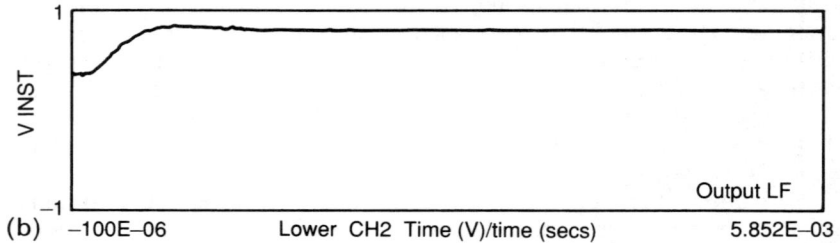

Figure 13.1 (a) Electrical input. (b) LF output

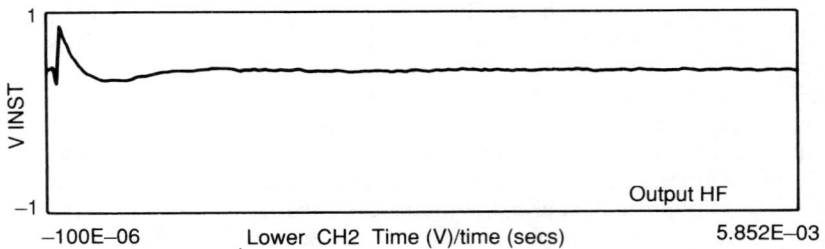

Figure 13.2 HF output

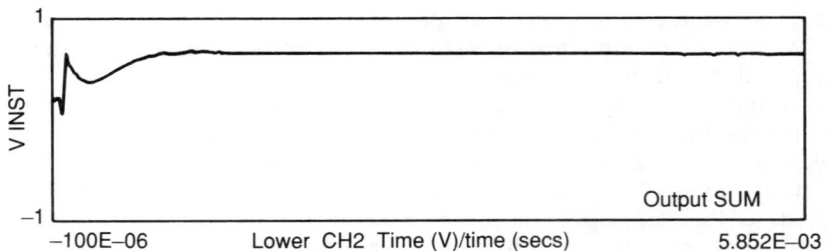

Figure 13.3 Summed output – in-phase

the two outputs shift in phase, one by $+90°$, the other by $-90°$, resulting in a $180°$ phase discrepancy at the crossover frequency. When summed, the $180°$ out of phase crossover point results in cancellation, and hence the 'frequency response' dip.

It matters naught whether this summation is made electrically at the crossover outputs, or acoustically in front of the loudspeakers; the effect is the same – cancellation.

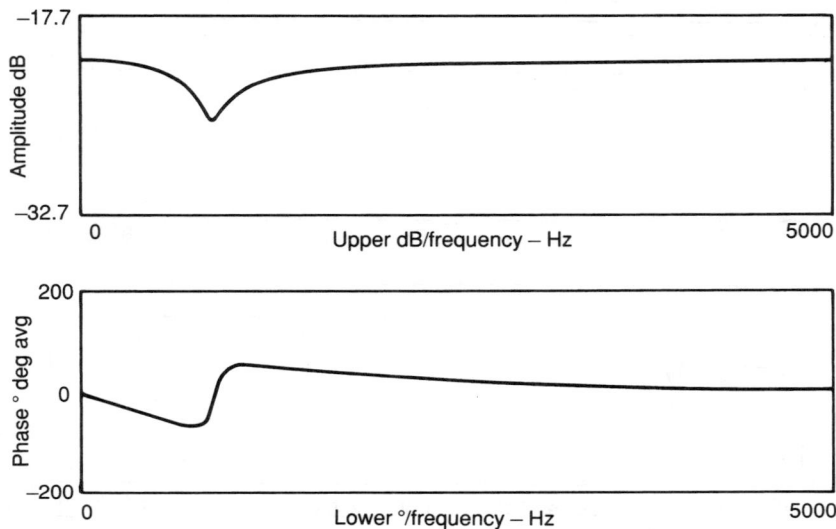

Figure 13.4 Summed output – in-phase

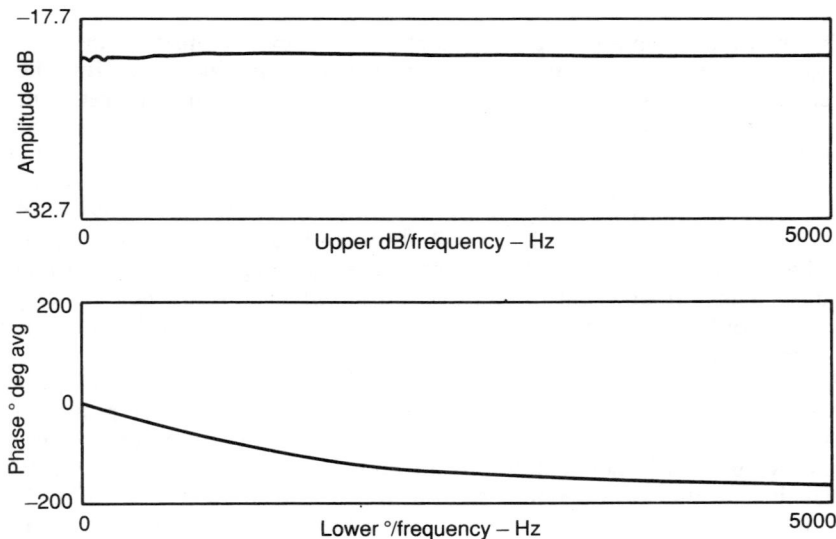

Figure 13.5 Summed output 180° inverted

Over a period of many years, hi-fi enthusiasts and studio personnel alike have become accustomed to looking for a 'flat frequency response' when assessing equipment from brochures or magazine reviews. Responding to this, the marketing departments of loudspeaker companies have often felt uneasy when confronted with a pressure amplitude response as shown in Fig. 13.4. Figure 13.5 is identical to Fig. 13.4 in all respects other than that one of the outputs has been reversed in phase. *Voilá*, the marketing department's dream come true – a 'flat frequency response'!

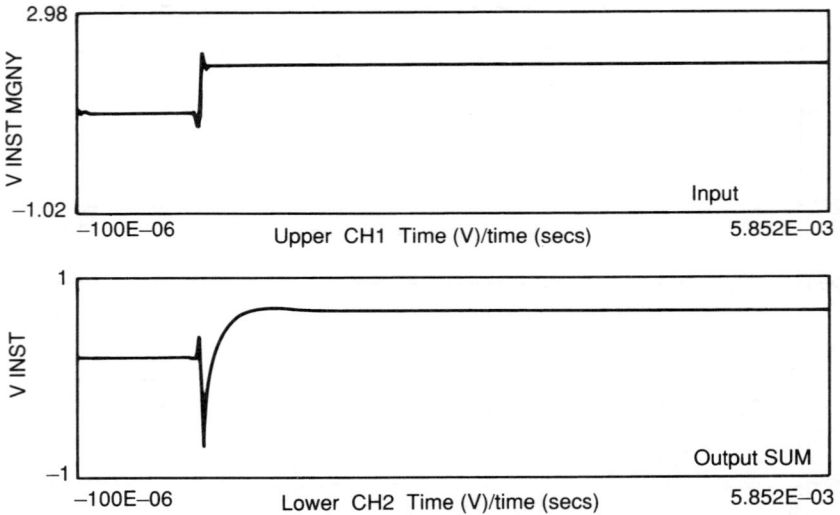

Figure 13.6 Summed output 180° inverted

Such a method of obtaining a marketable 'frequency response' is standard practice for a very great number of loudspeaker manufacturers. 'But the drivers are out of phase,' you may say. Indeed they are, but the argument goes that only by the time that they are well away from the crossover frequency could any cancellation be significant. One octave above the crossover point, the low frequency driver will be providing an output 15 to 18 dB below the corresponding output from the high frequency driver, and vice versa. As such, this could contribute to only a very small degree of cancellation, which is borne out by the flatness of the pressure amplitude response. They are only out of phase with each other at frequencies where, effectively, only one driver is operating. Therefore, it is deemed they are not out of phase *per se*.

Figure 13.6 shows the step response of a system connected according to this philosophy. It can be seen most clearly that as one driver responds to the input in one direction, the other driver responds in the opposite direction. Obviously with these conditions prevailing, we cannot even pretend to pay fealty to the aforementioned principles of absolute phase. This is straddling the phase fence in no uncertain terms. I just cannot accept that the step response of Fig. 13.6 is audibly the same as the step response of Fig. 13.3; but how does this work out in practice? Figure 13.7 is the step response measured at four feet distance from the front of a widely used small loudspeaker. The similarity to Fig. 13.6 is totally unmistakable, borne out by the 'frequency response' graph supplied with the units, which is very flat indeed. Many other loudspeakers are connected in a similar way, as it is a relatively standard practice for use with 12 dB/octave crossovers.

On numerous occasions when listening to studio monitor systems, people have commented to me about drums changing subjective pitch when switched between the main studio systems and the small monitors. The change in timbre can be partly attributed to the fact that minor disturbances

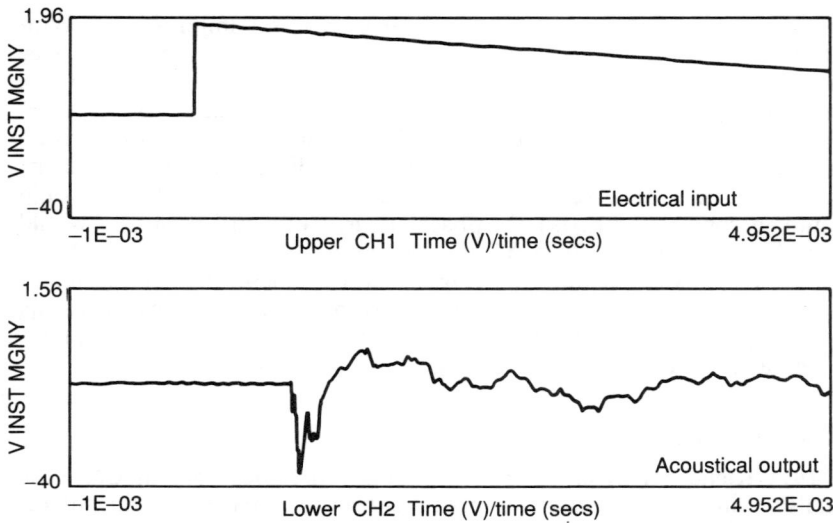

Figure 13.7 Small LS 12 dB/octave crossover

in frequency response are easily audible in a random signal such as untuned percussion or pink noise. There is, however, no shadow of a doubt that the phase considerations also play a major role. When phase is the prime causal factor, no amount whatsoever of 'frequency response' correction will bring the two systems into line.

The transient of Fig. 13.7 is quite patently not the transient of Fig. 13.3. The fact therefore that there is a change in subjective pitch when switching between systems is likely to be caused by the fact that we are not only switching between loudspeaker systems, but between the waveforms of Fig. 13.3 and Fig. 13.7. Many listening tests have been carried out with the high frequency drivers restored to their 'in phase' condition. By far the greatest number of people with whom I have conducted these tests preferred the 'in phase' connection, despite its 'frequency response' dip at the crossover frequency. The general impression was that the sound was more 'open', 'transparent' or 'natural', despite the response dip being audible; though by no means alarmingly so. Unfortunately you cannot publish a graph of what it *sounds* like!

We thus have a situation where the requirements for the integrities of pressure amplitude and phase are in total opposition. For conventional 12 dB/octave crossovers this must be the case. For passive crossovers, 12 dB/octave is a very useful compromise between acoustical performance, out-of-range driver protection – especially for tweeters – power loss/heat dissipation in the crossover, cost, and numerous other factors. Such crossovers are used because they are practical, and in many instances acceptable. The bi-amping of many small monitors with a suitable 24 dB/octave crossover usually produces a much more accurate system. However, such a device may not be a cherished, ubiquitous reference. No matter how 'improved' it may be, it may not be acceptable for some reference purposes because of its non-standard nature.

So what of the person who inadvertently replaces their JBL bass drivers with Gauss, and fails to realise the polarity reversal implications? With 6 dB per octave crossovers, it would be disastrous, though few large systems would be likely to use 6 dB/octave low frequency crossovers. Should the system employ a 24 dB/octave crossover, then less disaster. Near and away from the crossover point, the relative polarity of the drivers would be incorrect in terms of both pressure amplitude and phase, but the crossover region is relatively narrow, and the region of cancellation would be quite narrow. Nevertheless, the phase integrity of transient signals would be greatly compromised. With 12 dB/octave and 18 dB/octave crossovers, however, the answer is not so straightforward. These crossovers cannot possibly sum up to unity with any integrity of phase. It is a case of deciding your own priorities and using your own ears. Only one thing is certain – neither way round is absolutely correct.

13.5 Further points of debate

1 In 1971 Richard Heyser of the Jet Propulsion Laboratory at Cal Tech published in the *AES Journal* a far-sighted series of papers on the subject of arrival times of impulses. Decades later, I am surprised that his work is not more well known amongst studio staff. In the September 1982 edition of the same publication, Lipshitz, Pocock and Vanderkooy published a paper: 'On the Audibility of Midrange Phase Distortion in Audio Systems'. In their 'Summary and Overview', they stated that they did not understand why there were still reports appearing which state that the human ear is deaf to non-linear phase changes – I can only add to that my own incredulity that probably less than 1% of studio personnel with whom I discuss monitor systems can speak with any degree of lucidity on the concepts, and clearly audible effects, of steady state phase shift and impulse phase-slope characteristics.

Almost all-top line professional monitoring systems have an acceptably uniform pressure amplitude response. It must surely be obvious that one must look elsewhere to find the major sonic discrepancies between different systems.

2 Could it be that many previous attempts at assessing the general audibility of phase distortions have been carried out using equipment with inherent phase distorting properties? Such equipment could easily mask any potentially audible phase discrepancies being sought by the investigators. Probably the only readily available commercial loudspeaker system exhibiting truly excellent, on-axis, minimum phase characteristics, are the electrostatics. Though not suited to studio monitoring purposes, they have long held a reputation for 'clarity', 'sweetness', 'smoothness' and 'transparency'. These qualities can, almost without doubt, be attributed to the good axial linearity of both amplitude *and* phase response. The evidence has been before us for over 40 years.

3 In the light of the effect of conventional 12 dB/octave crossovers with respect to transient response, I have never understood how any company could preach adherence to the principles of 'absolute phase', then use a 12 dB/octave crossover in their system. Absolute phase with respect to which end of the spectrum – the high frequencies or the low frequencies? Certainly not both!

4 It may be worth pointing out that analogue magnetic tape recorders suffer from considerable phase distortions. Once again, this may have masked many prior attempts to determine the necessity for phase accuracy, especially as a magnetic tape 'master' was at the source of most reproduced audio signals. The subjective audibility of the phase accuracy of a good analogue tape recording is just about on the limit of acceptability. Consequently, each copy can be adjudged very similar to its predecessor. However, it is not so much the generation to generation differences which are noticeable, but the direct comparisons with two or three generations before, when differences really begin to show. The advent of digital mastering has now removed that mask, and has placed a new emphasis on phase response accuracy.

13.6 Extreme responses

I was recently asked to visit a studio where a music programme had just been mixed for television, but where the results had been generally considered to be light on bass. One of the owners asked me what I considered the problem to be, so I took along a few recordings which were well known to me, but nothing sounded *too* far from what I expected. We then listened to the mix in question, which had been sent to me a few days before, and which I *had* considered to be somewhat bass light. Sure enough, in the studio where it had been mixed, the overall tonal balance sounded fine, so this was no red herring. During the conversation, I was asked somewhat tentatively if it was possible for a room to be bass heavy on some music, and bass light on others, to which my answer was 'Yes'.

Figure 13.8 shows the responses of three loudspeakers in one control room, the solid line showing the typical response of a built-in monitor system with 15″ bass driver. The dotted and dashed plots are those of two different smaller systems being used on stands just behind the mixing console. The larger monitors, flush mounted in the front wall, show much greater linearity throughout the low frequencies, which is only to be expected as they can take full advantage of the large baffle provided by the front wall. Furthermore, they are not troubled by the effect of rear radiation striking the front wall and returning into the room, where the reflexions can cause irregularities in the total response.

Clearly, the response of neither of the smaller loudspeakers exactly matches the response of the larger system, but does this necessarily mean that the systems are incompatible? Well it is all a matter of experience, and how one interprets what one hears. The large system exhibits a reasonably uniform and linear response from about 50 Hz to 10 kHz. The smaller systems, partly due to their own frequency response characteristics, and partly due to their position in the room, show general tendencies towards a rise in the response at 125 Hz and a dip at 80 Hz, with the response returning to just below the '0' level at 50 Hz before falling off completely. At the high frequency end of the spectrum, the dotted curve closely follows the solid curve in general shape, but maintains its response for an extra third-octave or so. The dashed curve continues towards 20 kHz with a somewhat more 'flat' response.

The solid curve is typical of many good monitor systems, with a smooth

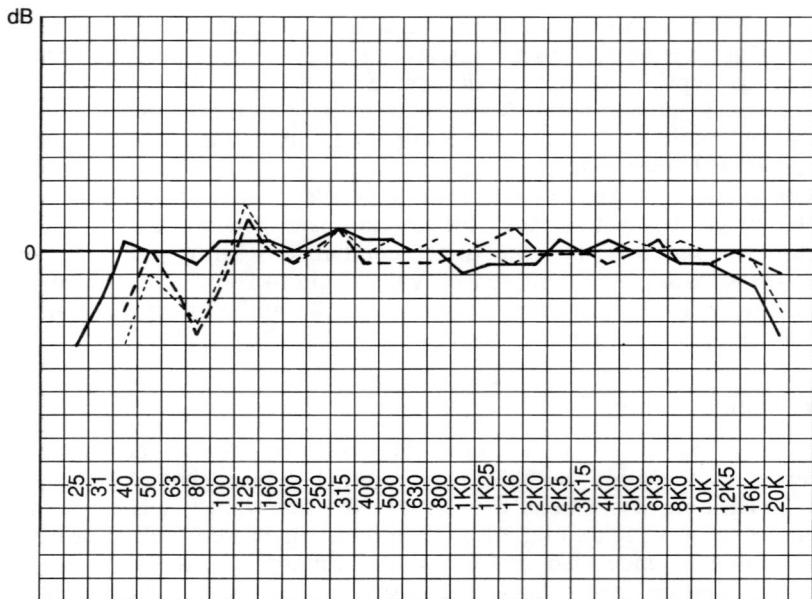

Figure 13.8 —— Response of a large, in-built monitor system; – – – and ········ responses of two different, small, free-standing loudspeakers in the same room as the large system

but gradual roll-off above 8 to 10 kHz, and below 50 Hz. The dotted curve is that of a small loudspeaker, designed by a well known manufacturer for studio use, and shows a reasonable similarity in the shape of the HF roll-off. The dashed curve is that of a domestic hi-fi loudspeaker of quite respectable quality. Purely in pressure amplitude response terms, they are generally much of a muchness from 160 Hz to 8 kHz, with their inevitable differences in tonality and transparency: balance-wise, engineers using the three systems tend to make acceptably similar judgements on all three systems in the above frequency range. But now let us look at judgements made at the low and high frequencies, and then at the whole frequency range.

13.6.1 Low frequency considerations

Figure 13.9(a) shows the spectral balance of a piece of music played on the large loudspeakers in the key of C major, with heavy fundamentals around the 63 Hz and 125 Hz bands. Figure 13.9(b) shows the spectral balance of the same piece of music, but this time transposed down into the key of E, with fundamentals in the 40 Hz and 80 Hz bands. When played through the in-built monitor system, the picture shifts to the left, but is generally similar in size and shape. This reflects the uniformity of the response of the large system in a good room. The two smaller loudspeakers both show reasonably similar response irregularities to each other, suggesting that much of their LF response is room and position related more than being due to drive unit problems. Both, incidentally, drop off the scale below 40 Hz. Figures 13.9(c) and (d) show the same piece of music, but this time played respectively in the key of C and E through the loudspeaker with the dotted plot in Fig. 13.8. (The response of the dashed plot is substantially similar.) This time, the spectrum when played in the key of C (Fig. 13.9(c)) has fundamentals which coincide with the response peak at 125 Hz on the dotted plot in Fig. 13.8. When playing the music in the key of E, the fundamentals coincide with the response dips at 40 and 80 Hz, and it can clearly be seen that unlike the comparison of the responses in Figs 13.9(a) and (b), the

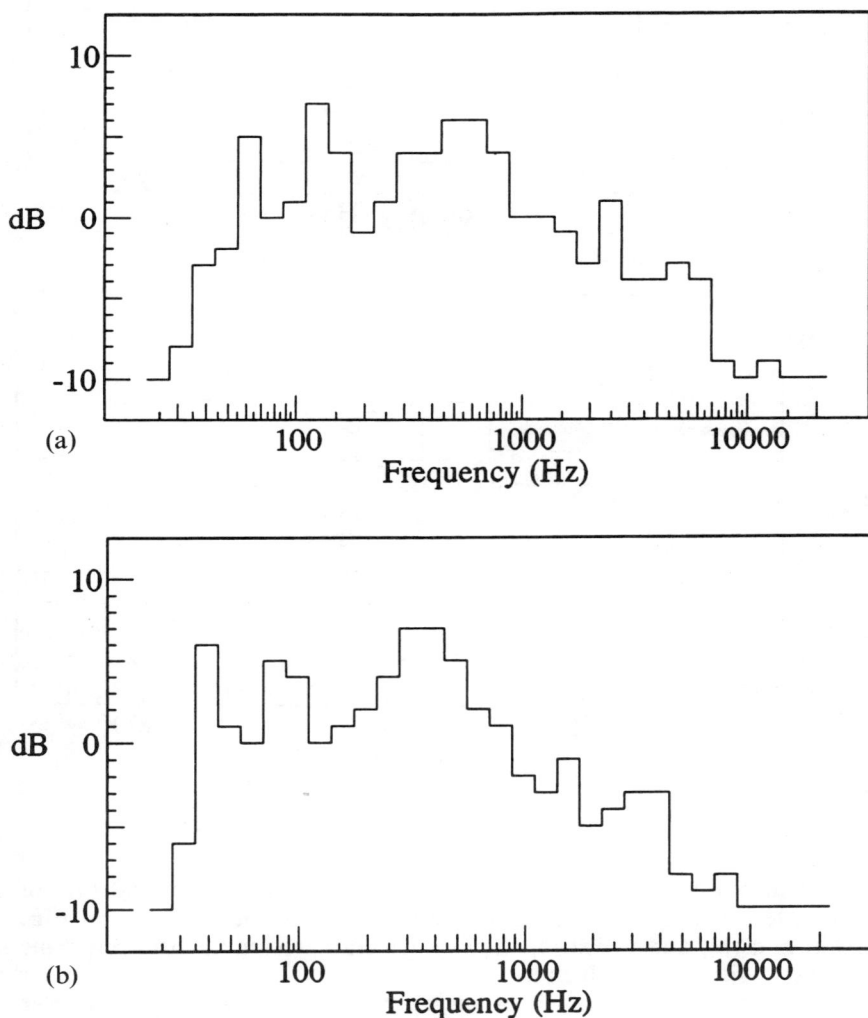

Figure 13.9 (a) Spectral balance of a piece of music played on large loudspeakers in the key of C_M; (b) spectral balance of the same piece of music played on large loudspeakers in the key of E; (c) spectral balance of the music of Fig. 13.9(a) played on 'dotted' loudspeaker in the key of C_M; (d) spectral balance of the same piece of music played on 'dotted' loudspeaker in the key of E

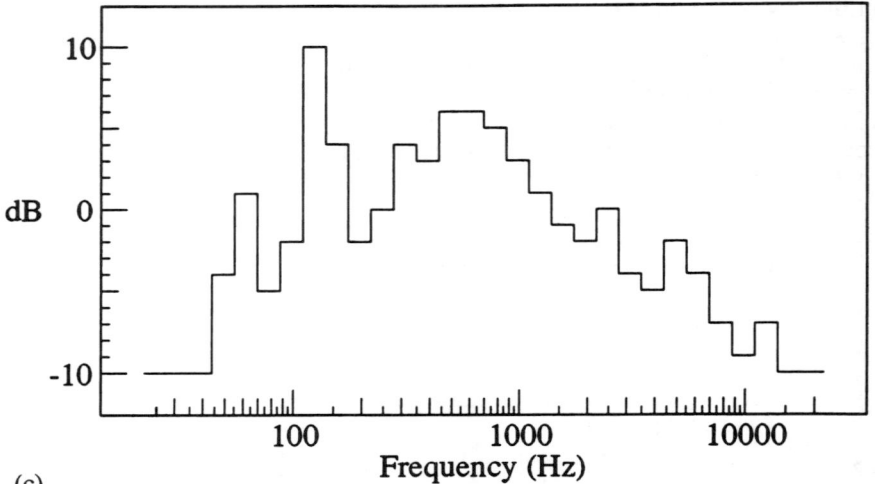

(c)

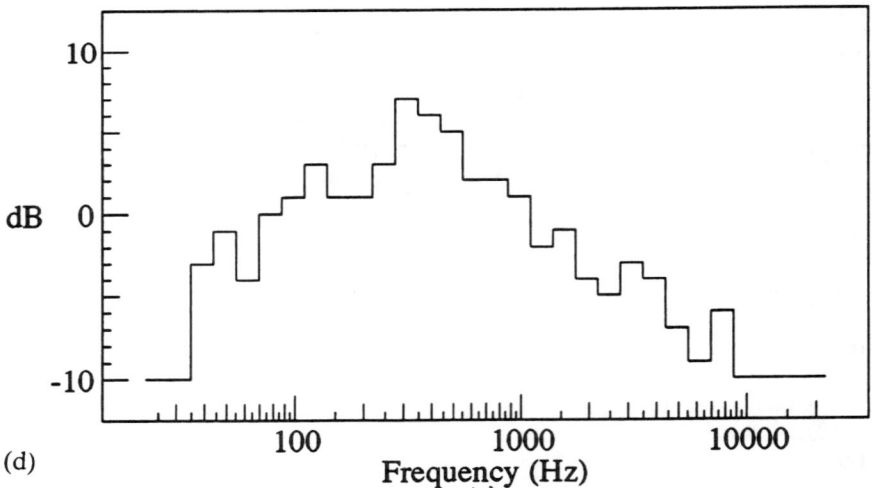

(d)

Figure 13.9 (*Continued*)

shape and power distribution of the responses in Figs 13.9(c) and (d) are quite different. Neither of them follow the shape of the responses in Figs 13.9(a) and (b), and whilst (c) shows a generally enhanced bass response, (d) shows a significant reduction.

The implication from this is that mixing on the loudspeakers with the solid curve in Fig. 13.8 would yield similar subjective balance judgements irrespective of the key in which the music was played. Mixing on the loudspeakers with the dotted and dashed plots would suggest different balances as being optimum, dependent on key. Mixing the music in the key of C would tend to produce a balance with less bass than when mixing the music in the key

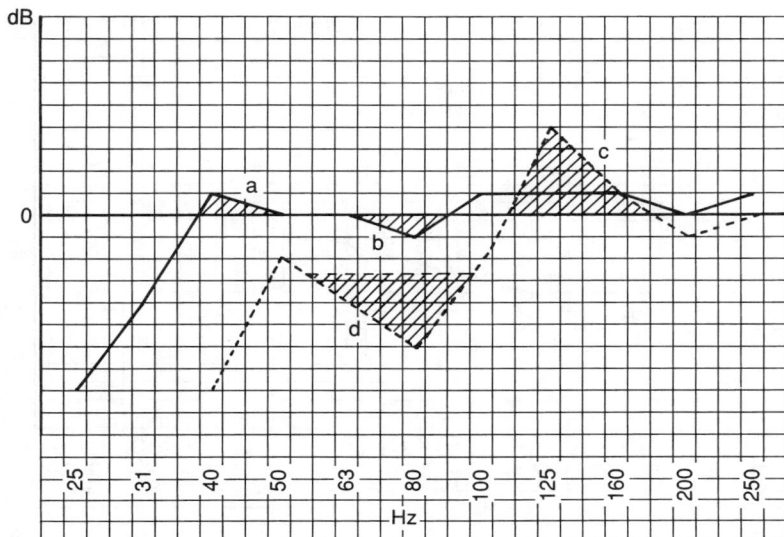

Figure 13.10 If the shaded section below the solid line a were to be used to fill in the dip b, then the whole of the solid line between 40 Hz and 250 Hz would be on or *above* the 0 dB line.

The shaded section c below the dotted line could be used to approximately fill in the dip d raising the dotted line to the dashed line.

Even so, the whole of the new dotted/dashed line would be on or *below* the 0 dB line, showing that the total power output between 40 Hz and 250 Hz was still much less than that of the loudspeaker depicted by the solid line

of E, where the reduction in the loudspeaker response would tend to cause the balance engineer to put more bass on to tape. So, it can be seen from the above description that the absolute degree of subjective bass lightness or heaviness can be dependent upon musical key, but the degree to which the mixes produced on the small loudspeakers would be considered bass light or heavy when auditioned on the large system contains even more variables.

Figure 13.10 shows two of the plots from Fig. 13.8, which are drawn on squared 'graph' paper. If we take all of the squares below 200 Hz, bounded by the dotted line and the '0 dB' line, which are above the 0 dB line, then use those squares to 'fill in' the dip around 80 Hz, it will yield a new dotted line which is at all points below the solid line. This indicates that, in absolute terms, when compared to the relatively similar mid-frequency levels, the large (solid line) loudspeakers produce much more relative output at low frequencies. Clearly, on pink noise, which has an equal power distribution in each third-octave band, the sound will be much more rich in bass when heard via the large loudspeakers. Not surprisingly, music having a generally well distributed amount of low frequency energy will consequently sound subjectively bass-light when heard on the small loudspeakers, compared to the same music heard at the same mid-range 'loudness' on the large monitors.

But, let us now consider music with very little content below 100 Hz. In the 100 Hz to 160 Hz region, the small loudspeakers, in the room where measured, actually have more squares above the 0 dB line than do the large loudspeakers. In this instance, the generally bass light small loudspeakers will possibly have

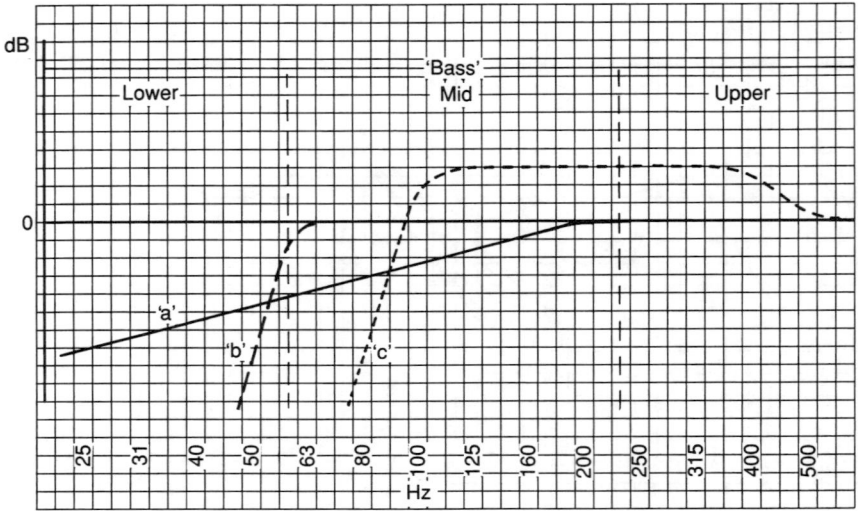

Figure 13.11 The plots in this figure show three very different bass responses, which despite imparting very different tonality or timbre to the music, nonetheless could under many circumstances all be perceived as having subjectively the same 'quantity' of bass, as related to mid and top, if their responses above 500 Hz were all the same. This would, however, be highly dependent upon the nature of the musical signal. In general, the rock world tends to favour something akin to the 'b' line, the classical world the 'a' line, and many cheap domestic systems the 'c' line.

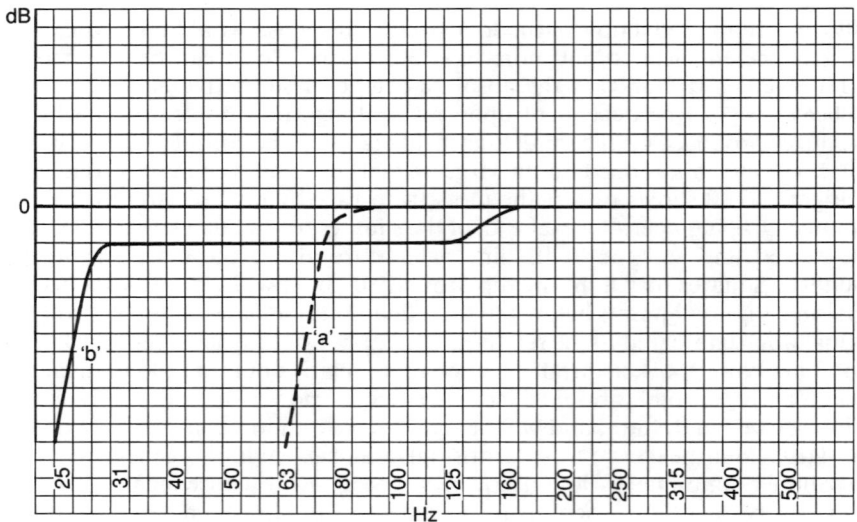

Figure 13.12 Plot 'a' is typical of many good quality small, free-standing loudspeakers, such as are frequently employed as close-field monitors, on top of or immediately behind the mixing console. Plot 'b' would be typical of a large, high quality, built-in monitor system. Clearly there will be more overall bass energy produced by 'b' on a musical signal possessing a wide-band low frequency content.

more subjective bass response than the large ones. Here, we would run the risk of balance judgements made on the small, 'bass light' loudspeakers, actually producing bass light mixes as compared to those done on the large system. A generally bass light monitor environment should produce a bass heavy mix, as a result of the engineer having to turn up the low frequencies more in order to correct for the lack of response from the loudspeakers, but above we have an instance where this would not necessarily be the case.

We tend to average out subjective frequency balance into roughly third-octave power bands, and often group these together into what we may refer to as bass, low-mid, mid, high-mid, top, or various other sub-divisions of subjective perception. In Fig. 13.11, we can see the bass divided into upper and lower regions. All three of the plots yield an average response of uniform bass energy from 20 Hz to 200 Hz when compared to the average mid-range level. Here, a bass response (a) which begins to tail off from the higher frequency, but which ultimately extends to a lower frequency, can yield a subjectively similar *quantity* of bass (though probably of different *quality*) to another unit (b) which remains 'linear' down to a lower frequency, but dies more quickly once the roll-off does begin. The third unit (c) has a boosted upper bass to help to subjectively compensate for the lack of lower bass response. Obviously, though, the subjective tonal quality of the low frequency response of loudspeaker (c) will by no means be as lifelike as that of (a) or (b) on any programme material with any appreciable amount of low bass. All three, however, on many musical mixes, may well yield acceptably equal quantitative balances of bass and mid-frequencies.

Given today's realities, much music is very likely to be mixed on small loudspeakers having a response more similar to that shown in Fig. 13.12, plot (a). Here the response is maintained as linear as possible down to a frequency limited by the size of cabinet. When compared to any of the plots of Fig. 13.11, this loudspeaker can be seen to have considerably less overall (20 Hz to 200 Hz) bass than any of the other units. Given the widespread use of such systems, there can be justification in certain circumstances to align any large loudspeaker systems which may regularly be used in conjunction with them, to a response more similar to that in Fig. 13.12, plot (b). Subjectively, on a wide range of music, the overall quantitative bass response of the systems of Figures 13.12(a) and (b) can frequently be very similar, though, obviously, compared to the more technically preferable absolute responses of Figures 13.11(a) and (b), the overall bass response of Fig. 13.12, plot (a) would be lacking. On the other hand, the larger systems often tend to be used louder, when the ear's own response begins to cause a subjective bass rise, hence possibly producing a similar mix on 13.11(a) when used quiet, and 13.12(b) when used loud.

Here we come across the horribly vexed question of whether we should be mixing for the 95% of people who will buy and enjoy the music on 'budget' domestic music systems, to the detriment of the achievable enjoyment for the people who purchase expensive hi-fi systems; or, whether we should mix to the best absolute standards, and the domestic listeners get what they pay for. There is no universally accepted answer to this dilemma, but there is a general trend for 'disposable' music to favour the former approach, and 'serious' music to follow the latter. (By 'serious' music, I am referring here to music which is intended to be timeless, as opposed to 'jump on the bandwagon' music, which will be largely forgotten by the following year. I am not referring to any particular musical genres.) When all of the above points are taken into consideration, it is little wonder that so much confusion and controversy still exists on the subject of monitoring.

13.6.2 High frequency considerations

At the high frequency extremes, particularly in the 'last' octave or so from, say, 8 kHz to 20 kHz, again different philosophies exist side by side. A large percentage of purpose designed studio monitor systems have intentional roll-offs above around 8 kHz, typically being 4 to 10 dB down at 20 kHz, following a gentle slope from 8 kHz. The solid and dotted plots of Fig. 13.8 were clearly designed in accordance with this philosophy. As mentioned earlier, loudspeakers designed for domestic hi-fi more usually have responses continuing more linearly to 20 kHz, but there are several reasons for the tendency for some large studio monitor systems to incorporate this high frequency roll-off. In the previous paragraph but one, I referred to the responses of Figures 13.11(a) and 13.12(b) yielding similar subjective levels of bass, when used at lower and higher sound levels respectively. Human beings tend to perceive more subjective low and high frequencies, proportionate to the mid-frequencies, when the overall level of loudness is increased. When large loudspeakers are used at moderately high levels, they *can* tend to become unnaturally 'toppy' in their tonal balance. Furthermore, high levels of high frequencies can become very tiring when a person is subjected to them for long periods of time. Many, many years of experience has led to a great number of experienced engineers opting for a gentle roll-off of the high frequency response of large studio monitoring loudspeaker systems, believing that such a curve yields the most natural results for the tapes leaving the studio.

There are, however, three further points of great relevance relating to the high frequency responses of loudspeakers. Firstly, the risks of a slightly dull monitor system producing slightly 'bright' mixes in the outside world is infinitely preferable to a slightly bright monitor system resulting in dull mixes in the shops. About this, there is almost universal agreement. Secondly, when domestic loudspeaker manufacturers produce loudspeakers for a mass market, there is still an alarmingly unrealistic number of potential customers who equate 'top' and 'brightness' with quality. This results in an absurd number of 'tizzy' domestic systems. Consequently, many domestic hi-fi loudspeakers, hi-jacked for monitoring, may possess a market-oriented response rather than one purposely intended for long-term monitoring, but which nonetheless may well be representative in many instances of the multitudinous other domestic systems. Thirdly, as Gilbert Briggs, of Wharfedale, pointed out in his books of the 1950s, a 'balanced' response around a centre frequency of 1 kHz or so tends to produce a more subjectively natural response. In other words, it is OK to go flat out to 20 kHz, if at the other extreme you are going down to 20 or 30 Hz; but if you are only going down linearly to 40 or 50 Hz, then at normal domestic listening levels, it may be prudent to begin to curtail the top end response above 12 or 15 kHz.

To extend linearly to 20 kHz whilst losing the low frequency response below 70 Hz may not only give a perception of a light bass, but also one of an excessive top, even though in absolute terms the top is not raised above the mid-frequency average. I feel that this latter point rarely receives its due attention, and many people, even of great experience, could do well to bear it in mind more regularly. Remember, all music is a perceptive experience, not an absolute one.

On the latter point of high frequency levels and the natural perception thereof, and giving due consideration to the fact that probably none of us possess the theoretically 'average' human response, there is one further point which I would like to consider. I have recently been asking a number of academics to look into this in audiological terms, as I am of the growing opinion that we do not perceive the intense and directional high frequency response from a loudspeaker in the same quantitative manner that we perceive it from a more distributed source.

For example, if we consider the sound emanating from an acoustic guitar or a cello, the sound in the top octaves will be being produced by many parts of the instrument, and certainly from many points along the strings. Once we collect the sum total of these high frequencies arriving at a microphone, or pair of microphones, then reproduce them via the tweeter of a loudspeaker system, they are re-radiated in a concentrated form, and the ear picks up the thin beam of high frequencies radiating in a direct line from tweeter to ear. Although another microphone placed at the position of the ear would measure the same quantity of high frequencies (whether receiving the sound from the instrument itself, or at the same level via a loudspeaker of linear response), I am convinced that the subjective high frequency level perceived by the ear is different for the two sources. Certainly the way in which the two sound sources would be collected by the pinnae (the outer ears) would inevitably be different, and it is a matter of fact that the perceived frequency balance of an incoming signal to the ear is direction dependent. I can only speak from my own experience at this juncture, but to me, in such tests, a slightly rolled-off response to high frequencies renders a typical loudspeaker more natural sounding when compared to many real instruments.

When all of the points discussed here are taken into account, it can readily be appreciated just how difficult it is to make any sweeping state-ments of what is absolutely right in monitoring. In reality, only when purpose-built monitors are positioned in skilfully acoustically engineered rooms can we begin to dwell very heavily on absolutes. In almost all other circumstances, we are in a mess of compromise, but this is where the skill and experience of engineers becomes of such great importance. It is only with a much fuller understanding of the pitfalls and problems that we have any hope of regularly producing 'balanced' results from a wide range of music, at least in anything less than optimum circumstances. Without that knowledge, we can only expect the talking at cross-purposes, and the 'false facts' that are so common in this rapidly expanding industry, to continue unabated.

Chapter 14

The sound of mixing consoles

Upon reading the title of this chapter, most people would probably presume that the main subject of this discussion was to be about the relative merits of the electrical signal paths of current popular mixing consoles. Not so; what we are about to look at is primarily the acoustic effect of the placement of consoles in control rooms. There are two major factors which cause problems, and many studio designers have long complained that console manufacturers seem to show little inclination towards addressing them. The first and probably most obvious problem is the physical shape of the mixing console and its effect on reflexions, and hence the imaging and definition of the monitor system and room.

The second problem is the quite alarming degree to which panels can resonate sympathetically with the music, adding undesirable colouration to the sound.

These problems are not trivial. Whilst console manufacturers cannot be expected to tailor the shape and size of consoles to each and every room, there are certain fundamentals which are so universally problematical that, by now, it would seem reasonable to expect that they should have already been addressed more thoroughly.

14.1 Console positioning and construction

The most significant of all physical problems exists where a large console possesses a deep, flat, vertical, resonant back. In many instances, the rear of the console receives virtually a full wavefront from the monitors, and almost invariably the rear of the consoles is hard, thus highly acoustically reflective. Sounds impinging upon this surface will reflect back towards the front wall, which is also frequently hard and reflective, often containing a large window. In turn the sound will then either return to the mixing console rear, to begin its journey to the front wall once again, or pass into the room. Chattering can begin between the two hard surfaces, setting up resonances which will colour the sound both in the frequency domain, where phase and amplitude will be disturbed, and also in the time domain.

With sound travelling at around 1000 feet per second, every foot which the reflected waves must travel before reaching the ear will cause that sound to be delayed by around 1 ms. Therefore, if the console was 6 feet from the front wall, a sound reflecting from the console rear, bouncing back to the front wall, then returning to the ear, would arrive at the engineer's ears around 12 ms after the arrival of the initial sound. The result of this is time smearing, in addition to the colouration produced by any modal chattering and resonant panel colouration.

What is more, the panel resonances will usually arrive at the ear via a non-direct path, reflecting off another surface. They will thus be perceived as delayed resonances. Delayed resonances are even less desirable than non-delayed resonances, as the temporal separation increases the ear's ability to detect them, a point clearly demonstrated in Chapter 12 in relation to horn reflexions. It does not even stop there; the fact that these resonant reflexions will bounce off a surface which is non co-located with the source of the drive signal means that they are also spacially separated as well as temporally.

It is difficult enough, even in very good control rooms, to support stable, clear, stereo images. When one compounds the issue with delayed, spacially separated, frequency dependent, amplitude and phase modified spurious sounds, then it will not be difficult to understand their potential for spoiling the clarity of the monitoring. All this because the console manufacturer did not realise the implications of the fact that their product was ultimately destined for work in a real control room, and not just to look pretty on an exhibition stand.

Most rooms will benefit considerably from the damping of the rear panels of the console with an automotive type panel damping material, together with a screen of 'Sonex' type foam wedges, preferably at least six inches deep, shielding the console from direct impact from the wavefront leaving the monitors. Depending upon whether the console rear needs ventilation, the absorber panel can either be attached directly to the console, or spaced off a few inches as a free-standing unit.

Consoles with very deep backs, especially those which go all the way down to the floor, are acoustic disaster areas. The manufacturers should be ashamed of their lack of awareness of the consoles' true circumstances of application. Full height console rears can form resonant cavities between the floor and the front wall of a room. They also block the path of the low frequency waves which should be allowed to pass freely under the console.

When monitors are mounted high up on the front wall of a studio, pointing down at a steep angle, there is a potential for the floor to reflect a wave back upwards on to the underside of a mixing console, and possibly back down to the floor before finally coming up once again towards the engineer's ears. Given this pathway, a considerable delay will be present between the direct and reflected waves. I am not suggesting that *this* problem is a function of poor mixing console design, but it is nonetheless a possible consequence of placing a mixing console in a room. I know of one designer who quite routinely places absorbent material below the mixing console in order to ameliorate this problem.

14.2 Siting of effects

Many consoles have built in or built on 'wings' for the mounting of effects. These wings are frequently fitted as a mechanical or electronic engineering exercise, rather than an acoustic one. Very great care should be taken to ensure that the wings are not sited such that resonant modes could be established either between adjacent wings, or between a wing and a wall. Neither should they be sited such that sound impinging upon the wings from the monitor system could be reflected into the critical listening area. Wings should be angled such that any reflected sound from the monitor system will pass away from the central listening area, and, if possible, into an absorbent area of the room where it will subsequently be lost. Furthermore, the top and bottom panels on the effects themselves should be checked for sympathetic vibration when certain musical notes are present.

Wings can be acoustically problematical, but with care they can be rendered all but neutral in their acoustic disturbance of the room. I certainly consider them infinitely preferable to the custom of mounting the effects in a long rack immediately behind the engineer's position, and usually angled upwards for a more clear view of the controls and labelling. Such racks place an almost perfect, large acoustical mirror behind the listening position, and if one strikes a line from the main monitors to the face of the effects, the reflected ray would come right back up to the engineer's ears with a delay of eight to twelve milliseconds or so. In many such rooms, one hears comments of a lack of distinction from the main monitors, with many people using close-field devices. Little wonder – such an effects rack is custom designed to wreck any sense of stereo imaging from the main monitor system. I do realise that the effects must be housed somewhere convenient, but when they are placed in such a way behind the console, the staff must accept that they have chosen to place an emphasis on the ergonomic operations of the electronics, which will significantly degrade the acoustics. If they have made that choice, then they cannot expect neutral monitoring, nor make any claims that 'accurate' monitoring could exist in such rooms.

14.3 Response disturbances

Figure 14.1 shows the before and after response of a room when a mixing console was placed in that room. The effect is clearly visible on the plot, and in all too many cases is clearly audible as well. I mentioned that exercises such as placing effects racks behind the console can drive people to rely more on the close-field monitors, but despite the fact that such monitoring can be less prone to some of the effects described above, the console design can still exert an influence over their response. One obvious problem would be resonating or rattling top panels on the console – particularly lightweight blanks where the console awaits the future fitment of further modules.

Figure 14.2 shows the response from a small loudspeaker, placed on the metre bridge of a console with a large, shallow angle top surface area, plus plenty of space round the knobs and a significant number of blank panels. An almost perfect reflexion, delayed by around one millisecond, can quite clearly be seen in the response plot. Fortunately, the ear is far less susceptible to confusion by vertical reflexions than by horizontal ones, but nonetheless, such reflexions should be avoided where possible. The console in Fig. 14.2 is generally well liked for its clean sounding electronics path, yet little attention seems to have been paid to its acoustic properties; in addition,

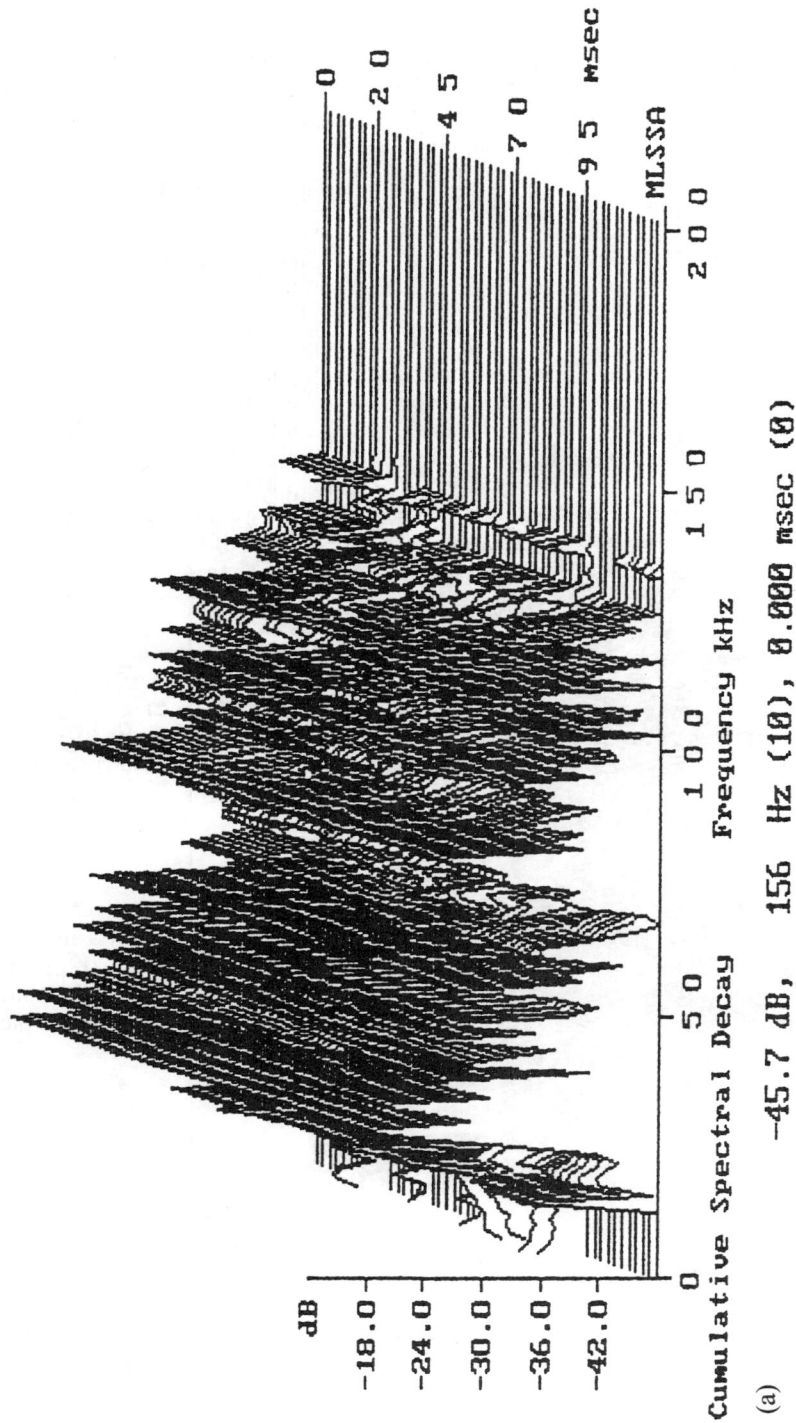

Figure 14.1 (a) Cumulative spectral density of empty room (position 1).

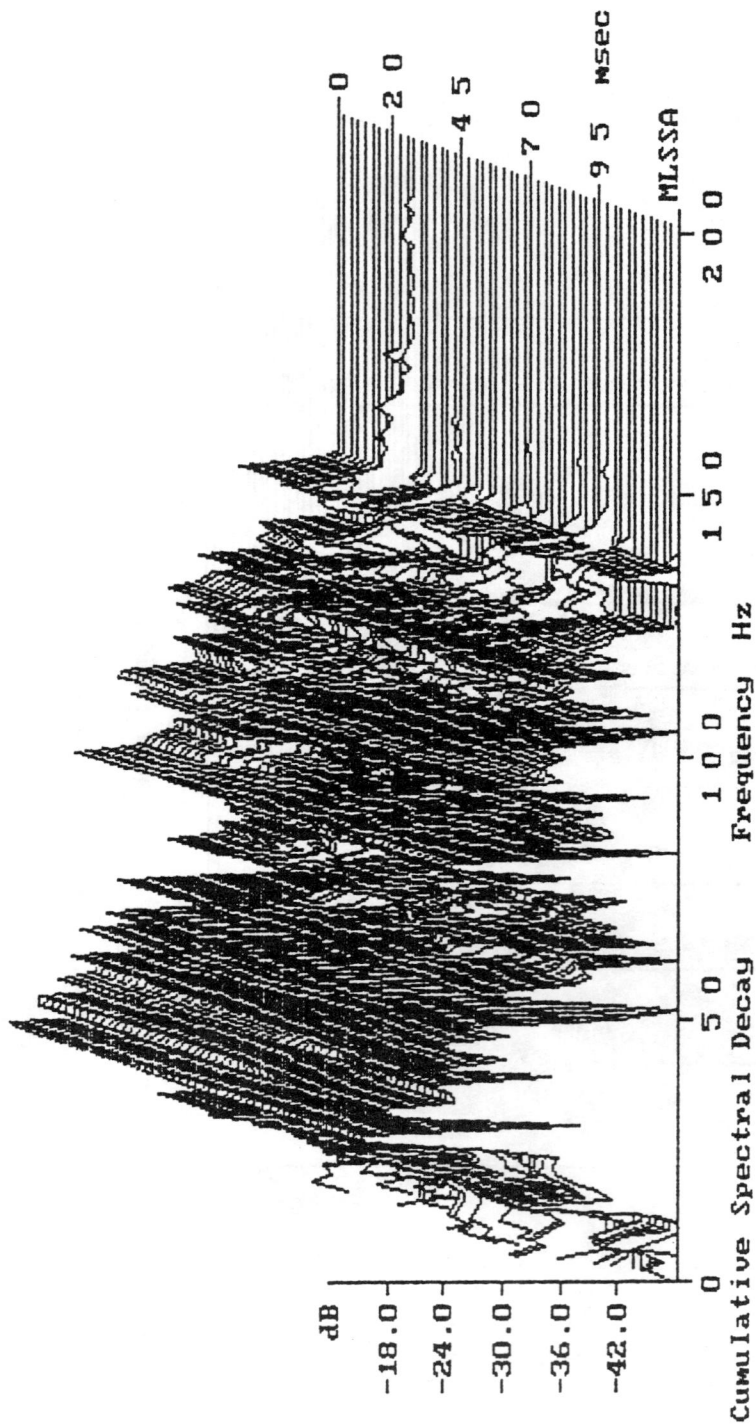

Figure 14.1 (b) mixing desk added (position 1)

Cumulative Spectral Decay

−38.8 dB, 50 Hz (32), 0.000 msec (0)

(b)

(a)   ELECTRICAL INPUT SIGNAL

LEVEL vs TIME

RESPONSE AT 1ft NO REFLEXIONS APPARENT

LEVEL vs TIME

RESPONSE AT 2ft CHARACTERISTIC DOUBLE TRACE

LEVEL vs TIME

RESPONSE AT 4ft EVEN GREATER DISTURBANCES IN TAIL

LEVEL vs TIME

(b)

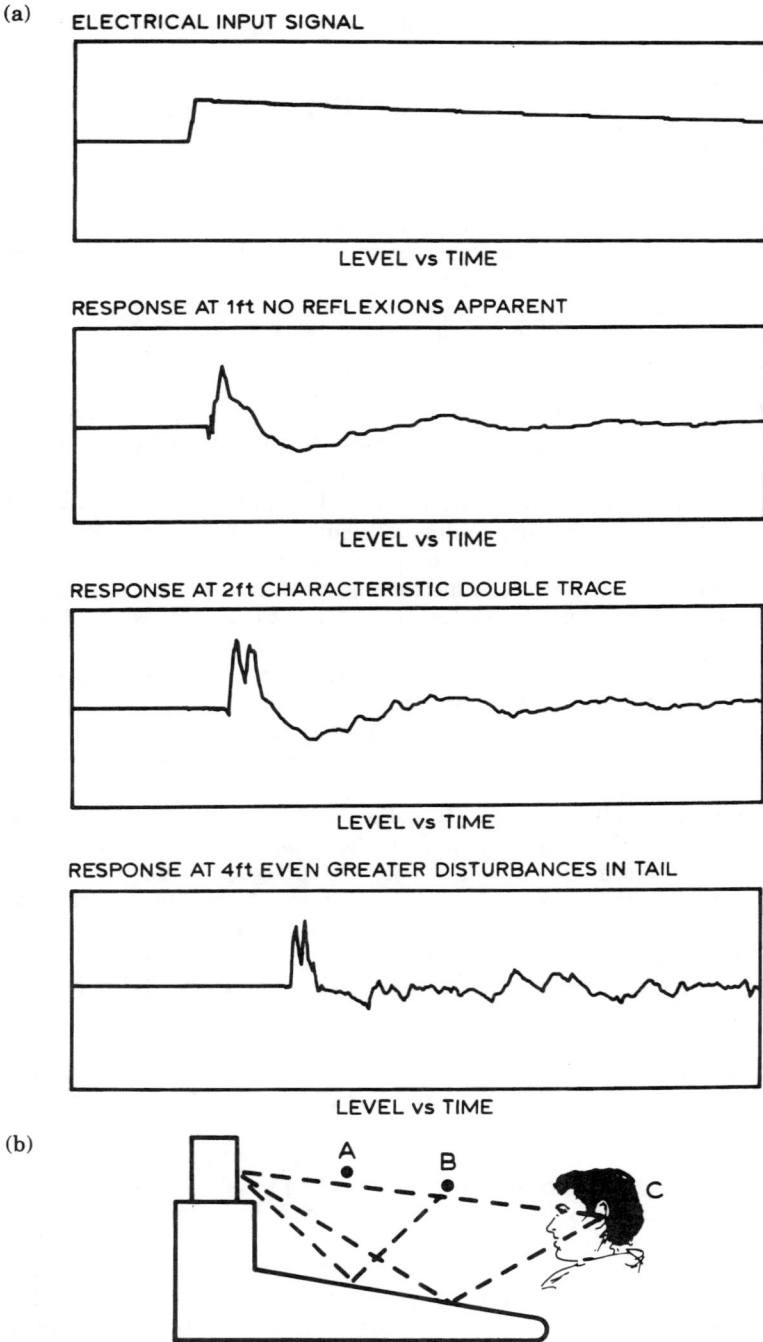

Figure 14.2 Close-field monitor response. (a) Effect of the reflexions from the top surface of a mixing console on the transient response of a small loudspeaker placed on the meter bridge (b) Loudspeaker directivity is too narrow to produce reflexions at position A. Reflexions are apparent at positions B and C with differing ratios of direct to reflected path lengths, hence they produce different composite transient waveforms as in (a). Positions A, B and C relate to the 1 ft, 2 ft and 4 ft plots in (b)

until treated, a 2 kHz ring was clearly audible from its panels upon excitation by a snare drum or similar transient signal, either directly or via the monitors.

As a general rule of thumb, predictability of final overall response is made easier when the console size is small compared to the room. Where the console is large in proportion to the room, and hence occupies a considerable percentage of the room volume, one is no longer really dealing with the room itself as an acoustical space. I cannot say that a large console in a small room cannot sound acceptable – one may get lucky, and some people do – but in general, other things being equal, the effect of a console is largely a function of the percentage of the space of any room which it occupies if predictability of response is to be maintained. From an acoustical viewpoint, the acoustically best console is no console.

14.4 General care and attention

I fully realise that console manufacturers exist in a highly competitive and cost conscious world, but I do wish that more of them would accept that the console is a part of a system, and not the all-important bit to which all other aspects of the studio must subjugate themselves. It would take only a little forethought and consultation to produce an ergonomically viable, acoustically streamlined console, which would cost little more to manufacture, yet could well give such consoles a significant extra sales boost. If one thinks about it, a console which does the least damage to the acoustics of a room will no doubt be deemed sonically superior to one of similar electronic clarity but greater acoustical disturbance.

When a well designed control room is first completed, with nothing installed but the monitor system, its sonic characteristics will almost always be heard at their best. As equipment is brought into the room, its acoustic neutrality will generally degrade. If sufficient ill-conceived and ill-sited equipment is installed in that room, then the room as used will probably bear no acoustic resemblance to the room as designed and constructed. Indeed, many well designed rooms are acoustically ruined by the careless installation of 'too much' reflective and resonant equipment. More care in the design and placement of such equipment will undoubtedly lead to better acoustic performance, and a better environment for all interested parties to achieve their desired goals.

Unfortunately, all those interested parties are not necessarily in close communication. It is quite remarkable that once a studio is up and running, the operating personnel frequently seem to forget all of the careful considerations about the overall attention to system detail. Equipment is moved around without thought, and people screw blank panels into holes which were to have housed equipment which was never purchased. Movement of equipment is a problem where overall symmetry is lost, creating differing reflexion patterns on the left and right hand sides of the engineers. The aforementioned blank panels can also be a problem, especially the larger ones, since these can ring and produce unnatural colouration unless suitably damped. Where such panels are either in the mainframe of the console or in the effects wings, they can be damped with automotive panel damping material, which is available from most of the larger car accessories shops.

Another alternative is 'Revac' or similar deadsheet, glued to the inside of the panel.

In the case of many of the large panels on the underside or rear of the consoles, such an application of damping material will usually noticeably improve the imaging and overall perceived neutrality of the monitoring. Unless specifically listened for, however, such resonances are not always obvious. They insidiously add their own character to the monitor output, usually in such a way that blurs the temporal and spacial response of the system, rather than adding any noticeable lumps into the perceived 'frequency response' of that system. It is often only when their unwanted contributions are removed that their significance can be readily noticed by its absence.

Most studio personnel have never had first hand experience of the comparison between a control room empty of all but its monitor system, and the same room when in a typical working condition. Most of those people would be quite alarmed upon realising the degree of acoustical degradation which usually takes place. In each and every studio, somebody should take responsibility for ensuring that any equipment installed in that studio is in itself acoustically neutral, and that its siting is consistent with good acoustical practices. Despite the way in which electronics seem to dominate the thinking of many people, the acoustic interactions in a working studio are far too complex for any electronic fix. There must be acoustical solutions to acoustical problems. Every so often, ships enter dry docks for the barnacles to be scraped from their hulls. Just as those barnacle build-ups reduce the ships' efficiency and performance, so the build-up of odds and ends in and around the mixing console can severely degrade the performance of a control room. A regular 'scraping off of the barnacles' can work wonders for control room performance, just as much as for ships. Untidy control rooms are not conducive to good monitoring conditions.

14.5 Meter bridge problems

When consoles are designed with large, flat topped surfaces for their meter bridges, then once again these seem to be a potential hazard to sonic neutrality. They are definitely not desirable from an acoustic viewpoint, not only because of their ability to reflect glancing sound waves towards the ears, but also because, in reality, they become home for anything and everything which seems to have nowhere else to go. Electric flowers which dance to the music seemed to be all the rage at one time in terms of desk top decorations, but cups, beer cans, extra pieces of hired-in, flight cased effects, and an entire array of different close-field monitor systems, are probably the rule rather than the exception where large, flat topped meter bridges exist. Often the proliferations of small loudspeaker systems are ostensibly down to the fact that various clients cannot 'get on' with the main monitors. Cannot get on with them? – I have known situations where they cannot even see them. I remember visiting one very well known studio in London where four out of the five drive units in each cabinet of a monitor system, for which they had paid just short of £20 000 ($30 000), were invisible from the engineer's position. They were entirely obscured by the plethora of loudspeakers and other oddments on the meter bridge.

14.6 Summary

I appreciate the dilemma here for the console manufacturers. Many people *like* flat top bridges; they are convenient. Is it the responsibility of the console manufacturers to save the studio personnel from themselves? *Per se*, no, but there are two points worth noting. Firstly, an angled surface is not only more desirable on purely acoustical grounds, but also more importantly, if cluttered bridges can be discouraged, then, ironically, the greater neutrality in the monitoring environment will probably be put down to an improved sonic performance and better reputation for the console itself; which it *is*, but not in the electronic performance to which most people would attribute the improvement.

It is surely thus in the interest of console manufacturers to look to the acoustic effects of the presence of consoles in control rooms. If the acoustic considerations are addressed from the beginning of the design stage, the cost implications are not great. It is largely a matter of the choice of geometry, and the attention to panel location and damping. Unfortunately, not all control rooms would necessarily exhibit sufficient sonic neutrality and imaging to render some of the more subtle aspects to be noticeable. However, in those environments which do offer such neutrality, it is all too frequently the presence of the mixing console which limits the realisation of the full sonic performance of the total control room system. At the other end of the scale, panel resonances have produced clearly audible degradation of monitoring performance even in some of the most rudimentary of control rooms. The problem of the acoustic design of consoles does warrant considerably more attention than it has customarily been given.

14.7 Monitor channel electronics

On completing control rooms, it has become somewhat customary for me to experience an extended period of listening to CDs and DATs which are well known to me, prior to the installation of the equipment in the room: except for the monitor system that is. It was by this means that I first became alerted, as did numerous other people, to the degradation of the room acoustics by the physical installation of the consoles. However, whereas without the console in the room, the CD or DAT machines were usually connected directly into the monitor system, via a simple volume control, the tendency was to route via the console stereo monitor returns once the console was installed.

In numerous instances, as this re-routing coincided with the console installation, it had not been immediately apparent that there was any more to it than a physical problem. In one instance, a power supply failure on a new console necessitated the reconnection of the CD player directly into the monitors, in order to demonstrate the room to some interested persons who had come to visit. Everybody involved with the studio had become accustomed to the sound over the previous few days, but there was suddenly universal agreement that the direct to monitor sound was an improvement. Once the power supply had been repaired, a series of tests were carried out listening direct from CD to monitor system, and then again via the console stereo returns. Via the console stereo returns, the stereo image closed in, some of the sonic transparency was lost, and the general fidelity was less than had been previously experienced.

Once alerted to this problem, we checked again on different makes of console, and it became apparent that many reputable consoles left much to be desired in terms of the sonic neutrality of their stereo monitor returns. In many instances, though, the difference could not be noticed when listening through small monitors such as NS10s, but on the higher resolution systems, the differences were in some instances alarming. Certainly we were faced with a lot of disappointed console owners.

In one instance, I decided to confront a manufacturer directly, and was surprised to receive a straight answer, without any defensiveness. It transpired that by using circuitry and components costing only £4 ($6) per channel, the problem could be solved, and this on a £50000 ($75000) console. I enquired as to why this improved circuitry was not standard, and was told that, firstly, in these cost conscious days, £4 is £4, and, secondly, that no customer had ever complained to them about the problem.

Surprisingly, even on some £250000 ($375000) consoles, this effect can still be noticed, which is worrying. In reality, the monitor chain begins in the monitor section of the console, so if things are less than optimum there, a limit is placed on the neutrality of the whole chain. Considering the fact that once pointed out, the differences are so clearly noticeable in good monitoring conditions, it seems surprising that we have not heard much more on this subject in the past; but I must emphasise that the degree of difference noticed is a function of the transparency and clarity of the monitoring conditions.

14.8 Irony

I was told not too long ago by a designer who had built three, more or less identical, rooms that acoustic and electronic sonic degradation do not always go hand in hand. Of the three rooms, one contained a Focusrite console, one a Neve, and the other an SSL. In a straw poll of recording engineers, although all three are excellent consoles, perhaps in terms of perceived sonic neutrality of their general electronic circuitry, they would be preferred in the order in which they are mentioned above. However, in the circumstances described, of the subjective overall neutrality when the three fully finished and equipped rooms were auditioned, the general preference of the rooms was in the reverse order. The SSL room, which might previously have been expected to run a close third, came out top on overall performance. The Focusrite, although renowned and revered for its electronic neutrality, was such an enormous physical presence in the room that the acoustic disturbance which it caused rendered the room 'Not worth measuring'. On balance of electronic and acoustic neutrality, the SSL actually beat the other two. To go to so much trouble to achieve an electronically pure signal path, then to build it into an acoustic monstrosity, seems total folly. An analogy would be to fine tune a racing car to be significantly faster than its competitors, then send it out on to the track with a dirty windscreen. Its true potential would be unlikely to be realised, as the driver would be unable to see clearly where he or she was going.

I am afraid that, at the moment, console manufacturers in general seem to be paying scant attention to the serious matters of the interface of their products with the rest of a working studio system. Perhaps more people should complain to them.

Chapter 15

Control rooms

15.1 Diversity of forms

Recording studio control rooms have become almost as disparate as the domestic listening environments in which the end products will ultimately be enjoyed. Many complaints are now being heard from the hi-fi fraternity that the variability in the spectral balance of modern recordings is beginning to become unacceptable. There is probably a wider range of monitor systems now in commercial use than at any time in the past, and furthermore the acoustical philosophies behind the acoustic design of the control rooms themselves are beginning to become polarised into what are effectively areas of mutual exclusivity. Almost inevitably there is a grey area in-between, and design concepts exist which aim to straddle the divide. However, gone now are the days when the established 'text book' concepts of control rooms led to one general trend in design formats. There are a very great number of 'organically developed' control rooms which have grown out of largely untreated rooms and an ever expanding inventory of equipment. On the other hand, there are the specifically designed control rooms which generally now fall into three generic categories: absorbent, scattering and diffusive.

The evolution of control rooms has seen progress from something tanta-mount to no more than a booth, to the present day situation whereby, in many circumstances, control rooms are the main location for the actual performance. As we have already discussed in Chapter 2, there are two very distinct recording processes, one involving the acoustical pick-up of an event performed in a different space, and the other the purely electronic or electrical synthesis of music, usually initially performed within the actual control room. Much static equipment such as tape recorders have in many instances now been relegated to a machine room; particularly to help reduce the disturbance created by their mechanical noise, but also to make space. The new demands on space are made by ever larger recording consoles, ever increasing numbers of effects, and by the very great quantity of space which is needed for effectively moving the entire musical ensemble into the control room. All too frequently, the purpose-built studio areas are no more than a convenient store for flight cases and superfluous equipment.

15.1.1 Changing trends

Whilst the objectives of control rooms for studios which are generally to be used to record classical/acoustic music have changed little, the objectives of control rooms for electronic recording are somewhat confused. The term 'control room' may now be something of a misnomer: 'general purpose room' may be more apt. The rooms are to be the performance space during the recording process, hence need to be large enough to accommodate all required personnel and machines. During mixdown, the rooms frequently still house enormous amounts of equipment, as in many cases instruments are no longer committed to tape but are MIDI controlled and time code triggered directly into the mix. During recording, the monitoring system may be used to create the required excitement for the musical performance, yet, when mixing, perhaps a more 'definitive' monitoring environment is required. One major problem in all of this is that the mountains of equipment imported for each session can be sufficient in size and variable in nature as to render control of the acoustic circumstances, at best, arbitrary.

I no longer believe that any one type of conventional control room can be optimally suited to all types of music. I suppose that it is true to say that no absolute agreement on control room priorities ever did exist, but now the requirements of the classical/acoustic and rock/electronic camps are polarising. A distinct gap now seems to exist between the two sets of requirements. The criteria for which side of the divide a recording process will lean are whether the music ever existed as one performance in real space, or whether the music was synthetically generated. In the former instance, real acoustic reflexions exist, the perception of which rely largely on inter-aural time and phase differences as well as amplitude differences, whilst in the latter case, positional information is almost entirely in the amplitude domain. Chapter 2 discussed at length the differing and often conflicting priorities for the realistic perception of each of these domains.

15.1.2 Inherent conflicts

Although the advent of time domain pan pots will soon create another aspect for consideration, a degree of mutual exclusivity will still apply to optimisation criteria for the best perception environments for the two aural mechanisms for detection of source direction. In a 1988 paper by Toole and Olive[1] they stated, '. . . in the making of recordings, the monitoring environment should have some important acoustical similarities to the intended listening environment. There is the further implication that the necessary similarities, in this respect at least, might be satisfactorily achieved electronically, in the form of synthesised reverberation.' They went on to say: 'An amount of equalisation appropriate for one listening condition may not be equally appreciated in another. It is one more source of variability in the trouble-prone record/reproduce cycle.' Toole and Olive were identifying problems in the chain from the musician to the record purchasing listeners, yet, in one way, the requirements to fulfil these objectives are almost contradictory.

In the first statement, they were referring in their paper to the detection of timbral subtleties. Dependent upon the signal being transient or steady in nature, reverberation will either assist or detract from the respective abilities to detect those timbral characteristics. In the second statement, the implication was that any equalisation which may be added in the mixing process will be dependent upon the combination of transient/steady state nature of the musical signal, and the degree of reflexion/reverberation in the mixing room. In other words, changing the acoustics of a control room by piling in variable and arbitrary amounts of keyboards and effects racks will change the day-to-day assessment of the appropriate amount of equalisation to be applied to a sound, but the effect cannot be quantified because it is in turn dependent upon the transient or steady nature of the musical signal. It is not possible to use any rule of thumb such as, 'when there is a lot of equipment around, I use a bit less equalisation than I think I ought to'. The variables are so circumstance dependent that no general rule can apply, and this is the cause of one control room compatibility problem.

15.1.3 Addressing the dilemmas

Obviously, in the early days of recording, all music was of acoustic origin. Control rooms developed in a way which gave the recording staff the 'clearest' picture of what was happening. Control room monitor systems were generally the 'best' loudspeaker/amplifier systems available, and hi-fi enthusiasts sought to follow suit. The rapid expansion of generally available hi-fi in the 1970s led to the growing use of 'domestic' references in the studios, and from that, the 'near-field' approach to monitoring was born, when it was realised by studio staff that close range monitoring was less affected by room interaction. By around 1980, the practical requirements of the working studio had held back the progress on control room fidelity to such a degree that for general use, domestic hi-fi was frequently 'higher-fi' than the control room monitor systems. In order to hear all of the relevant detail which could probably have been noticed in the home, whilst still being held back by the constraints of control room practical requirements, new approaches to control room design were developed. Unfortunately, the optimisation for certain criteria in these new rooms could only be achieved by some sacrifices being made to their suitability for general musical usage. Concurrent with this fork in the design road, totally new music began to grow around the new technology, taking into account the strengths and weaknesses of each genre, therefore effectively locking-in each design concept to types of music born out of those differing rooms. Self perpetuating cycles had begun.

15.1.4 Separate ways

By the beginning of the 1990s, three distinct approaches to control room design had become apparent. The extension of the main line development from the classical approach was the scattering type of room. In these rooms, a certain amount of 'life' was retained in the room, with distinct reflexions being maintained, but the energy in the most objectionable room modes was scattered or broken up into less obtrusive smaller modes, largely

by the use of geometry. Selective absorption was used to help to keep the
frequency/reverberation time parameters within generally accepted 'desirable' limits. By the mid 1980s, Tom Hidley in particular had begun building
rooms which were almost anechoic from the point of view of the monitoring.
These rooms were capable of supporting very powerful phantom images
from highly unstable, amplitude panned stereophonic sources. Diffusive
rooms were also beginning to emerge, frequently using patent diffusers in
an attempt to straddle the fence which was beginning to divide the other
approaches. Fence straddling can of course be either very comfortable or
excruciatingly painful, dependent somewhat on how one lands! An intermediate approach was the 'Live End, Dead End' concept which brought a
more diffused reverberation from the rear of the room, leaving clean the
'first pass' of the wave-front from the monitors.

It is probably worth looking at the fundamentally differing approaches in
slightly more detail to try to understand just what each was trying to
achieve, and how the designers were trying to achieve their aims. To avoid
appearing to criticise anybody else's approach, I shall first outline the
general approach which I used until 1990.

15.2 Reflective/scattering rooms

Having been brought up through the time-honoured, conventional recording processes, I suppose that my own concept of a control room was also
fairly conventional. Until recently, the classical requirements for a control
room were neutral, full-range monitoring; sufficient space to house the
mixing console, the tape machines, outboard equipment and recording
personnel; and also to provide a pleasant environment, conducive to the
mood of the work in progress. The operational variables were few, possibly
the number of people in the control room would vary, but it would vary
around a fairly consistent average. The control room was a constant; a
fixed, known reference – that was its very function. It was the point from
which to make all decisions, and was the reference to which all decisions on
sound quality were based. Within the constraints of any given building,
dimensions could be optimised for the purposes of recording and monitoring. Internal shapes and geometrical designs could also be chosen to give
the best internal acoustics for those same purposes. The shift from the
concept of an idealised monitoring environment to a multi-functional
recording and mixing room has obviously not changed the laws of physics
which govern the acoustic design. The change has, however, very markedly
shifted the compromise points in ways which now put many more variables
into the equations for 'good' control room design.

15.2.1 Modal distribution

For a long time, I adhered to the general concept of a control room being
an attempt to mimic a typical domestic environment, but without any of
the unduly predominant room modes found in most domestic rooms. The
usual way of scattering each of the room modes over a wider frequency
range is to angle as many surfaces as possible, in order to avoid the parallel
surfaces which reinforce the more dominant axial and tangential room

modes. Where the distance between two parallel surfaces gives rise to a mode whose wavelength is a function of that distance, the parallel surfaces will allow a mode to reinforce itself on each subsequent 'bounce' between those two surfaces. By angling all of the surfaces and avoiding parallels, each subsequent bounce encounters a different distance between the surfaces, hence the Q of the modes are reduced, broadening the energy spread towards a point of overlap into a more uniform overall room response. In general, it is the strongly reinforced modes between the parallel surfaces which become most noticeable as room colouration.

Parallel or angled, the total energy in the modes will be the same – it is just that in the parallel walled rooms, the energy is concentrated at certain frequencies, hence is usually more objectionable. So this concept involves maintaining the reverberation times of the contentiously non-existent 'average domestic listening room', whilst attempting to remove any 'untypical' predominating modes. I say 'untypical' here in the context of the fact that although domestic rooms do typically have predominating modes, they tend to differ in frequency from room to room, hence a predominant mode in the control room would be typical of only a very small proportion of domestic rooms. It therefore seems prudent to reduce any such modes to the general level of the other modes.

At low frequencies, the angled walls are usually insufficiently large or angled to have any effect, therefore absorbent trapping is used to deal, as far as reasonably possible, with the low frequency modes. Usually, such modes cannot easily be dealt with domestically, although three-piece suites do go some way towards the break-up of the lower modes. In situations where the reverberation times of a control room are undesirably long, mid and high frequency reverberation times are dealt with by means of surface materials. Control room reverberation times usually tend towards the low side of the usual domestic range, the interesting point here being that they have been tending towards a steady decline, particularly in rock and electronic music studios, as time has gone on.

15.2.2 Principal design aims

With the above philosophy, what we are generally trying to achieve is a control room in which one could reasonably represent the general acoustic of a domestic room. In other words, if blindfolded, one would be aware of being in neither a hall, a box room, nor an anechoic chamber. Such a concept still works well today for classical type recordings of acoustic ensembles, especially when the complement of personnel and equipment within the room is kept to the levels roughly similar to those in the rooms of the not too distant past. The compromises which have crept into such designs have been rooted in the need for more floor space, as ever increasing amounts of personnel and equipment have invaded the control rooms; and also a generally lower than previous optimum reverberation time.

The reason for the lower reverberation times is partially to compensate for the highly reflective nature of much of the aforementioned equipment, but also due to the fact that such equipment is normally associated with electronically generated music, panned in the mix solely via left/right, amplitude differentiating pan pots. As we have discussed in previous chapters, such amplitude panned images can be somewhat indistinct in reflective environments, so my approach was to keep lateral reflexions well separated in terms of time and amplitude from the direct, axial wavefront leaving the loudspeakers. This meant mounting the loudspeakers roughly one-third of the way along the length of the long wall of the room, and using a front wall which was quite absorbent in the mid/high frequency band. Compared to the direct wave, lateral reflexions would be required to travel a considerable distance to and from the side walls in order to reach the ears of the listeners. They would thus be lower in level and more separated in time. Any higher frequency reflexions from equipment within the room, which may return in the direction of the monitor wall, would be absorbed to avoid confusion and smearing of the direct wave, as would be the case if a hard monitor wall reflected them back into the room.

I still held to the principle that as spaciousness was a function of lateral reflexions, those reflexions *must* come from the sides. Reflexions from the forward direction served only to time smear and timbrally colour the 'true' sound. Vertical reflexions were relatively innocuous as we are so used to floor reflexions that evolution has led us to largely ignore them. (In evolutionary, survival advantage terms, it is somewhat more advantageous to be aware of the absence of a floor rather than the presence of one.) Reflexions and reverberation from the rear can also be generally beneficial in terms of spaciousness, so long as no 'chatter' develops from the rear reflexions returning to, and subsequently being reflected from, the front.

15.2.3 Inherent dilemmas

Within this design approach, a conundrum began to surface, based largely on expectations from the past clashing with the requirements of the present. Throughout the 1980s, an ever increasing proportion of recorded music was from electronically generated sources. The control rooms when empty were acoustically quite different from the same rooms when loaded with equipment, and the unpredictability of the equipment load made room optimisation difficult. Unduly deadening the room broke the link with the past practices, hence people frequently came into such deader than usual rooms and felt unsure. Low frequency reverberation times were largely unaffected by the influx of equipment, so such a room which acoustically compensated for the vanloads of equipment present in the recording process frequently showed a reverb time tilt towards the low frequency end of the spectrum when in a relatively empty, mixdown mode. In order to take down the low frequency reverb time, based on the questionable assumption that the subtleties of the mixdown acoustic were more important that the acoustics during the recording process, some rooms began to be more heavily 'trapped' at low frequencies, leading to a general reduction in *overall* reverberation times. Reverb times were ratcheting downwards, step by step.

Subjectively, such rooms still looked as though they ought to be more reverberant, and I well remember a period of a few years where the term 'over trapped' was applied to many rooms. Effectively, engineers and producers were not hearing as much low frequency reflected energy as they had come to expect, so the rooms were deemed subjectively to be bass light, with the resulting tendency towards bass-heavy mixes leaving those rooms.

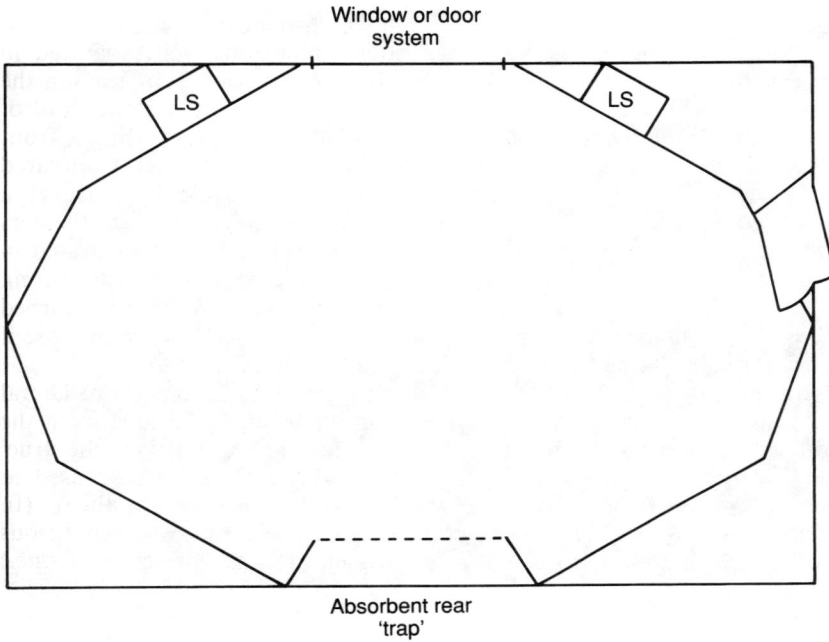

Figure 15.1 Geometrical plan of typical reflective/scattering room. Frequently used with considerable LF trapping

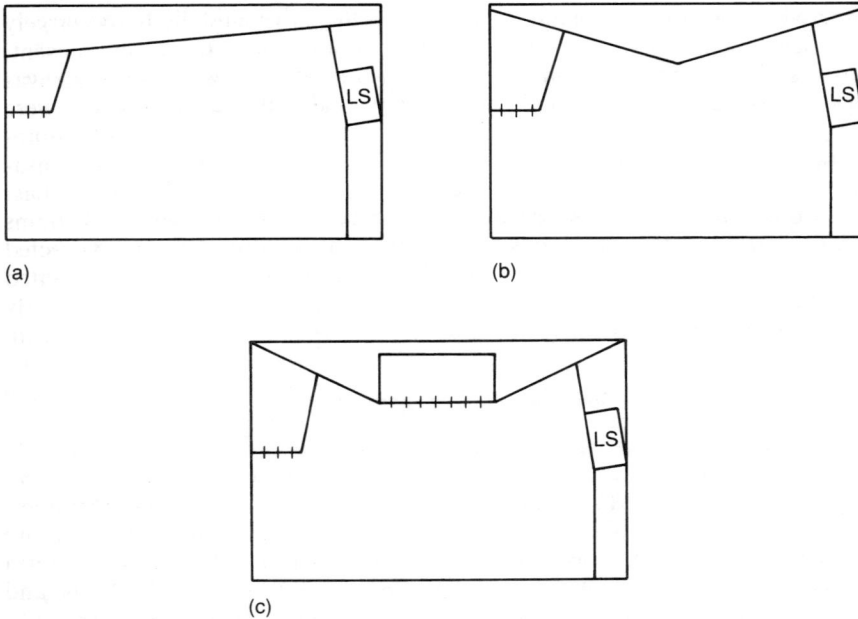

Figure 15.2 Side elevations showing three typical ceiling geometries of rooms as shown in Fig. 15.1. Typical trap entrances shown

Figure 15.3 Typical framing of a control room

By the middle of the 1980s, partly as a function of past experiences, and partly in response to the changing expectations in terms of what was required of a control room, some designers began to use loudspeaker directivity control to attempt to split the subjective mid/high frequency reverberation times with respect to either the monitors, or to the general conversation within the room. Figures 15.1 and 15.2 show a typical geometry of such a scattering room, with the typical framing of such a room being shown in Fig. 15.3. Experience had dictated that, certainly for most of the people for whom I was building studios, a relatively 'domestic' acoustic was deemed most suitable for a control room. The changing perspective, however, was based upon the ever increasing percentage of recorded music being from positionally unstable synthesised sources, and also to deliver more subjective punch. Both of the latter effects were becoming highly fashionable and were actually developing around the strengths and weaknesses of contemporary monitoring acoustics.

15.2.4 Alternative variations on a similar theme

Other people were using different approaches to address the changing requirements of electronic music. For example, some designers used the 'Live End, Dead End' (LEDE) concept whereby the front half of the room was very dead, with all the life of the room being to the rear. Such a concept allowed a very clean first pass of the wavefront from the monitor system to the ears, creating a 'reflexion-free zone' around the mixing position, with the life being added from the rear of the room. Various approaches were all largely aimed at better support for the stereo phantom images, and the perception of less 'coloured' transients, whilst allowing an overall reverb time in the control room which met conventional expectations. These approaches were not so much based on 'first principles' but were developments which sought to provide an improved acoustic for the new music, whilst maintaining many of the attributes of the older generation of rooms. Indeed, the LEDE principle has been used successfully in the design of hundreds of control rooms around the world.

Recording studio owners had been loath to specialise, as obviously, from a business point of view, few studio owners would wish to tell a potential client that the room was not intended for that client's particular type of music. The search for the ultimate control room which was all things to all people was still very much in full swing; the reality, however, was becoming quite a different matter. As the general drift of room design tended to larger and drier monitoring environments, a large proportion of the classical/acoustic fraternity were staying with the older style of reflective/scattering rooms and wide directivity monitoring. They were not just ultra-conservative die-hards, they were far too experienced and skilled for that. Something was evidently amiss from their point of view as far as many of the more modern rooms were concerned.

15.3 Diffusive rooms

A further development in the search for a better compromise was the introduction of diffusers into control rooms. Much of the spaciousness produced by lateral reflexions, so well liked by the classical people, is a function of distinct specular reflexions in the reverberance of a room. Too much reflexion all but destroys the stereo imaging of amplitude panned electronic music, but too little gives what some people would deem to be an unnaturally low reverb time, with consequent implications for erroneous choice of timbral assessment when equalising or mixing any given signal. Diffusers break up the reflexions in such a way that no distinct specular reflexions can be observed. By such means, a small number of distinct, lateral, specular reflexions can be selectively introduced by the addition of reflective surfaces into a relatively dead room. The reverb time can then be increased by the installation of diffusers, which add to the overall reverberant energy but without introducing any specular reflexions which could compromise the stereo imaging.

The commercial diffusers, such as those marketed as RPG (reflection phase grating) use series of slots, or pits, of differing length, based on a mathematical prime number, quadratic residue sequence, to break a wavefront into an exceedingly large, uncorrelated reflexion of energy, hence removing any individual, specular reflexions from the reflected energy. Much capital has been made of this technique by Dr Peter D'Antonio, at RPG Diffusor Systems Inc., in the USA. The mathematical principles were based on the work of Manfred Schroeder, one of the world's great theoretical acousticians.

Thought of in different terms, if a sugar lump were to be placed in the bottom of a glass, representing an unwanted room resonance, there are effectively three ways of restoring the bottom surface of the glass to a relatively level condition. The 'absorbent' approach would be to remove the sugar lump. The scattering approach would be to smash the sugar lump

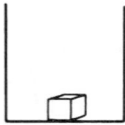

Sugar lump on bottom of glass
representing an unwanted room resonance –
an irregularity on the flat surface

Absorbent room removed the lump
completely

Scattering room breaks the lump into
individual crystals, converting the
single large irregularity into very many
minor ones. The base level is raised by
the total volume of the sugar lump (the
equivalent of the resonant energy)

Diffusive room pulverises the sugar into
minute homogeneous particles, also
raising the base level but producing a
uniform surface with no significant
grains (specular reflexions)

Figure 15.4 Absorption, scattering, and diffusion

into its consistent crystalline granules. The diffusive approach would be to pulverise the sugar into a smooth, fine powder from which no grains could be detected. The analogy goes further: the surface of the bottom of the glass would be raised in the two latter cases (Fig. 15.4). Similarly, the level of the overall sound would be higher in the scattering and diffusive rooms as compared to the absorbent room: effectively, the rooms would be subjectively louder.

Most rooms using the diffusive technique are constructed using patented diffusers, which phase-scatter the incident wave and re-radiate the energy as a reverberation rather than with echoes. They effectively have a reflexion-free zone for some milliseconds after the direct wave has passed, then the diffusive reverberation is returned to the listening area, but free from individual reflexions. The technique is relatively new, and I am still undecided in some of my own opinions. I appreciate that it provides many 'technical' solutions to the problems, but, psycho-acoustically, it may turn out to be neither as 'absolute' as the absorbent rooms nor as 'representative' as the scattering rooms. Those who like them will continue to use them; we cannot dictate what is 'right' in such a subjective area. I do, however, see much promise for the acoustical variability of such diffusers in the studios themselves, where the recording acoustics can be pleasingly enhanced. One of my main reservations is that most people do not as yet have diffusers in their conventional home environments. It is true that few people live in anechoic rooms either, but the anechoic room is essentially monitoring what is on tape, without attempting to compensate for the average room. The close-field systems in the local environment of the mixing console will be used as a more 'domestic' reference.

15.3.1 Brushwood room

I have often wondered just what the effect would be of building a diffusive control room having a solid floor and a sturdy front wall to house the monitors, the other walls and ceiling being of open frame construction covered in chicken wire and thin fabric. The frames would be built leaving a six foot gap to the solid walls and ceiling. The gaps would then be entirely filled with logs and brushwood, randomly scattered and well mixed up, with the diameter of the branches varying from six inches to a quarter of an inch or less, and lengths from inches up to several feet. I am sure that the fire hazard could be overcome by an adequate supply of automatic inert gas discharge devices; alternatively, synthetic, fireproof 'logs' and 'branches' could be used. It could be an interesting approach. Panels of appropriate size and texture could be added to the side walls for the introduction of any desired lateral specular reflexions. I am sure that some people would like it very much, whilst others would possibly find it less desirable than other approaches. I have in fact heard from the East that depths of broken glass have been used acoustically for diffusion.

The problem with all compromises is that different people choose a different order of priorities. Furthermore, many of the subtleties of technically addressing a problem can lead to remarkably similar measured solutions: but all too often, despite the objective similarity, the subjective differences are far from subtle. As yet, we have no means of predicting many of the subjective subtleties of objective change. Some of these concepts will be discussed further in the following chapter.

15.4 Absorbent rooms

While the arguments continue, the conclusion as to precisely what constitutes or represents the average domestic listening room seems to be no nearer. Indeed, it is highly contentious as to whether that average would have any practical meaning even if it *were* realised. If even 10% of the population lived in 'average' circumstances, then that would still leave a large majority of the music buying public totally unrepresented by the 'average' reference rooms. There is still a strong body of opinion that believes studio control rooms and monitor systems should accurately portray what is on the tape, and should not attempt to confuse the issue by attempting to mimic the wide-ranging ambiguities of domestic listening conditions.

After four or five years out of the business, around 1984, Tom Hidley, the founder of Westlake Audio and Eastlake Audio, returned to studio building, announcing his totally new approach to control room design. He had, amongst other things, taken his ideas on bass-trapping to an extreme point of all-trapping, and he exercised ruthless control over the room acoustics and monitoring. There was a limit to how far he would bend to a client's whims as the new approach was to be no compromise to anything which would disturb the stereo localisation, or uniformity of response.

Hidley has promoted his 1980s control room concept as the 'Non-Environment Environment'. Certainly, under these conditions, fewer overall performance compromises need to be made. The concept seeks to emulate, indoors, the acoustics of monitoring outdoors, but without the obvious drawbacks of wind, rain, cold and extraneous noises. At first sight, large anechoic chambers would seem the closest practical approximation, but even here problems arise. To achieve reverb times of 0.1 sec or below at 30 Hz, a room of say 30 metres cubed with absorbent wedges of 3 metres length would be a reasonable starting point. Reasonable in theory, that is; in practice it is obviously out of the question. There is, however, another problem. When we are outdoors, environmental noise and visual cues render the relatively anechoic environment to be natural, but the low ambient noise level and visual constrictions of an anechoic chamber can be most disconcerting to many individuals. It can feel both unnatural and disturbing in a way that such an environment would not be conducive to the mood for recording or mixing.

15.4.1 Differential RT

The 'monitor dead' 'non-environment' approach is one solution to the problem, allowing for reflexions from the floor and front wall of the room, with the remaining surfaces being highly absorbent down to low frequencies. Very rarely, even outdoors, do we walk on anechoic surfaces, so a floor reflexion can be perceived to be very natural. Reflexions for the natural acoustic presented to any people speaking in these control rooms are further enhanced by the hard surfaces of the equipment in the rooms, together with glass or other hard surfaces on the front wall only. Looking from the monitor loudspeakers, however, we can see no reflective paths to the listening position. The mixing console blocks to a large degree any direct paths from the floor reflexion to the listener's ear. The rear of the mixing console itself, or any other pieces of hard equipment, will be treated with at least six inches of material such as 'Sonex' absorbent, in order to prevent any reflective path back to the acoustically reflective front wall. In effect, the monitors face into an anechoic termination, whilst the personnel in the room hear an acceptable amount of 'life' from their own speech and actions. The rooms thus sound quite natural to be in, but are effectively anechoic to monitor in.

Rooms of this nature therefore have reverberation times which are entirely dependent upon the position of the source of the sound. I am using the term 'reverberation time' somewhat loosely here: possibly 'echo time' would be more apt, as reverberation never exists in these rooms. No single reverberation time figure can in any way describe the performance of such rooms. Many people find these rooms very easy to work in, and also find that the mixes which they produce travel well. In other words, they find that the mixes show few surprises when played in a wide range of domestic listening conditions. For those people, *no* average seems to be the best average: compromise to one and you immediately compromise to all. I do not claim that these rooms are a universal panacea, as in anything so

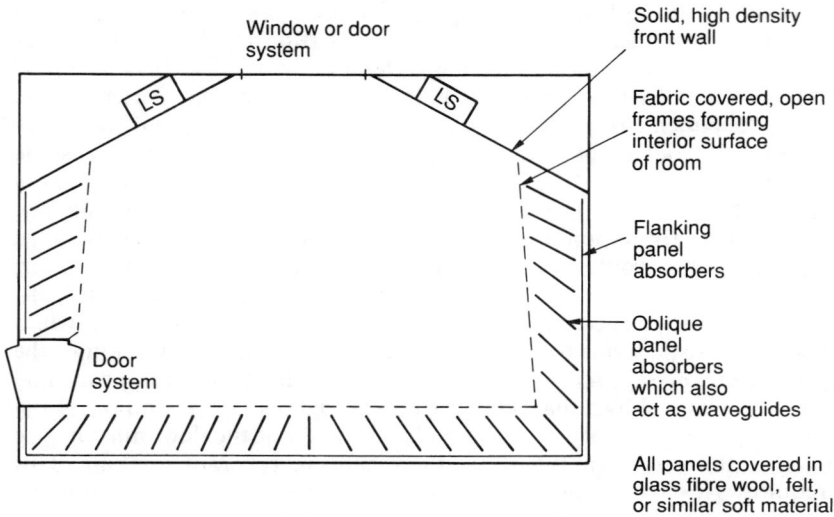

Window or door system

Solid, high density front wall

LS

LS

Fabric covered, open frames forming interior surface of room

Flanking panel absorbers

Oblique panel absorbers which also act as waveguides

Door system

All panels covered in glass fibre wool, felt, or similar soft material

Figure 15.5 Plan of typical absorbent room

Fabric covered open frame ceiling

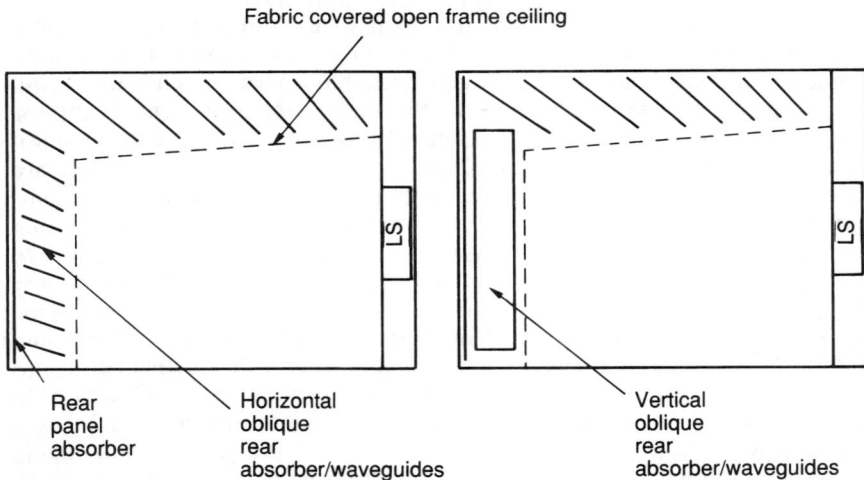

LS

LS

Rear panel absorber

Horizontal oblique rear absorber/waveguides

Vertical oblique rear absorber/waveguides

Figure 15.6 Side elevations showing ceiling absorbers and two different typical rear absorber arrangements

subjective as studio acoustics that is surely not possible, but this philosophy does seem to have a growing army of followers.

15.4.2 Size requirements

One drawback to the wider use of the absorbent type of design was that they did tend to require a rather large shell, in comparison to the final working area within the finished room. A typical construction is shown in Figs 15.5 and 15.6. Except for the front wall and the floor, all surfaces are covered by around three feet depth of angled panel absorbers, behind which are continuously joined hanging panels running the full length of each wall, almost from floor to ceiling. These flanking absorbers are quite effective down to frequencies whose half wavelengths are equivalent to the length of the panel. From this it is obvious that maintaining control to very low frequencies inevitably involves rather large rooms. There are also some audiological perception problems which seem to dictate half wavelength paths which equate to the length of the room, so for a subjectively, acceptably smooth response, one would be looking for a room with an absolute minimum front to back dimension of 14 ft at 40 Hz, 18 ft at 30 Hz, 28 ft at 20 Hz, and so forth. Practical rooms tend to be around 16 ft, 24 ft, and 32 ft for those respective frequencies. In the latter instance, that would translate in realistic terms to an empty shell, prior to the construction of the control room, of, say, 45 ft by 35 ft, by 15 ft high. A room approaching such dimensions is shown diagrammatically in Figures 15.7(a) and (b).

Another prerequisite of this room philosophy is that the floor must not be resonant in the audio bandwidth, as there is no reverberant masking. This can require as a starting point a concrete base, floated on rubber, pneumatic, coil spring/hydraulic, or some other form of suspension, with a total system resonance of 10 Hz or less. The final solution could be a concrete slab of 8 to 12 inches in thickness for conventional audio frequency rooms, and more in the case of the infrasonic rooms. Clearly such an approach to room design would not be suited to a construction on the fourth floor of a timber framed building! Ideally, the largest possible space around the room perimeter is also desirable. Despite the elaborate, absorbent 'trapping' systems, the low frequencies still penetrate, bounce off the structural walls, penetrate the trapping once again, and return to the room environment, albeit severely attenuated. An attenuation through the trap of 24 dB or more can be achieved down to the half wave-path of the trap design, with a minimum of about 22 dB being required to prevent any perceivable pressure change being audibly detectable in the response curve. Even shorter low frequency reverberation time can be achieved if the closest perimeter wall is itself diaphragmatic. A timber framed structure clad in foam rubber and acoustic deadsheets would be a worthwhile addition, with a further surrounding air space before the structural walls are encountered. Once again, however, both cost and shell size are increased – not to mention weight!

Recently, I have been able to shrink Hidley's approach, by using highly damped membrane absorbers. These use felt covered 'kinetic barrier' mats, stretched over wooden stud partitions, the other side of the cavity being covered with conventional plasterboard/deadsheet/plasterboard sandwiches. These materials, such as Noisetec PKB2 and LA10, have been successfully allowing the application of the overall non-environment philosophy to rooms of conventional size and even less. Indeed, some very respectably neutral control rooms have now been built in shells as small as 12 ft × 10 ft × 8 ft high.

15.4.3 Monitor considerations

From the point of view of monitoring system design, the 'non-environment' rooms are something of a mixed blessing. Choice of actual drivers can be

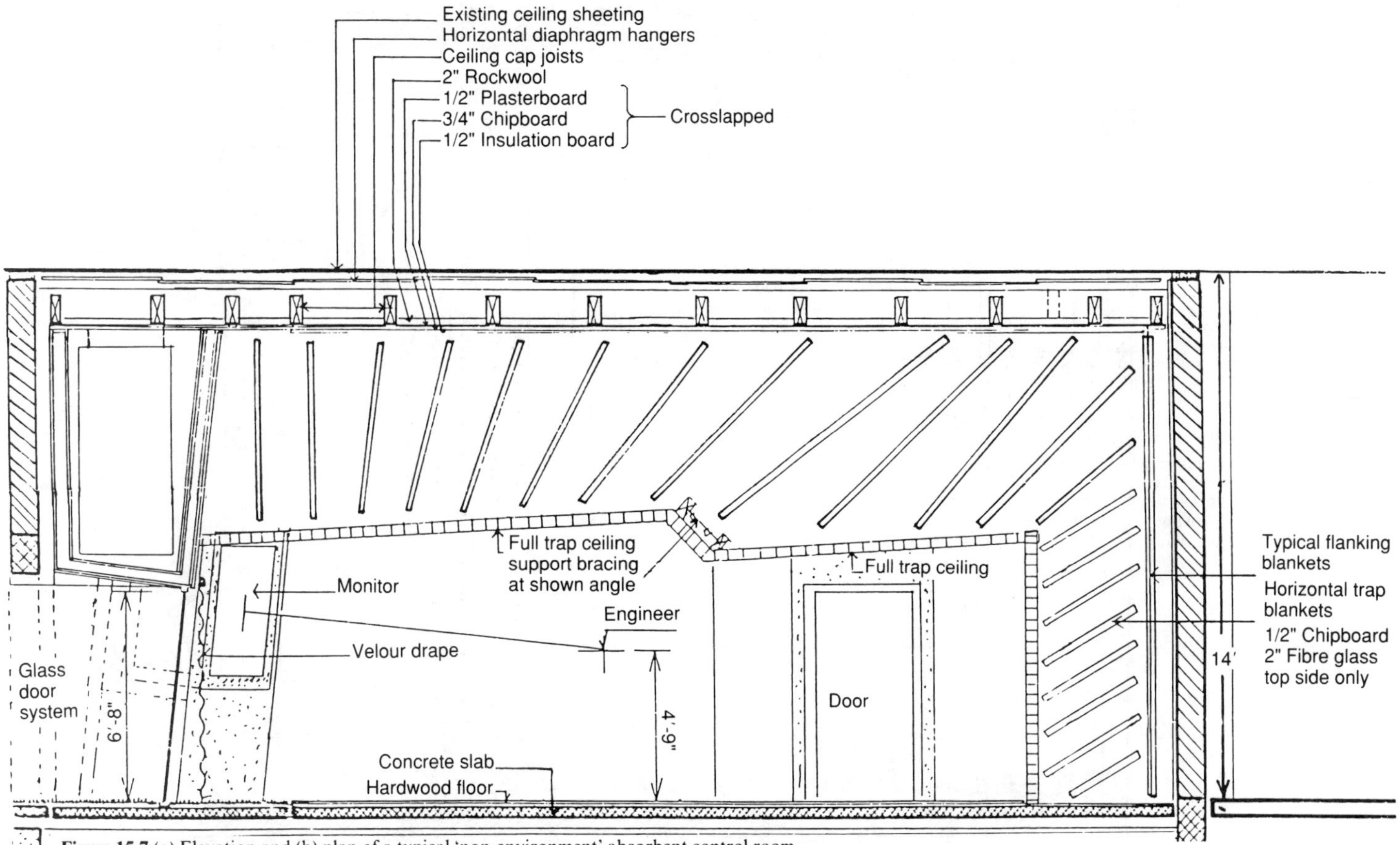

Existing ceiling sheeting
Horizontal diaphragm hangers
Ceiling cap joists
2" Rockwool
1/2" Plasterboard
3/4" Chipboard
1/2" Insulation board
Crosslapped
Full trap ceiling support bracing at shown angle
Full trap ceiling
Monitor
Engineer
Velour drape
Glass door system
6'-8"
4'-9"
Door
Concrete slab
Hardwood floor
Typical flanking blankets
Horizontal trap blankets
1/2" Chipboard
2" Fibre glass top side only
14'

Figure 15.7 (a) Elevation and (b) plan of a typical 'non-environment' absorbent control room (courtesy of Tom Hidley)

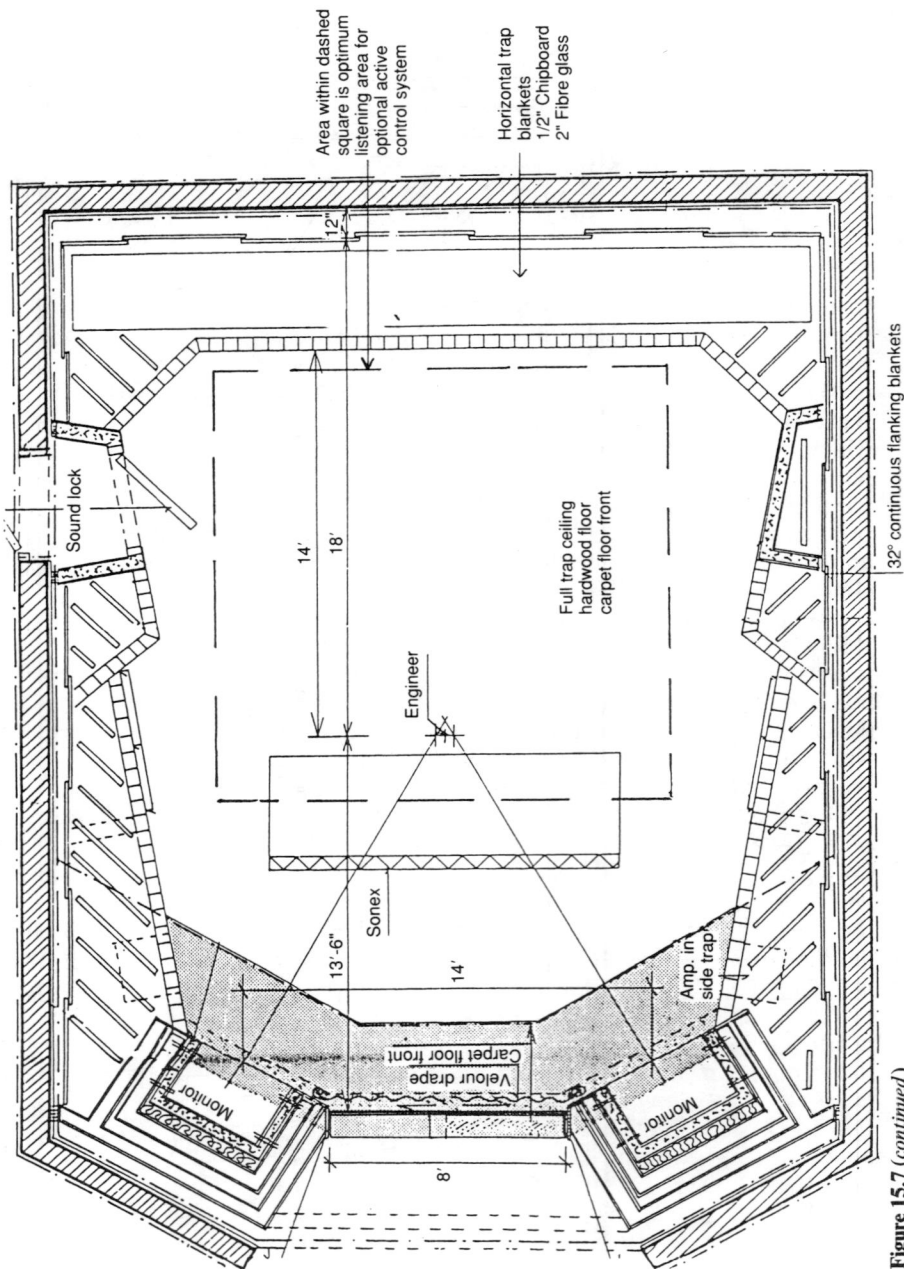

Figure 15.7 (*continued*)

Area within dashed square is optimum listening area for optional active control system

Horizontal trap blankets
1/2" Chipboard
2" Fibre glass

32° continuous flanking blankets

Sound lock

12"

14'

18'

Full trap ceiling
hardwood floor
carpet floor front

Engineer

Sonex

13'-6"

14'

Velour drape
Carpet floor front

Amp. in side trap

Monitor

Monitor

8'

made more readily on the basis of sonic characteristics, without as much attention having to be paid to the directivity indices. If listening is carried out virtually on axis, what happens 60° off axis is of little consequence as it will not be reflected back into the general listening area of the room. Axial impulse response assumes a much higher priority over total power response than would be the case in a more reverberant room. I personally believe that this will ultimately lead to more definitive, repeatable, and accurate monitoring. A monitor system designed for such purposes is described in Chapter 19.

15.4.4 Subjective perception

In terms of perception, the 'non-environment' rooms do have differences from conventional 'scatter the modes' rooms. Phase accuracy within a signal appears to be much more perceptible, as the characteristics are not masked by the random phase of the reflexions from the reverberant field. Back in the 1950s, Manfred Schroeder produced a 'phase organ' upon which he could play tunes. The 'organ' consisted of a pulse train having 31 harmonics from 100 Hz to 3 kHz, all with zero relative phase angle. By varying the phase, and phase alone, of certain harmonics, notes were produced which were clearly audible above the 100 Hz pulse train buzz: the amplitudes of the harmonics were in no way varied. The 'organ' was very effective on headphones or in anechoic rooms; however, in reverberant conditions, no tones were audible, just a more reverberant buzz. Schroeder's 'Models of Hearing' from the Proceedings of the IEEE 1974 makes very interesting reading (see also Fig. 6.2). Digital audio recording preserves much more phase coherence than analogue recording, so, relatively suddenly, phase/impulse accuracy is needing to be addressed in a much deeper way than was previously the case; hence the sudden upsurge in interest in the more phase preserving 'monitor dead' 'non-environment' rooms.

One point which had been raised on numerous occasions was the problem of subjective levels of artificial reverberation committed to tape in the different kinds of rooms. 'Surely,' people said, 'in an absorbent room you will put more reverberation on to the mix.' Well, that does not appear to be the case. In the 'dead' room, the ear seems to be more sensitive to the amount of added reverberation, such that a less reverberant overall sound is accepted as the norm when mixing. In a more reverberant control room, reverberation is added to the mix to such a degree as is necessary to make it noticeable above the naturally reverberant room. It appears that it is the differential which we perceive, as we soon adjust to the environmental levels. If there is any difference at all, the tendency is to use less reverberation in the monitor dead room, as the freedom from room RT overhang allows one to hear more detail in any artificial reverberation which may be added. The reverberation which may be added artificially is in any case usually much longer than that of any respectable control room, so it is distinguishable from the ambient reverberation of the room.

15.5 Jacks of all trades, or masters of one?

Obviously, in real life, control rooms do not just fall neatly into the black, white and mid-grey of the three approaches described – there are rooms of all shades in between. The principles of scattering the modes, diffusing the reflexions, or absorbing the reflexions, are, however, the three constituent building blocks from which all other rooms are derived.

To recapitulate, many classical and acoustic recordists rely heavily on the spaciousness of lateral reflexions to present a realistically natural reproduction of recorded music containing a considerable proportion of inter-aural time/phase cues. Pinpoint positional accuracy is not in the nature of the original performance, so the spaciousness for definitive positioning trade-off seems wholly justifiable. The order of the day would seem to be spacious control rooms with a somewhat live acoustic, especially from the sides, and relatively constant directivity, wide dispersion-angle monitors. As we are seeking to attempt to reproduce an actual performance of an event, there is no reason to use monitoring levels above those experienced at the real event. The SPL requirements from the monitors are further eased by the 'help' from the reverberant loudness of the room. Plenty of high top end also seems to be desirable for classical work, possibly because much of the sense of space comes from the higher harmonics. Whilst timbral neutrality is a fundamental goal of *all* monitor systems, it is particularly important in these circumstances, as firstly, one always has a real life comparison, and secondly, it is more easily achieved in systems of moderate SPL requirements.

Electronic music has no 'original' for comparison, it is heard first through the monitors of mixdown. Stereo imagery and pinpoint positional accuracy are fundamentals of the art form. Highly transient sounds, often of extremely 'unnatural' frequency range, can make punishing demands on monitor systems, especially when played against a realistic backdrop of a drum kit at a natural level. Undoubtedly, such music is best supported by an absorbent approach to control room monitoring. The deadness of the acoustic produces no confusing reflexions, but, on the other hand, neither does it help the loudness. Given the lack of help from the room, together with the higher required SPL in the room for much electronic/rock music, the one metre axial output from the monitor systems may be required to produce 10 to 15 dB more than monitor systems for the reflective or diffusive rooms.

There are powerful and different philosophical arguments applying to monitor system bass response. One can either maintain it as flat as possible for as far as possible then let it drop abruptly, or one can let it roll off gently, beginning at a higher frequency than the previous case, but possibly extending lower in frequency for the 10 dB down point. Subjectively, the difference is considerable: if one looks at the Fletcher–Munson curves for equal loudness, although at 1 kHz 10 dB is considered to subjectively double loudness, at frequencies below 100 Hz a mere 4 dB can double loudness. The differing low frequency philosophies for the monitors can obviously have drastic repercussions on the subjective bass balance and tonal character when mixing. The Fletcher–Munson curves are shown in Fig. 3.1.

15.5.1 Proposal for best compromise

With current technology, specialisation of control room monitoring and acoustics, with regard to the type of music most likely to be recorded in that control room, would seem to still be with us if the best results are to be hoped for. The subjective requirements for the different types of music, and indeed for the hearing mechanisms associated with those different types of music, more or less dictate that such should be so. If one were to subjectively rate monitor system/control room combinations on a scale of A to F, then an A/A for classical/electronic is all but impossible to achieve with current technology. The way that things are going, an A/C or C/A is probably a better compromise, dependent upon what constitutes the majority of any particular studio's clientele, as opposed to a B/B compromise. There are studios such as broadcast or jingle studios where a B/B compromise may be a necessity – as is the case with mobile recording trucks which perform a wide range of duties – but as mobiles begin from a disadvantaged set of dimensions, a C/C may even be a laudable achievement.

The only way which I can currently see any hope of an A/A would be to use an absorbent room, as for rock/electronic music, but with delayed and frequency contoured small monitors placed in the side walls, enabling the introduction of lateral reflexions for classical monitoring. The high frequency roll-off and precise delay would take some experienced assessment on initial set-up, but would, once set, hopefully be left at a relatively constant setting. Remember, those lateral reflexions *must* come from the sides, they cannot be incorporated into the frontal signal if a natural perception is desired. Following on from this, though, that would suggest that only surround sound systems in the home, probably with multi-channel recordings, will be the way of the future. Many of the things required in a classical-music control room will currently not be found in the record buyers' homes.

Further to this, there are other requirements for the classical/electronic specifications. Most classical recording personnel seem to require more high top on their monitors. The rock/electronic personnel often use systems which are around 1 dB down at 8 to 10 kHz and, say, 3 dB down at 15 kHz. Whilst one would normally expect a reduced top end response on the monitors to produce a top-heavy tape, a system at high level with a flat top end response can produce listening fatigue such that a disproportionate amount of top is subjectively lost by the ears. The subsequent compensation can produce a top-heavy tape from the monitors with *most* top. The 1 down at 10, 3 down at 15 curve seems to have been adopted by much of the industry as an unwritten standard. This once again relates also to the Fletcher–Munson curves (see also Section 13.6.2).

15.5.2 Dual monitors

Certainly, on grounds of timbral neutrality, different monitor systems tend to be preferred by the different camps. Here, however, it may be possible to mount two monitor systems side by side, blanking off the one not in use. If the rock/electronic monitors were positioned furthest apart, the classical/acoustic monitors would not unduly suffer by being slightly closer together, as the side-wall mounted, lateral reflexion monitors would add a sense of spaciousness which may be compromised by the closer spacing of the main classical/acoustic monitors. These things do not sit happily with me, but this situation does seem to reflect the current state of the art.

Chapter 16

Control room specifications – a theoretical nightmare

I remember reading a specification for a new television studio in a European capital city. The specification quoted design reverberation times in octave bands, which were to be shown mathematically when the plans were submitted with the tenders. It stated that the reverberation times were to be calculated according to the Sabine Formula, yet, in the very same sentence, went on to say that they were aware that the Sabine Formula did not necessarily have direct relevance to the actual achieved results. I was told by many people in the television industry in that country that sound was not of great importance; the emphasis was on visual effects. On the other hand, I do know of a few television companies who have gone to very great lengths to provide the finest sound quality. Overall, though, within the international industry, sound is still not very high in the order of priorities of many multi-media companies.

16.1 Underlying necessities for specifications

There are two very obvious answers to why this state of affairs exists, and a third which is less obvious, but casts its shadow very heavily over the whole concept of acoustic specifications in general. The first reason is that the overwhelming majority of people watch television on sets with very poor audio reproduction systems, and whilst it is true that a growing number of people are buying 'hi-fi' video systems, the overall percentage is still very small. The second reason is that the human visual sense is so dominant that when hearing and seeing at the same time, much more sensory emphasis is on the visual material. This reason has for a long time held back the pressure for television sound to improve. On the other hand, when most people are presented with a first class sound to accompany the vision, they usually consider the experience worthwhile. Commercially, however, there are still too few of them who would then go out and buy a 'hi-fi' television system, to make the financiers of the TV industry consider any significant progress a matter of urgency. It is this which leads to the third point: who is in control?

There are the 'visionaries' (no pun intended) of the industry who look to the future of more 'hi-fi' television, and commit now to the installation of sound studios of the first order. They will indeed be considered visionaries if they have at their disposal the means to make the finest quality sound when the expected demand materialises; as their companies will be the first to be able to take advantage of the new boom. Acousticians, technicians and musicians may also consider them visionaries come what may, but if the boom does not materialise, then their accountants, chair-persons, bank managers and lawyers may not be so enthusiastic. The third important point is just that – the real power in the television world often lies well outside the jurisdiction of professional audio people.

Television studios cost an enormous amount of money to set up, and when such projects are undertaken, no one person usually has the cash in their pockets. Almost without exception there is involvement from banks, institutions, shareholders or even governments who are looking at the venture in mainly financial terms. Large amounts of money usually come with many legal 'strings' and require business plans to be submitted in such a way that everybody involved has the 'protection' of an agreed set of figures and specifications. The intention is that there should be no arbitrary nature in the assessment of the finished project if it can be shown to meet the paper specifications. Under such circumstances, the provision of special-ist sound control room acoustics which cannot be specified in subjective detail, or given a repeatable 'quality rating', tend not to figure in many projects. That is unless a very influential person in the power structure has the experience and credibility to press for higher priority to be given to sound; and also to ask the financiers to back a 'knowledge' of what will be 'special' without provable subjective specifications.

In the specialist sound recording industry or the music industry in general, it is also true that large sums of money may be involved, but their history has been more a story of the development of previous successes rather than written specifications. This is partly as a function of evolution, partly because many stupendously successful recordings have been made in low budget studios, and also partly because the hierarchy of the music recording industry is more 'into' the artistry of the music. For many of their contemporaries in radio and television, football matches, drama, documentaries and political events are responsible for the majority of their work. For these latter events, current television sound is usually deemed 'good enough'. Even I must admit that whilst improvements in the sound could rarely be a bad thing, I have often become very excited about the nail-biting finish to a cricket match, which may have been broadcast from Australia via telephone lines with appalling sound quality. If it had sounded as natural as a man sat speaking in front of me, the commentary may even have lost some of its sense of urgency and distance. I also wonder if Neil Armstrong's moon landing would have been quite so dramatic without the noises, bleeps and the strangled sound from the microphone in his helmet.

Of course the sound personnel in reputable broadcast companies strive to achieve the best results, but it is so much more 'structured' in general than is the music business, where companies will gamble more readily with designs of a more innovative nature. The divide between broadcast and record companies as discussed here is really one of priorities. The music people have to gamble more because they face greater competition in terms of their original, creative music output, whereas the broadcasters largely deal with music which has already been 'created' and is merely (or not so merely) being reinterpreted.

16.2 Subjective/objective correlation

What I am trying to highlight by all of this is that a strange inversion exists, whereby in most instances, the greater importance that the sound assumes, the less possible it becomes to specify acoustic parameters in meaningful terms. The further one progresses down the road to subjective acoustics, the further one gets from hard, provable facts. In the first sentence of this discussion, I related to a design proposal which had to be provable, so even in an unprovable area, the consultants had to find something close which could be used as a provable guide. They opted for the Sabine Formula, which requires a diffuse, reverberant sound field, such as can be approximated to in some large halls or reverberation chambers. When one introduces absorbers, however, the genuinely diffusive nature of the sound field is lost, and the Sabine Formula becomes at best approximate, and at worst, very misleading. Indeed, a 100% diffuse sound field could not exist, as it implies a totally random energy flow with a net energy flow of zero. In reality, there is always an energy flow away from the source of the sound. Nevertheless, in a highly reverberant space, a good approximation is achieved via the Sabine Formula.

In small acoustic spaces, however, the Sabine formulations begin to fail because of the inevitably higher levels of clearly definable reflexions or echoes. Also, anything introduced into the room, such as a carpet or a human being, will have a much greater percentage influence on the acoustics of a small space than on that of a larger one. For example, let us consider the reverberation time for a live room of a studio, primarily designed for recording drums. Given the size of a typical live room of, say, two to three thousand cubic feet, whatever specifications the bare room may meet, the introduction of people, instruments, rugs to prevent the drums slipping, holdalls full of drum sticks and dusters, and many other small items, will change the sound in an unpredictable way; sometimes greatly changing the reverberation time. The reverberation will also be less diffuse, because the items introduced will cause local areas of absorption. *True* reverberation is independent of room position.

Furthermore, five live rooms built to realistically achievable identical reverberation time characteristics, especially in terms of RT60 alone, if built one of wood, one of smooth stone, one of concrete, one of rough stone, and one of plaster will all have radically different timbral characteristics due to the materials of construction, yet all can have similar performance in terms of conventionally written specifications. In such instances, a room may meet *absolutely* a written specification, yet be deemed to be entirely sonically disappointing when compared to a room of identical conventional written specifications but built of different materials.

Many designers use controlled specular reflexions, or multiple echoes, to achieve or at least to bolster the reverberation. The principle is similar to the way that old tape echo machines used to synthesise a sort of reverberation by processes of multiple repeats from a number of replay heads and appropriate feedback loops. When this synthetic reverberation is mixed with the true reverberation, it is not easy to find any readily accessible or easily understandable process of enumeration which could adequately specify the perceived sonic performance in such a way that, from the

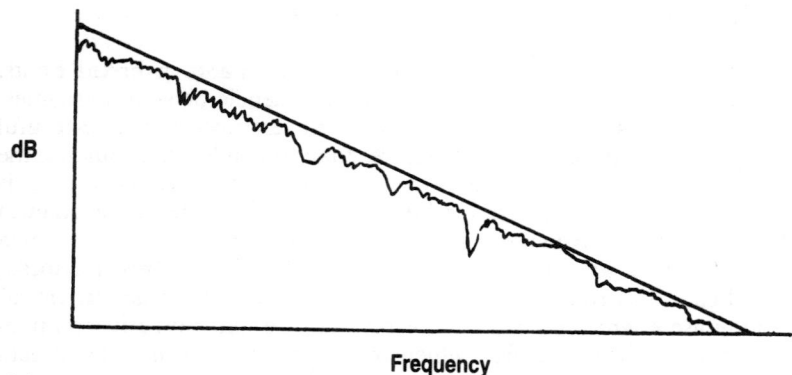

(a)

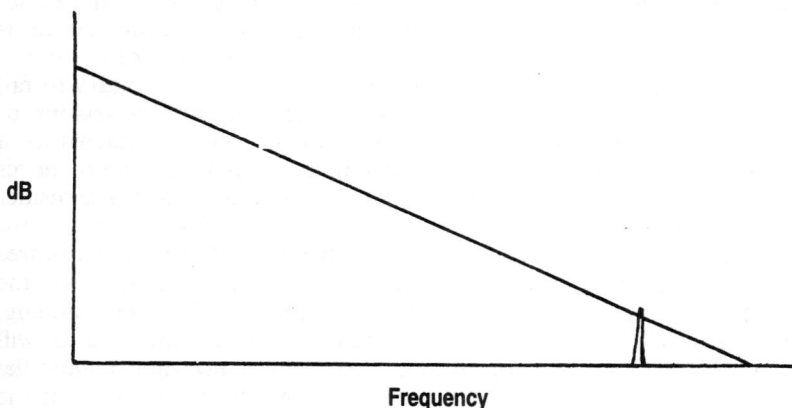

(b)

Figure 16.1 The above plots represent sections of a noise-criteria boundary by which the noise represented in (a) would be numerically acceptable but the noise in (b) would not. Clearly, the noise in (a) would contain vastly greater energy than in (b), and in many circumstances (b) would be more easily masked in virtually all real situations. (b) would be greatly preferable to (a). Too strict an adherence to statistics can create absurdity!

written specifications alone, an identical sounding room could be built. RT60 measurements also only show the time for the reverberation to decay to a point 60 dB below the initial sound pressure levels. Whilst it is true that, in most genuinely reverberant spaces, this decay is relatively uniform, in more complicated or less homogeneous rooms the RT20 or RT40 projections can be very different between rooms with identical RT60 specifications. The energy/time curves of the different rooms can be very dissimilar indeed (see Chapter 17).

On the subject of sound isolation, similar caveats exist. Again in the more specification conscious broadcast industries, they often work to noise criteria (NC) figures. The NC give a plot of an envelope of maximum levels of noise against frequency. These must not be exceeded if the specifications are to be met, yet from Fig. 16.1 it can be seen that if a single spot frequency in a relatively unimportant part of the spectrum exceeds the NC, the result is deemed unacceptable. If, on the other hand, a broad band signal remains $\frac{1}{2}$ dB below the NC curve at all frequencies, the specification is met. In most circumstances, the much greater energy of the sound in the second case would be more objectionable than the relatively innocuous failings of the spot frequency, but to rectify this failing, the cost of the building could be greatly increased if the NC figures were adhered to rigidly – even though, in practice, perhaps no problem existed.

Where NC figures are used for justification of data compression techniques in digital recording systems, the converse can be true, where a broadband noise slightly exceeding the NC can be less obtrusive than a spot frequency in a sensitive area which by half a dB does *not* exceed the NC, but conditions here vary with masking effects. Simplicity does not rule in these areas.

Experienced designers can take into account all of the many factors in the equations to provide sensible results. Obviously one option would be to lower the NC in the former example, but that could impose enormous cost increases whether a practical problem existed or not. On the other hand, reputable designers would not be satisfied with a job which met written specification yet failed to satisfy subjectively. A problem often exists when acoustically non-specialised architects specify to 'known' criteria without referring to specialised acousticians. Under these circumstances, errors of judgement can easily be made in terms of subjective results, despite acoustic specifications being satisfied.

Subjective acoustics is so much like an iceberg, with 90% or thereabouts not being obvious. Of what *is* visible of an iceberg, a change of 50% may seem a significant change, but there are two things to bear in mind. A change to 50% of what is *visible* would be a change of only 5% to the whole, but any such change would also alter the balance, and hence the angle of flotation, thus revealing sections which were not previously visible, and submerging parts which were formerly above the surface. Without a precise knowledge of what lies *below* the surface, no change can be entirely predictable. Such is the capricious nature of acoustics: the deeper one gets, the deeper one needs to get.

In an address to the 72nd AES conference in Anaheim in 1982, Ted Uzzle summed things up beautifully by saying: 'No sound system, no sound product, no acoustic environment can be designed by a calculator. Nor a computer, nor a cardboard slide-rule, nor a Ouija board. There are no step by step instructions a designer can follow; that is like Isaac Newton going to the library and asking for a book on gravity. Design work can only be done by designers, each with his own hierarchy of priorities and criteria. His three most important tools are knowledge, experience, and good judgement.'

Quoting Lettinger from his 1981 book *Studio Acoustics* (Chemical Publishing Co., New York): 'Nothing is gained by specifying the noise level limits in a room using an NC curve. According to this method, a noise is not acceptable when any part of the spectrum exceeds the limiting curve, no matter how narrow the frequency band which surpasses it is. But then, a noise is also unacceptable with a spectrum equal to that of the pertinent

NC curves but which slightly transcends these curves. Yet the two noises, carrying the same rating number, may differ widely in their A-weighted sound levels.' (A-weighting is used to relate measured sound levels more closely to perceived sound levels.)

There is nothing new about any of this, but it is surprising how many people in the recording industry are still unfamiliar with these aspects of design. Back in 1963, Schultz in his *Journal of the Audio Engineering Society paper* (Vol II, pages 307–317) 'Problems in the Measurement of Reverberation Time' stated: 'In a large room, if one has a sound source whose power output is known, one can determine the amount of absorption in the room by measuring the average pressure throughout the room. This total absorption can then be used to calculate the reverberation time from the Sabine Formula. This method fails badly in a small room, however, where a large part of the spectrum of interest lies in a frequency range where the resonant modes do not overlap, but may be isolated. In this case, the microphone, instead of responding to a random sound field (as required for the validity of the theory on which these methods depend), will delineate a transfer function of the room. It does not provide a valid measurement of the RT in the room.'

16.3 Control room evolution

When it comes to designing a control room, the problems described above almost entirely preclude designs based on any rigid formulations. Control rooms began in the first days of electric recording as 'control booths' – very small rooms with the monitoring consisting of a very basic loudspeaker. Even fifty years on, around 1970, control rooms were still largely of what would today be considered to be very rudimentary designs. Any acoustic treatment consisted of Helmholtz resonators and absorbent panels, and, if one was lucky, some geometrical 'control'. One was *very* lucky if any two locations in the room sounded similar.

Although almost every control room for record production was by then stereo, there were still mono disc cutting rooms and mono broadcast control rooms. Few if any control rooms had been designed with stereo as a prime consideration. Most were updates of old mono rooms, and many still had four-speaker monitoring from the original multi-track days of four track. By four-speaker monitoring, I do not mean the quadraphonic layout of the mid-seventies, but four in a row across the front of the room – one to each track of the four-track tape recorder! Eight and sixteen tracks were later routed through the same four monitors via switching matrices.

The main problem as everything began to require stereo mixing was that the rooms were often inadequate for good imaging or frequency balance. 'Good' rooms which many people liked became renowned, though many of their designers could not pinpoint exactly why one room 'worked' whilst another generally failed to please. There were still very few genuinely good rooms even by the time that 24-track machines began to appear in earnest in 1973; but around this time, Tom Hidley, then at Westlake Audio in Los Angeles, began to make international waves with a concept of control room design philosophy which was always recognisable. It was intended to produce repeatable results around the world, but the claims were too far exaggerated, and people soon began to complain that work done in one room of Westlake design did not necessarily sound sufficiently similar in another.

It was, however, a very bold step by Hidley, and though he did not achieve his goals in the 1970s, he changed the face of recording studio design throughout the world. He relinquished his shares in Westlake, moved to Switzerland, and started Eastlake Audio; he had moved east. In 1979 he passed Eastlake to David Hawkins who had been his representative in the UK, and retired to Hawaii. Since his return to studio design in the mid-1980s, he has returned to Switzerland.

What Hidley did not know, and *could* not have known at the time (mid-1970s), were the implications of the then yet to be properly defined Chaos principle. There were just too many small changes from room to room which would conspire to produce radically different sounds. What he and most other studio designers were then using as their major reference was one-third octave spectrum analysis with graphic equalisers on the monitors. Most current studio designers now accept that third-octave analysis and monitor equalisation was a disaster, but Hidley had conceived a system of control room design and monitoring which, possibly for the first time on a large scale, was enabling wideband flat amplitude responses without excessive use of equalisation. It was certainly a beginning, and whilst by no means perfect, some of his rooms of twenty years ago are still in front line use by major record companies: a testament to his efforts.

The problems which fooled most people at the time stemmed from the concept that between reverberation time and the pressure amplitude response around the position of the engineer, the performance of a room could be specified. As we now know all too well, if a monitor system is installed in a room, it can be the culmination of the minor bumps on the bumps of the response, caused by complex internal reflexion patterns, which dictate the overall character. A third-octave equaliser was far too crude a tool to do anything but move the bumps around.

The reverberation time would affect the steady state response of the room, but the direct signal would only be affected by the monitor system and its method of mounting. Each would affect the other in terms of perceived responses, but differently on steady state and transient signals. I remember in 1974 when I first approached Tom Hidley about building a studio for Virgin Records, for whom I was technical director at the time. We had previously been using JBL and Tannoy monitors in our existing studio, and when I went to Los Angeles to meet him, he told me that the Westlake systems which I had heard were using JBL drivers. To me, in those days, JBL monitors sounded hard, or so I thought. I asked if he had any alternatives in mind, and he suggested that we used Gauss drivers for the low and mid-frequencies. The room using the Gauss drivers, once completed, measured identical in third-octave, pink noise terms to the JBL room in Los Angeles, yet the two sounded very different, and could not be equalised to sound similar.

What was going through our minds in those days was short of concepts of phase response. Though many papers had been published indicating its importance, classical acoustics was still largely adhering to Helmholtz' philosophy of phase being imperceptible by humans, which we now know not to be so. But even then, the very fact that I had asked for Gauss instead of JBL drive units in an equalised room should have illuminated gigantic beacons in our minds. If the Gauss and JBL *could* be equalised to sound alike, then no preference should exist. If they could *not* be equalised to sound alike, then something was missing from our concepts of achieving similarity in 'repeatable' room designs via third-octave equalisers. I remember Tom Hidley saying to me in the late 1970s, 'It's phase distortion; that's what produces fatigue and inconsistency!'

Looking back, he was of course right, though what could have been done about it at the time is uncertain. *That* phase distortion was of course that of a non-minimum phase, uncorrectable type, produced by inappropriate reflexion patterns. Since then, however, as the general requirements for control rooms have multiplied, so have the different approaches of different designers, most of which are attempting to produce a relatively neutral acoustic. By definition, neutral implies that it neither adds nor detracts anything, hence travelling from one neutral control room to another, even of different design, should ensure a consistent sound character; at least if using similar monitor systems. Such is of course still not the general case; in reality, so many rooms sound so different, and even identical rooms with different equipment in different locations *can* sound very different. This was one of the major reasons for the widespread adoption of 'near-field' – or more properly 'close-field' – monitors. (Near-field is an acoustical term for the zone immediately in front of a sound source where the air acts as a 'lumped element' and the sound does not decay by the usual 6 dB with doubling of distance.) With the small monitor so close to the listeners, the room plays less of a role in the perception of this sound, but, necessarily, desk reflexions and reduced frequency range mean that there is a price to pay for their use.

16.4 Acoustic control

There have been many approaches to the problem of linearising the responses of control rooms by acoustical means. The first attempts were based on 'tuning' the room by means of resonators and absorber to produce a uniform reverberation time with frequency. The problem with this approach was that even if different rooms of different designs achieved not only a uniform response with frequency, but the same measured response with frequency, for example 0.4 sec from 50 Hz to 10 kHz, then as we discussed earlier in the context of live rooms built from different materials, how they were built and what they were built of could substantially change their perceived sonic performance. What is more, if they had either different monitor systems with different directivity characteristics, or different areas of different tuning or absorption, or both, then once again they would be perceived to be very different, though all meeting the generally laid out specifications. This was discussed in the opening paragraph with regard to the broadcast industry working to predetermined specifications. It necessitates each company's own set of 'golden ears' having to make their individual assessment of fine tuning.

Another approach towards linearising the performance of a room is to use a technique of geometrical control, providing many non-parallel surfaces and selected areas of absorption and reflexion. In a rectangular room there are three types of resonant modes: axial, between parallel walls and parallel to four other walls; tangential, travelling round four walls and parallel to the other two; and oblique, travelling round all walls (including floor and ceiling) and parallel to none. The axial modes are the strongest in terms of energy. The oblique and tangential modes lose energy more quickly, both by striking more surfaces per given distance travelled, and because surfaces tend to absorb better when the incident wavefront strikes them obliquely, rather than at right angles.

In such rectangular rooms, the spread of the modes *can* be changed, generally levelling the response to some degree; but at low frequencies, control geometry angles cannot be made large enough, so low frequency control is again achieved by absorption, often by 'bass traps' placed strategically at the suitable points of problem modes. Such was the early Westlake approach which produced some very pleasant rooms to work in – but identical as intended, they were not. They could never be, because the specifications could not take into account the effect of differing shapes, sizes, decorations and equipment. They could, and were, built to be capable of providing uniform third-octave pressure amplitude responses, but the reflective patterns, the phase, and hence transient responses, were different, so they could never be properly assessed until each one was built and heard in use.

The ear has an ability to detect as separate events the early and late arrival of sounds. If a sufficient time delay exists between the direct sound from the loudspeakers and the first reflexions from the room, prior to any general 'pseudo-reverberation', then ears, especially 'trained' ears, can usually soon begin to 'lock' on to the incident sound, and allow the brain to consider separately the room. This relies on the 'Haas Effect', named after its first proponent. Taking into account this phenomena were the 'Live End, Dead End' rooms, as proposed by Don Davis in the late 1970s, having the front half of the room made very acoustically dead and the rear half of the room made reverberant to the desired degree, sometimes by means of Schroeder diffusers.

The concept prevented early reflexions from reaching the mixing position. The 'first pass' of the incident wave from the loudspeakers was thus uncontaminated by related reflexions for a sufficiently long period of time for the ear to clearly differentiate the direct sound of the loudspeaker from the effects of the room. Once again, different monitor systems produced different results, so consistency depended upon both the monitor system for the direct sound, and the design of the rear of the room for the reverberant sound. Again, there are many people who like these rooms and find them easy to work in, and other people who do not like them so much. As with other approaches, they are not consistent between measured and perceived subjective performance, though it must be said that this need not be a problem for engineers familiar with the rooms, who know them well, and can achieve consistent results.

The seemingly obvious choices, if the rooms are to be entirely consistent, are to standardise on room shape, size, equipment location and design, but this would be totally impractical considering different requirements of different people, and also the suitability of available property. Apart from that, subjective agreement of design would never be achieved in such an individualistic industry. The second option is the anechoic chamber, as anechoic is anechoic irrespective of shape, but such rooms are not very pleasant to be in. Indeed, some people find them very disturbing, as all normal auditory cues on wall positions are missing; you do not hear the room which your eyes see. Such rooms also run against the long held concepts of rooms bearing some general similarity to the 'average' domestic room, though by now it is widely agreed that any average is itself so unrepresentative of the majority of rooms, especially from country to country, that little relevance exists between the concept and the reality.

16.5 Zero × error = zero

When Tom Hidley returned to studio design around 1984 and moved back to Switzerland, he came back determined to achieve the room-to-room consistency which had eluded him in his first decade and a half of trying. It does not require many reflexions to make a person feel comfortable in an anechoic chamber – only two or three will soon restore a sense of ease. On the other hand, as shown by the Schroeder 'phase organ' experiment (see Chapter 6), it also only takes a few reflexions to mask much low level detail and phase information. Hidley thus concluded that by making the front wall and the floor to be the only reflective parts of the room, other than the equipment such as the mixing console, the remainder of the room could be made absorbent to as low a frequency as possible without producing any unpleasant characteristics from a human point of view. By making the front wall the only vertical hard, reflective surface, it provided a very effective monitor baffle extension, but itself could not produce any reflexions of the music, as the loudspeakers were pointing directly *away* from the only hard wall.

Any reflexions from the floor would have to arrive at the listener in the same vertical plane as the direct sound, so no disturbance of the stereo image could be expected. In practice, a soft back to the mixing console effectively stops all but the low frequency floor reflexions from reaching the engineer's ears. When one adds to this concept the use of only one type of monitor system, it is easy to see how a very great degree of consistency could be achieved from room to room. The only significant room-to-room differences are to the sound of people speaking in the rooms (dependent upon the distances to the reflective surfaces), and the responses below 40 Hz or so, which are dependent upon the amount of space available for low frequency absorption. This approach has probably come the closest yet to achieving absolute room to room consistency.

After becoming familiar with Hidley's new concepts, I co-sponsored with him some further research done by Brazilian acoustician Luis Soares at the Institute of Sound and Vibration Research at Southampton University, England. It was an attempt to find mechanisms by which to achieve similar absorption in smaller rooms, or more effective absorption of lower frequencies in larger rooms. Hidley himself refuses to drop below certain room sizes, as he is rightly very conscious of the previous criticisms of his claims to uniformity of listening conditions for the pre-1980 rooms, but his concepts are so powerful that I have now used them to very great effect, even in some *very* small rooms.

The rooms are, however, very demanding of their monitor systems, usually requiring specifically designed units. They soak up almost all of the incident wave, so very little reinforcement of loudness is available from these rooms. Furthermore, any off-axis irregularities of the monitor system are heard off axis with ruthless accuracy. Effectively, the rooms need high power, smooth, wideband, coincident source monitor systems. Such systems are not easy to locate commercially, so most come specifically with the rooms.

It is something of an irony that these rooms, which have begun to achieve consistency on a subjective basis, have only done so by having a dual specification. They are 'monitor dead'; in other words, the monitors drive into a largely anechoic termination, hence the rooms have no specification in terms of reverberation time as they are approximating to a free field. There is, however, a second specification which could relate to the direction and distance of echoes or specular reflexions from a sound source inside the room, such as from the listening position, but this would relate only to the general ambience from the point of view of persons within the room. It would have no bearing on the monitoring, except in some extreme psychological concept of 'comfort factors'. Not surprisingly Tom Hidley referred to these 'monitor dead' rooms as 'non-environment environments'. You feel as though you are inside a room, yet you listen to the music as though you are in a field. The logic behind this is that even if small changes in terms of specifications can have what would appear to be disproportionately large subjective repercussions, then the only way to make the practical percentage differences approach zero is to make the *specification* approach zero. A ± 20% error on two dead rooms is still two dead rooms!

From a subjective point of view, there are no absolute rights and wrongs in terms of control room acoustics, but although some people have made great recordings in difficult acoustics, that is still no reason to justify the haphazard approach to many designs. In reality, though, many such difficult rooms have either been 'home-built' by people who *thought* that they knew more than they did, or designed by designers operating under constraints from clients who introduced 'strange' sets of working priorities. Sometimes, such approaches have resulted in rooms of excellent performance, but well designed rooms which were designed with good knowledge and sound principles stand a better chance of success.

One of the major reasons for the reluctance of many studio owners to believe what the 'experts' tell them is that too many of them have expected too much from technical specifications. They have shied away from what they saw as a black art of acoustics when no definitive relationship between subjective and measured results could be clearly demonstrated. Hopefully, from the previous pages, a more clear understanding can be achieved about the degree of complexity involved in acoustic interactions, and Ted Uzzle's comment should be taken as the golden rule, to the effect that there is absolutely no substitute for experience in such a capricious 'science'.

16.6 Is it time to abandon the domestic listening room concept in control room acoustics?

Physically, domestic listening rooms usually bear scant resemblance to by far the greatest majority of sound control rooms, whether those control rooms be in film, radio, television or for pure music recording. Although many films are now released on a video format for home viewing, the big impact is still usually from initial release on the silver screens, and it is the box office performance which usually dictates the success or failure of a big production. Consequently, many film studio sound control rooms or audio sweetening rooms do try to simulate the effect of a large auditorium, and, indeed, some of the multi-channel surround sound formats are solely dedicated to the cinema world. Such formats require very special recording and mixing facilities in order to achieve the desired effects. Mixing for a stereo video release is a separate exercise, where the earthquake effects of the cinemas can be trimmed to reasonable domestic levels. In the original cinema formats, they would probably only serve to produce unwanted distortions in the home.

Television and radio also have a specialised and relatively captive market. By far the majority of listeners receive what is usually only a one-off broadcast, via rather arbitrary audio amplifiers and loudspeakers. Almost invariably the sound reproduction systems are much too small for realistic full range performance, but through them the essence of the programmes must be clearly intelligible if ratings are not to fall off rapidly. Reference at the mixing stage to typical reproduction systems is in these instances a totally worthy exercise. Thus the television, radio and film industries have dedicated staff working to relatively clear goals. They know their aims and they know their audiences. The record industry, on the other hand, is usually less formal, less structured, and much less certain about the listening conditions of its primary markets. In line with most other studio designers, in the past I have held to the general principle of recording studio control rooms conforming to some degree of domestic listening conditions. However, as time has progressed and listening conditions have diverged, I now wonder whether the goal was ever either achievable or realistic. What is more, attempts to adhere to this philosophy may now be doing more harm than good.

In small European countries such as Greece or Portugal, the total populations of the countries are exceeded by some of the world's individual cities. Especially where those countries have languages which are not widely spoken elsewhere, or there is no great export of their musical culture, it is reasonable for their recording studios to pay at least some attention to the 'average' listening conditions of their home markets. On the other hand, in a country such as the UK which produces an enormous amount of music for worldwide consumption, it is surely better to use control rooms which allow monitoring of the fullest range and the greatest overall neutrality. In Europe alone, the range of domestic acoustics is so wide that any statistical average of domestic listening room acoustics would only relate to a very small, single figure percentage of listening rooms as a whole. Under these circumstances, the implications of average are taken to represent a bulk of the listening population, whereas the reality is that it relates only to a tiny minority.

Insulated Scandinavian rooms with excellent thermal, and hence often acoustic, insulation are an acoustic world away from the mainly stone and plaster rooms of Spain, often with their hollow bricks which are almost transparent to low frequencies. The heavily carpeted, draped, and softly furnished rooms found in the UK bear little if any relationship to the rooms of either Sweden or Spain, and if one takes into account many of the very small domestic rooms in Japan, or bamboo houses of the Far East, where quite surprisingly large quantities of British recorded music are sold, the 'average' becomes even less meaningful. The disparate acoustics of the above examples, not only from furnishings, but also from their sizes, materials of construction and types of construction, bear very little similarity to one another. Furthermore, it cannot even be said that their shapes are generally rectangular. To the eye, they may be, but their acoustic shapes can vary enormously, where walls can be acoustically apparent only at certain frequencies. When one adds to this problem the fact that so many people now listen to music in cars, or on headphones where no conventional room exists at all, then the concept of having a meaningful average to aim for becomes totally redundant.

Given all of the failed attempts in the past to achieve a degree of commonality in control room acoustics, at least to the extent that mixes should be perceived to be largely similar from one control room to another, one must now question whether it is really worth the effort, given that the end result has so little bearing on reality. If it was generally adopted policy that control rooms should be largely dead to the monitors, and that internal acoustics were for the benefit of those speaking in the rooms, then adherence to this concept of reproducing a worthless average of domesticity could be dropped altogether. The benefits from such a policy would be very greatly increased commonality between control rooms, especially from different designers; much less room performance disturbance from the influx of varying quantities of equipment; improved phase and transient perception; an excellent relationship with the very large percentage of listeners in cars and on headphones; and an ability to hear loudspeaker and monitor system differences and problems for precisely what they are. Small loudspeakers could still be used near the mixing console, which give a more domestic feel to the music, irrespective of any reasonable room in which they may be placed. The large monitors could again then be used as more 'accurate' full range systems which related more to the sound of full range, true hi-fi equipment; and also to what had actually been recorded!

In the context of the very serious domestic listener who has spent a great deal of money on a high quality hi-fi system, in many ways compromising the control room performance to an unjustifiable mean standard actually works against the needs of the audiophile. Surely it is better that the control room be optimised for the necessary perception of sonic detail. If a music lover truly wants to perceive what is recorded, and to add a controlled, if desirable, amount of room acoustic to the reproduction, then it should be incumbent upon the music lover to create a good set of ambient acoustics for the reproduction: this way everybody should benefit.

If these policies were adhered to, it is very highly unlikely that anybody listening at home below true hi-fi standards would notice any degradation in performance. Indeed the very converse is true – more consistency in the control rooms should lead to more consistency in the tonal balance of the discs or tapes on general sale. For far too long we have been attempting to chase a mythological domestic reality, and the results in most cases have been detrimental. It is now high time that such misconceptions were allowed to fade.

I was personally very reluctant to drop my own belief in some domestic similarity factor, but the fact exists that in five years of building rooms which are dead as far as the monitoring is concerned, the general acceptance of the mixes from those rooms has been excellent. There have been fewer uncertainties about the general balance of mixes when taken to a wide range of domestic locations than from the rooms where some compromises to the 'average' domestic situation had been allowed for in the design. Despite some of the new rooms having stone floors and stone front walls, whilst others have wooden floors and wooden front walls, the general sound of the monitoring, with similar systems, has been very close indeed. All of this backs up what Tom Hidley has been saying in recent years since his adoption of this policy in 1984; and he above all others is super-sensitive to criticisms of room-to-room inconsistencies after his claims for the Westlake/Eastlake rooms of the 1970s. It seems that for as long as one tries to juggle with the parameters to be more domestically representative, the goal remains nebulous. Ironically, experience has dictated that when one removes those parameters from the monitoring equation, the goal crystallises in front of one's ears. The following two chapters will take these concepts and conflicts further.

Chapter 17

Non-environment control rooms – their origins and acceptance *vis-à-vis* conventional control rooms

17.1 Origins of the 'trapping' system and 'non-environment' rooms

The concept of 'monitor dead' or 'non-environment' rooms was discussed in Chapters 15 and 16. Suggestions that 'bass traps' had no mathematically proven acoustic basis have been rife for years, but recent research carried out by Luis Soares has begun to throw more light on the subject. The fact that they do work is patently obvious to all who have ears to listen, but one problem with their academic acceptance has been their empirical origins and complex nature of operation. Within the recording industry, the term 'bass trap' can usually be traced back to Tom Hidley in the late 1960s.

In the 1950s, Hidley worked for JBL in Los Angeles, and on one particular occasion a loudspeaker was taken into a listening/testing room which was known to have low frequency problems. Upon setting the system up, it was noticed that the LF response was smoother than usual, and this was initially put down to the new loudspeaker design. When the particular loudspeaker was subsequently auditioned elsewhere, the benefits were no longer apparent, so other loudspeakers were taken into the test room, where once again a smoother than previous LF response was noticed. Something had clearly happened to the test room, but the only change to the room was that a grouping of blackboards and screens had been moved into the room for temporary storage. When these were removed, the bass problems returned, so Hidley asked his colleague Bart Locanthi what was happening. Locanthi, whose knowledge of acoustics was at that time significantly greater than Hidley's, replied that the boards were acting as traps – the low frequencies effectively went in but did not appreciably reappear.

Some years later, Tom Hidley was working in New York, when the Record Plant (NY) asked him to look at some LF problems in one of their rooms. He remembered the 'traps' from his JBL days and decided to try a system of angled, free hanging baffles in a giant contrivance on wheels. It proved too heavy to be movable, but nonetheless, once in position, it dramatically improved the acoustics of the room. Such 'traps' and Tom Hidley became almost synonymous over the next twenty years. After a brief retirement from studio design in the early 1980s, Hidley returned with a new 'all-trap' approach around 1984.

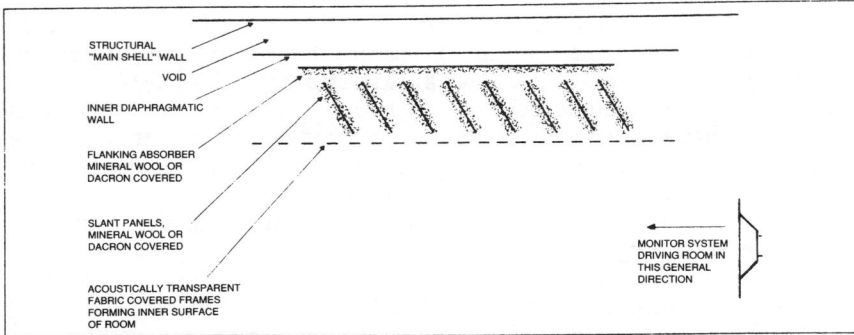

Figure 17.1 Plan view of full development of a Hidley style wall trap

The trapping system has developed over the years to a high degree of effectiveness and predictability, but although the free-hanging, fluff-covered baffles look to be simplicity itself, the acoustic goings-on which enable them to be effective have proved highly complex. When the traps form the bulk of a room, they act as absorbers, diffusers and waveguides, reducing very significantly the broadband energy which can return to the listening area after the first pass from the monitors. The empirical evolution of trap design has passed through many phases on its way to current thinking. During this process, many designers have used the concepts 'parrot fashion', frequently achieving a success rate greater than would be expected from mere fluke by virtue of the fact that the traps work in such multi-functional ways. On the other hand, the inappropriate use of such systems has also led some designers whose applications have been unsuccessful to suggest that the whole concept is flawed, and inappropriate for their studio design application.

In their current forms, the 'non-environment' rooms originating from Tom Hidley's mid-1980s ideas are highly effective in the control of low frequency reverberation times. Whilst the fine detail of the construction concepts still requires a degree of practical experience in their fine tuning, the basic concepts are now quite well understood. In essence, the rooms which are heavily trapped show modal characteristics which are, firstly, typical of rooms which are around 20% physically larger, and, secondly, lower in level and much broader than the modes of a similar, untrapped room.

17.2 Theory of operation

Investigation to date suggests the following processes to be at work. Figure 17.1 shows a section of a typical trapped wall. The flanking panel which hangs parallel to the wall is extremely important in terms of the effectiveness of the overall system. The 'chip cutter', slant panels in front act partially as waveguides, bending the low frequency incident waves to cause them to strike the flanking panels at an angle that renders the absorption more effective than would be the case for a too shallow or too direct strike. This is akin to the use of wedges in anechoic chambers, where the wavefronts largely strike the absorbent materials in a gradual manner. Three feet of foam wedges, while containing less absorbent material than a 3 ft thick, solid foam block, nonetheless is more effective in terms of acoustic absorption. Again, the destructive power of a wave striking a sheer cliff is greater than that same wave could achieve when rolling up a sloping beach. It is thus important that the slant panels are orientated such that they capture the wavefront at an optimum angle to steer that wave towards the flanking absorbers at an angle of maximum absorption.

The waveguide effect can clearly be shown by hanging 'baffles' consisting only of dense mineral wool or similar absorbent in the position of the normal slant panels. When this is done, the absorption is greatly reduced at low frequencies. Inserting a thin solid panel within the absorbent baffle will begin to improve the low frequency absorption, which will continue to improve as the solid panel is thickened, thus becoming less transparent to the low frequencies. Once the panel thickness becomes sufficient for the waveguide effect to be significant, then no further increase in thickness will show any benefit in terms of low frequency control. Indeed, as far as the low frequencies are concerned, the solid panels alone will show a more marked improvement in the performance of the trap than would be the case for the absorbent (mineral wool or whatever) panel alone.

When the panels are formed from a combination of solid core and absorbent covering, the absorbent covering has entirely different modes of operation in terms of the low frequency, and the mid/high frequency absorption. At middle and high frequencies, the absorption is a function of density, porosity and thickness, and is entirely conventional in operation. At low frequencies, where the wave is directed between the panels, the wavefront entering the slant array will follow the waveguide panels. The individual spaces between the panels, the ceiling and the floor will form duct absorbers. The sections of the wavefront passing immediately adjacent to the surface of the panels will have to 'drag' their way through possibly several feet of absorbent. Figure 17.2, in exaggerated form, shows how the wavefront will be distorted in shape as the absorbent slows down and reduces the amplitude of the sections of the wave which are forced to pass through the absorbent material. Bearing in mind the complex path which the wavefront must follow in order to re-enter the listening room, especially in the light of the effect of the absorption of the flanking panels and a certain degree of absorption in the slant panels themselves, it is not too hard to see the potential for reflexion suppression.

The effective low frequency limit of the trapping is partially a function of the size of the flanking panels, where the largest dimension of the panel determines the half wavelength of the lowest frequency which can be effectively absorbed. The room design itself also has a bearing upon the overall operation. Were the room to be considered a duct, then an absorber placed in that duct can be expected to achieve a certain degree of absorption. It is a well known acoustical principle that an absorber placed at the end of a side branch off that duct can achieve greater absorption than when placed directly in the duct itself. If one imagines the slant panels as producing a series of side branches off the main duct (the room), then the greater effectiveness of the flanking absorbers when placed behind the slant panels can be more readily understood. Some of the complexity of the systems can now be seen, as things are happening on several different scales simultaneously.

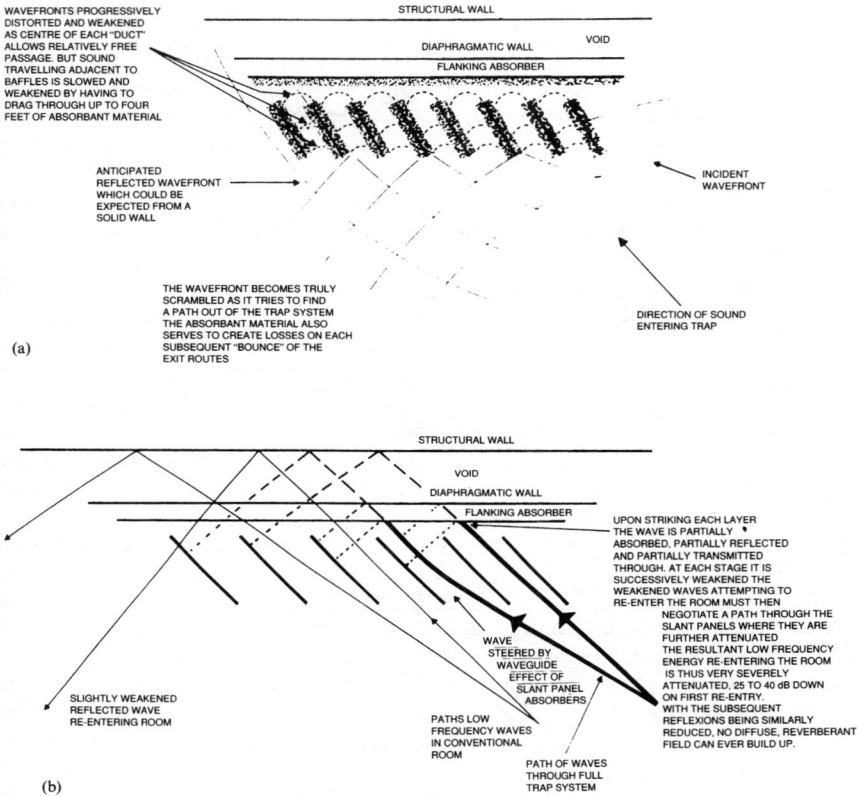

Figure 17.2 (a) Mid/high frequency absorption in a typical trap system; **(b)** low frequency absorption in a typical trap system

When these traps are enclosed within a diaphragmatic shell, the effect on low frequency reverberation time is even more noticeable. Reverberation is in fact a misnomer under such circumstances, as no diffuse field ever develops. Individual reflexions decay before any diffuse field can be realised. Such a diaphragmatic shell would typically consist of a $4'' \times 2''$ timber frame, boarded on one side with a plasterboard/insulation board/plasterboard covering, somewhat similar to the old BBC 'Camden Partitioning'. Other combinations can be used, however, with plasticised deadsheets replacing the insulation board, or deadsheet/felt membrane absorbers being used for the internal skins. The stud cavities themselves can be filled with absorbent material, and varied in depth to achieve maximal absorption over different frequency bands.

Where a significant gap can be left between this shell and an outer, sound containment wall, the low frequencies are even further controlled. The low frequencies will see the inner wall as being relatively transparent and hence will see the larger room of the sound containment shell. Some further attenuation will take place as the LF travels once each way through the wall, subsequently reducing yet again the LF energy returning into the

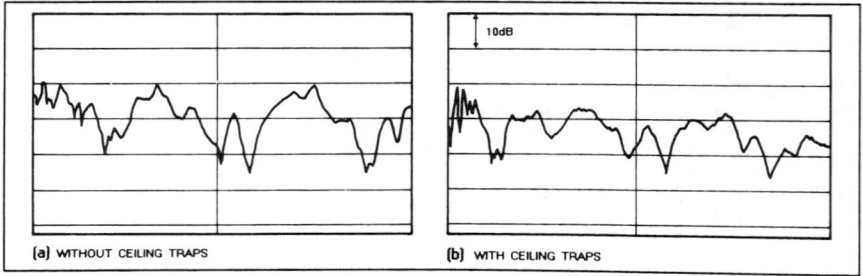

(a) WITHOUT CEILING TRAPS   (b) WITH CEILING TRAPS

Figure 17.3 Effect of ceiling traps on modal pattern of room

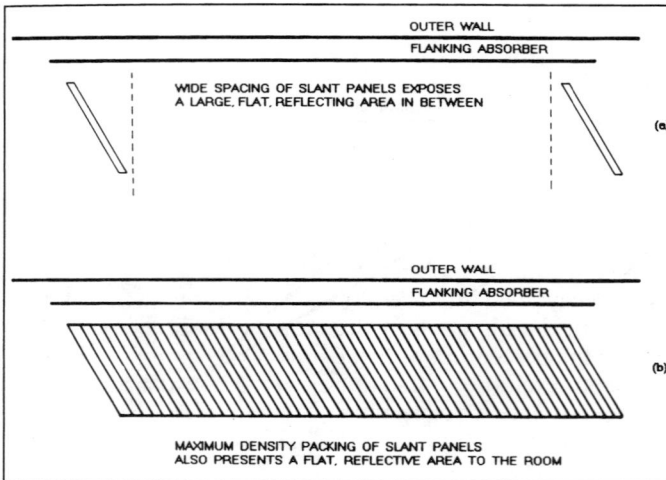

Figure 17.4 Effects of extremes of slant panel spacing

room. As can be seen, the different mechanisms keep nibbling away at the potential reverberant energy, gradually taking it down to insignificant levels.

A further aspect of such systems is that sound passing over the series of side branches can be slowed down by the highly dispersive nature of the multiple slant panels and the gaps in between. This is another means by which the room appears to be acoustically larger than its actual physical size. Typical modal patterns are shown in Fig. 17.3(a), with Fig. 17.3(b) showing the very even overall response after the addition of full ceiling traps; in this case the ceiling baffles were 24 feet long. Experience has shown that for the greatest sonic 'spaciousness', the slant panels should be in the order of 12″ to 18″ apart. Both extremes of spacing would yield flat walls, as one at each end would expose the flanking panel to only random incidence absorption, while too many panels, taken to the extreme, would yield a solid mass of panels (Fig. 17.4). Two feet to four feet appear to be the optimum range for overall trap depth, as below two feet, audible effectiveness drops off rapidly, whereas over four feet, further increases produce little significant effect, and would generally be considered wasteful of both materials and available floor space.

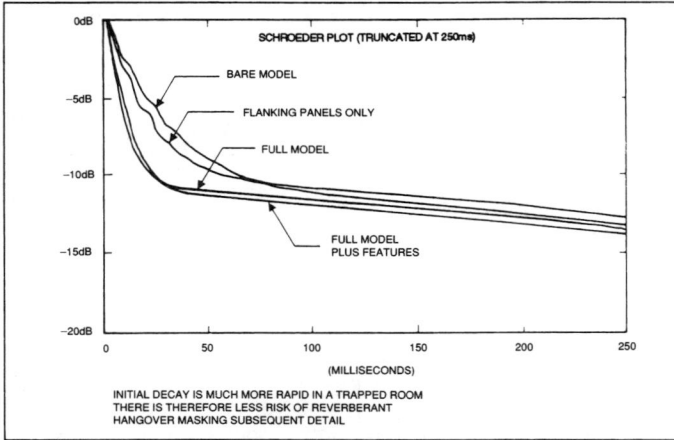

Figure 17.5 Shroeder plots of impulse decay curves of a modal room as trapping system is installed step-by-step (Courtesy Luis Soares)

Figure 17.5 shows a Schroeder plot of a typical decay curve for such a room. As can be seen, unlike a conventional room with a linear reverberation decay, the non-environment rooms lose their energy very rapidly in the initial stages of their decay. The rapid removal of energy, particularly when the room is excited from the direction of the monitor loudspeakers, allows much more 'space' for the perception of fine detail in the sound immediately following any transient excitation. Such a decay curve renders normal reverberation time measurement to be all but meaningless, as we are no longer dealing with a 'room' in any acoustically conventional sense of the word. Figure 17.6 shows the decay tail of a 20 Hz highpass filtered step function for the first critical 20 ms after excitation from a well designed monitor system. The lack of resonant/reverberant overhang is clearly apparent, rendering insignificant the amount of masking energy available to muddy or smear the audible clarity of the monitor response. Strictly in terms of definition, imaging, general clarity, and the overall ability to show fine detail, when equipped with suitable monitor systems, such rooms are appearing to achieve results which have hitherto been rarely realised.

17.3 Aims, priorities and early reactions

Ever since the early days of Westlake Audio, Tom Hidley had a goal of achieving a commonality of control room performance from room to room and country to country. Looking back on it, given the variability in shapes, sizes and installed equipment from one room to another, the goal was probably unachievable given the technology of the day. I, myself, in 1970 built a super-dead control room for a client who agreed with the general idea. Again, because of the dead acoustic giving no help to the monitor loudness, I had to install four specially designed, electronically crossed over loudspeakers, using 18″ bass drivers of relatively high efficiency. I liked the monitoring – indeed, many people liked the monitoring – but the room as a whole was not well received and was rebuilt within months on more conventional

RESPONSE OF LOUDSPEAKER (a) AT 1 METRE
SHOWING A COUPLE OF CYCLES OF LOW LEVEL
RESONANT HANGOVER (IN THE LOUDSPEAKER)

RESPONSE IN ROOM AT LISTENING POSITION, SHOWING
VERY LOW LEVELS OF REVERBERANT OR REFLECTIVE
HANGOVER THE DECAY TAIL CONSISTS MAINLY OF THE
ODD COUPLE OF CYCLES AT 20Hz VISIBLE ON PLOT (a)

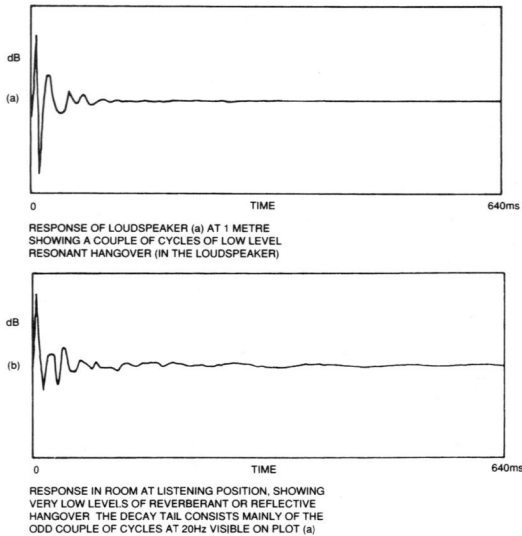

Figure 17.6 Impulse response of monitor system and room at Liverpool Music House (0 to 200 Hz)

lines, though the studio recording areas remained the same for almost twenty years. The super-dead room, at Majestic in London, was an early attempt to remove the room from the monitoring equation. Had I only realised then what I know now, the addition of a hard front wall surface and a hard, instead of carpeted, floor would have rendered this room an early version of something remarkably similar to some of this current thinking.

17.4 Practical realisations

After some of Luis Soares' one-tenth scale modelling of some of these described techniques at the ISVR, I built a modified 'full scale model' at the Liverpool Music House (LMH) and also incorporated a new monitor system utilising Keith Holland's newly developed axisymmetric horn, which was optimally matched to a TAD TD 2001 compression driver. The whole place was one giant test rig.

The new, large, highly absorbent rooms, however, would give no help in terms of loudness to the loudspeaker output. The LMH control room was in a shell of around 600 ft^2, and 15 ft high. We realised that high SPL monitors would be necessary, and a good step function response would require a minimum number of crossover points if amplitude and phase were to be maximally linearised. At the time, a superior horn and driver system appeared to be our only hope of achieving these goals on a reliable basis. In order to support such transient accuracy at the listening position, especially in terms of the reduction of the masking of further detail in the transient tail, a relatively dead room, even at low frequencies, was a further, seemingly mandatory requirement.

Without accuracy of both amplitude *and* phase responses, there is no

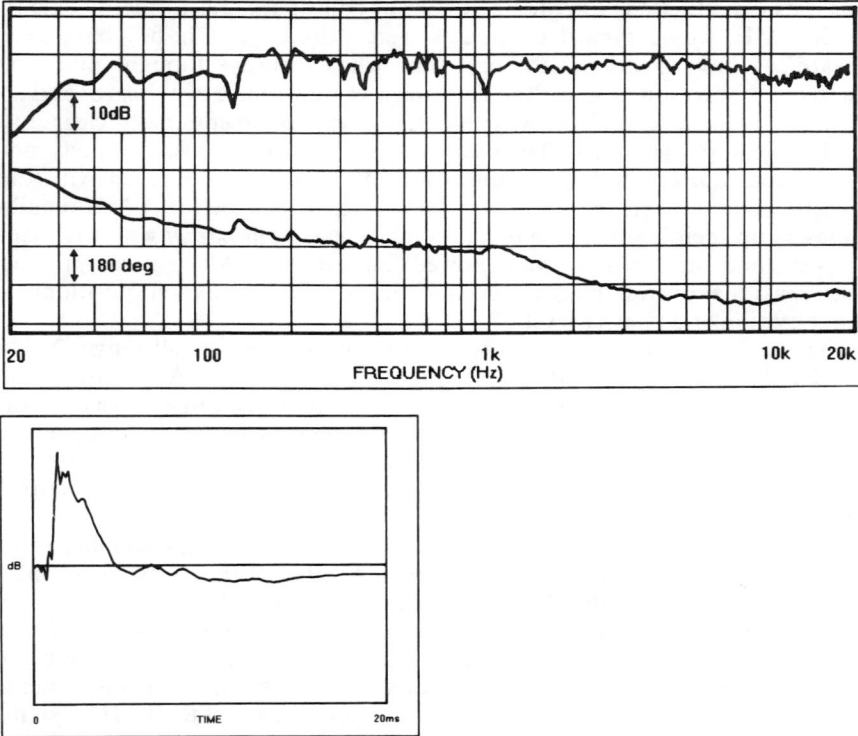

Figure 17.7 **(a)** 2-way monitor measured on-axis at 2 metres in situ in Liverpool Music House. **(b)** Decay tail of step function – first 20 ms at Liverpool Music House (20 Hz and 20 kHz filtered)

hope of any system following a square wave or a step function. In order to construct the LMH before the conclusion of the research projects of both Luis Soares and Keith Holland, I cannot deny having to borrow a number of Hidley's more proven techniques for the 'full scale model' at the LMH. The results are shown in Fig. 17.7. Plot (a) shows the amplitude and phase responses of the initially installed system with a temporary 20 Hz high pass filter, while (b) shows the step function response. Both of these measurements were taken via multi-point fast Fourier transforms, in the room. There is no smoothing or third-octave averaging – they are raw plots. Everybody involved was delighted with the performance of the room and monitor system, both sonically and in terms of measurements – the two do not always correspond. The new horn was a revelation, maintaining its response with the TD 2001 to around 22 kHz and yielding an exceptionally smooth directivity. The horn performance details had been first announced and published at the November 1990 IOA 'Reproduced Sound' conference.

17.5 The acid test

Certain aspects of control room assessment are very highly personal, so to try to avoid being carried away by the initial reactions of those involved with its construction, I invited a stream of colleagues and former clients to visit LMH. Some typical comments were: 'the biggest hi-fi I have ever heard', 'I have heard for the first time what I always thought that true stereo ought to sound like', 'the best imaging that I have ever heard' and so forth. A general consensus was that the bass was exceptionally clear and tight; the top was sweet, flat, clear and smooth, and most definitely not archetypal horn sounding. I could not have been more pleased with the reactions of the visitors. In fact it was six weeks before LMH finally hooked up their Yamaha NS10s; everybody had been happy with the main system, and tapes taken away showed few surprises. Mixing was proving easy, as the overall clarity allowed clear-cut decisions on positioning, equalisation and relative level.

The new 'non-environment' rooms provided in effect a full range, 20 Hz to 20 kHz large scale, close-field monitoring situation. Apart from the aspect of a more general reference to the frequency range of a typical, domestic system, the large monitors would seem to be generally restating their claim to be the main reference for a mix. Four years later, in 1995, I am continuing to use this concept, and the reactions to the LMH control room have proved to be typical.

It was a few weeks after the initial auditioning, however, that I began to receive some unexpected comments from a number of the people who I had taken to the LMH to the effect of: 'It's excellent, though I still like my own studio; but if you can give my studio some of the properties of LMH, then I would like to speak to you about it.' Unfortunately, they were asking for some mutually exclusive characteristics such as a warm, wrapped around, low frequency character, with the definition and clarity of the LMH system. Obviously, if the all-encompassing, warm, low frequency response was a function of the low frequency reverberation characteristics of their room, then it could not be achieved in conjunction with the clarity of the LMH system, as the clarity of that system had been achieved by the effective removal of any low frequency room reverberation. Clearly this was the conundrum that the LEDE and diffusive rooms were attempting to address, but their problem was one of room-to-room-to-outside-world compatibility. The non-environment approach was achieving the much-sought consistency, and was readily able to produce compatible mixes for the domestic market, but clearly some people missed the 'intimacy' of rooms of certain other types.

In Chapter 3, I discussed the concept that given the weaknesses in the electro-mechanical monitoring systems, it was not unreasonable to expect exponents of very differing types of music to opt for different systems. In Chapter 15, I expanded this further by suggesting that no single control room design could necessarily provide optimum conditions for music of both acoustic and electronic origins. The chapter concluded with a proposal for a room of a 'non-environment' type, with dual monitoring and an artificial reverberation system distributed along the side walls which could be switched in at will. I did not fully appreciate just how much of a necessity that may be until I received the delayed responses from the LMH auditioners. On the other hand, of the next ten clients for whom I built such LMH style rooms, after becoming familiar with them, especially in their ability to highlight fine detail, no owner had expressed a wish to choose any other design concept for future rooms. The owner of the second room of this type which I built ordered two more the following year. Two of their engineers left to start up business on their own, and again ordered a similar style room.

17.6 What *are* our objectives?

For much of the time, the control rooms are now the performing studios, and it is the 'vibe value' of the conventional monitor system in a conventional control room which has now become such an established part of the performing side of a recording process. A neutral environment is by no means always desirable in a performing room. Possibly the concept of a multi-monitored, optionally artificially reverberated room as discussed in Chapter 15 may well be the only way out if one room is intended to cater for all tastes.

Returning to the subject of the purpose of a control room, if we restrict ourselves initially to the classical concept of a reference room, then amongst other things, one of the assessment aims will be to check the suitability of the mixes for domestic consumption. The question would seem to be, if no one type of room can be representative of all typical listening environments, then which design concept will produce the best end results on the types of listening systems for which those rooms were *not* optimised? Personally, for such assessment work, I prefer the new rooms. The new non-environment rooms, with suitable monitor systems, are almost certainly capable of greater definition of fine detail, due to their greatly reduced masking of low level sounds by the reverberant hang-over from any immediately preceding high level signals. Undoubtedly, there are many people who enjoy the detail and clarity of these rooms.

The difficulty of working to standards in such a subjective area is rather similar to comparing Mike Tyson with Muhammed Ali. Tyson's strength usually overcame opponents with overwhelming power, whilst Ali's successes were based on speed and accuracy. As both became World Champions, can it be said that either approach was right or wrong? I think not. Taking things to an extreme, if a studio with poor monitoring was consistently producing big selling recordings, could it be considered to be a poor studio? It is not beyond the realms of reason that the success could be down to the effect of a couple of stunning lady members of staff spurring on the bands to new heights of performance. Yes, it certainly *can* come down to such non-engineering criteria, but while we cannot define it, then nor can we completely deny it.

I have now built three rooms in Liverpool, only a very short distance from each other. There is an old-style room with an old-style monitor system, a new-style room with an old-style monitor system, and a new-style room with a new-style monitor system. They were in fact built in that order. All of the owners like their own studios very much. They have developed partisan clientele who opt for whichever studio provides them with their specific needs; the clients settle on whichever studio works best for them with their music.

When a client now asks me which approach to recommend, if I have a good prior knowledge of the type of music which is likely to predominate I can make suggestions, but I still prefer sending the client, with his or her respective clients, to listen to the various approaches before final discussions take place. The pro-conventional room lobby cite the warmth, the power, and the intimacy as pro's for their rooms. They complain about lack of intimacy in the non-environment rooms – as if the music were a separate happening in which they were only observers and not fully involved. The pro non-environment followers claim superior imaging, definition, clarity, ease of decision-making, and general 'accuracy' for their rooms.

Having experience of working in many of the previously discussed types of room, and considering all that has been discussed in Chapters 15 to 17, I shall sum up in the following section some of my current thinking and personal tendencies towards control room design. It was first published as a paper, presented to the 'Reproduced Sound 10' conference of the Institute of Acoustics in the UK, in November 1994, and was co-written by Dr Keith Holland, Tom Hidley and myself (entirely by post as we were living in three different countries). It may serve here to reinforce some of what has been said, and also to address the concept of reverberant masking of fine detail, which with digital recording techniques is becoming increasingly important.

17.7 Control room reverberation is unwanted noise

It has been customary practice in most situations to produce control rooms with reverberation times approximating to, or slightly less than, some statistical 'average' domestic listening room. Numerous methods have been used in attempts to remove peaks and dips in the reverberation times of the rooms which create idiosyncratic colouration of the monitoring conditions. This has led to musical balances which may not 'travel well' when reproduced in other rooms. Despite fifty years of efforts to produce rooms which subjectively sound both consistent between themselves and 'typical' of the outside world, the general state of inter-room compatibility is still not good. This paper argues that only by minimising control room monitoring RT can greater commonality be achieved; but this requires the abandonment of any adherence to the old policy of mimicking any perceived domestic 'norm'.

17.7.1 Sources of inconsistency

There are only two sources of inconsistency between control rooms or any other listening rooms: the monitoring systems and the room acoustics. In conjunction with each other, however, this combination, even when ostensibly well designed, has conspired to produce some alarming degrees of difference in subjective sound quality. In fact, the difficulties in producing rooms with a consistent sound quality, irrespective of shape and to some degree size, has put constraints and pressures on the designers of loudspeakers, searching for a way to make the loudspeakers themselves less subject to perturbations in the responses caused by the different rooms in which they may be used. To be fair to the loudspeaker manufacturers, the room problems should be sorted out by the room designers, at least in professional circumstances, but the consistency problem, particularly at low frequencies, has been difficult to resolve.

The modal patterns of conventional rooms are very complex. Even two rooms dimensionally identical can show great deviations in their low frequency reverberation times, not only according to the materials and thickness of their walls, but also to what may exist on the other side of the walls. Control rooms constructed of dry wall partitioning are particularly prone to having their low frequency characteristics dominated by the dimensions of the more massive brick or concrete structures in which they are assembled. The resolution of the calculations of low frequency modes within such rooms is not trivial, and is further modified by anything else which may be constructed in the greater space.

Tiring of such inter-room inconsistency, producers and recording engineers have relied to a very great extent on the close field or 'near-field' loudspeakers, placed on or close to the mixing console. At less than the critical distance at which the room begins to dominate in the perceived response of the loudspeaker, recording staff have been more content with what they heard from the same loudspeakers in different rooms. Once a particular set of loudspeakers had been chosen (varying from individual to individual in terms of which ones each person felt produced the most consistent mixes and sounds, when related to the outside world), each producer and/or engineer would frequently continue using that type until something even better for them became available.

The small loudspeakers had a further advantage of relating more closely to the frequency range of the majority of domestic reproduction systems. Whilst this is indeed a worthy reference to conduct, as nobody would wish to short-change or disappoint the majority of the listening public, too great a reliance on the small monitors frequently left one or two octaves of low frequencies completely unmonitored, and much low level, fine detail could not be detected. In some instances, this fine detail could have been used to great effect if only it could have been heard, but in other cases the fine detail included noises, harmonic clashes, the operation of gates or compressors, or a host of other artifacts of the recording. Currently, these are frequently noticed first only by the music lovers who choose to spend a great deal of money on their domestic hi-fi and CD collection. That such a situation exists does not bring any great honours to our profession. Quality control is *our* responsibility, not that of the retail purchaser of the finished product.

17.7.2 Conventional design concepts

There have been a number of attempts to smooth out the modal responses of rooms in a search to provide a reverberation time/frequency performance which could closely approximate to the specified requirements, but without the irregularities which contribute to the undesirable, individualistic sound of so many rooms. To mention a few, there have been damped Helmholtz resonators, tuned to absorb undesirable frequencies which may predominate a room's RT. At best, the results are only approximate. There has been the 'Live End, Dead End' approach of Davis, Wrighton and Berger, which sought by means of an absorbent front half of a room to leave a suitable time interval between the incident wave's first pass of the ear, and the subsequent 'life' or pseudo-reverberation produced in the rear half of the room. By this means, reverberation was perceived *as* reverberation (I use the term loosely here) without colouring the direct sound from the monitors. A further means of achieving a more uniform RT with frequency in these rooms is by means of quadratic residue diffusers, which by suitable place-ment in a room can achieve a high degree of very diffusive reflexions, free of the discrete and possibly unwanted specular reflexions which can predom-inate at certain frequencies in conventional rooms.

Amongst others, there is the Geddes approach of the steeply double sloped wall, sloping in the vertical and horizontal directions, designed to disrupt the stronger axial and tangential modes in order to drive all modes into the less regular and more easily controlled oblique form. In addition to the above, there is the multi-faceted approach, spreading the modes into the broadest overlap. Such rooms have frequently been used in conjunction with monitors whose directivity was tailored to the rooms, such that 'patchworks' of reflective and absorbent surfaces could be positioned to yield an overall desirable RT/frequency characteristic. These rooms usually incorporated 'bass traps', low frequency absorbers of considerable volume, to deal with the less directional frequencies below 300 Hz or so.

In the hands of capable designers having a comprehensive knowledge of the underlying reasons behind each approach, there have been differing degrees of successes and failures. Sometimes luck has lent a hand, whereby the personnel of a certain studio had a natural empathy with the techniques used, or at least soon learned to adapt to their characteristics; but luck could sometimes work in the reverse sense. A major problem which arose was that whilst each technique could be used to approximate to one given specific RT goal, although these could be measured to correspond very well with their target figures, subjectively, they all sounded different. What is more, none of them produced their pseudo-reverberation in a way that was typical of real domestic conditions. Only the RT figures matched well; the perception was entirely different.

17.7.3 What domestic RT?

In itself, the concept of an average domestic reverberation time is a highly dubious target to aim for. It seems to be locked in to our thinking, probably from times when music was more ethnically localised, and when the sound reproduction equipment and the housing of the majority of probable purchasers of the music were likely to be more similar than they are today. In recent years, some well-known recording artistes have been selling in the order of ten million records a year, worldwide. The internal acoustic of an all concrete and brick, Southern European house is very different from either a granite built Scottish house, a Far Eastern wood or bamboo house, or a heavily furnished Californian apartment, yet all must now be considered as quite probable final destinations for much of today's internationally marketed music. All purchasers pay roughly the same prices for their recordings; so all could claim the right to expect *their* domestic conditions to be equally considered, which is clearly impracticable. What is more, there is a very great percentage of the music buying public who now predominantly listen either in motor cars or on headphones, and, for these people, reverberation, pseudo-reverberation or even significant reflexions hardly exist. So, why do we still produce control rooms with any significant degree of reverberation at all?

17.7.4 Alternatives

Undoubtedly, anechoic rooms are not very pleasant environments in which to spend very much time. Indeed, many people find them unnatural, disturbing, and, in some cases, even stress-inducing. Especially when loud music is reproduced under such conditions, there is cause to believe that it triggers a 'fear' response in many people, when the dynamic range exhibited between the silence and the music causes their brains to warn them of impending danger. In nature, such wide dynamic ranges are usually only experienced in times of natural disaster, where great energies are being released. On the other hand, anechoic conditions are highly repeatable in any given frequency range, which is why they have been so widely used for comparative and absolute measurements. Under anechoic conditions there is a tendency for loudspeakers with the best amplitude/phase compromises, and hence the best transient responses, to show a degree of reality which is noticeable to a greater extent than from any loudspeakers in conventional rooms. Shroeder's 'phase organ' experiment suggests that this would be so, as the perception of the 'tones' he produced by varying the phase relationships of harmonics in a pulse train under anechoic conditions was totally lost once reflective surfaces were introduced into the listening environment. Certainly anechoic conditions provide the best circumstances for detecting fine detail in the sound. Such conditions enable pin-point stereo imaging, and a tightness and reality in the overall perception, but it is largely their unpleasantness in the human comfort sense that has prevented them from becoming the natural quality control and monitoring environments. In anechoic conditions, the problems of creating loudspeakers with smooth responses at extreme off-axis angles is alleviated by the fact that 'out of the normal listening area' anomalies are absorbed, and hence do not return to the listening position.

17.7.5 Practical solution

It has been becoming apparent to a growing number of people that the optimum monitoring conditions for current recording requirements would be an anechoic room in which human beings would feel comfortable. The achievement of this requires a room with a dual acoustic, one for the monitoring and one for the people working within it. If the front wall is made to be both massive and reflective, and the main monitor system is flush mounted in this wall, then the wall itself provides an excellent baffle extension for the radiation of low frequencies, but cannot be 'seen' in ray-diagram terms by the monitor loudspeakers themselves. By making the wall suitably irregular on its surface, very pleasant specular reflexions can be returned to the persons within the room in response to their speech and actions, but nothing of the musical signal from the loudspeakers can be returned to the front wall, and hence back to the listener, as all other wall and ceiling surfaces are made as non-reflective as possible over as wide a frequency range as possible.

To further brighten the room, a reflective floor can be used, which helps to reduce the directionality of the 'front wall only' approach. The subjective neutrality of the floor reflexions is best described if we consider a large anechoic box, in which we fix securely a loudspeaker and a measuring microphone. Were listeners to enter the box (standing vertically with heads adjacent to the microphone), listen carefully to the sound, then listen for any perceivable changes when a hard, reflective surface was placed along one side, they would surely notice both a change in tonality (by reflective colouration) and a possible change or loss of precision in the horizontal position of the apparent source of the sound. In each case, before and after the introduction of the reflective surface, the response at the listening position could be measured and plotted via the microphone.

Let us now consider the situation if we were to rotate the box through 90°, by rolling it over on to one side such that the reflective surface, which we had placed alongside one wall, was now on the floor. As the microphone, loudspeaker and reflective surface had all been fixed in position before the rotation, then their relative positions would be fixed, so a further measurement taken in this position would be absolutely identical to the one taken previously. On the other hand, if a listener were to re-enter the box, with his or her head in a similar position as before (though of course now rotated through 90° relative to the previous standing position in the box) a very different perception of the sound would be noticed, as the reflexions would now be arriving at the ears from below and not from one side. Colouration would tend to reduce, back towards that of the anechoic conditions. Furthermore, the horizontal positioning would return to pinpoint accuracy, and little if any disturbance would be noticed in the vertical positioning.

The reason for this is that our ears, or at least those of by far the greatest majority of people, are enormously more sensitive to position and colouration by disturbances in the horizontal plane than in the vertical plane. Being restricted to largely two-dimensional movement over the surface of the earth, in fact usually moving very little during our lives in terms of distance from a floor, it is not surprising that evolution has tended in this direction. Until the introduction of the air-launched bomb and the landmine, humans have had no real threat to their existence from predators attacking from above or below.

17.7.6 Subjective perception in low-RT control rooms

When listening in stereo in a control room designed according to these principles, and especially when an absorbent screen is placed at the back of the mixing console, then exceptionally uncoloured sound, with precise stereo imaging, can be experienced over a large area of such rooms; dependent upon the suitability of the monitor system of course. Enough rooms built using the above principles are now in existence worldwide to be able to give a clear description of the general trends and tendencies in the way that they are used, and the results which they produce. The principle, to the best of our knowledge, was first introduced by one of the authors (Hidley) in the mid-1980s in his 'non-environment' control rooms, in which the monitors were intended to respond subjectively as closely as possible to driving into 2 pi space. By its very nature, the concept is highly tolerant of shape and size variations, as anechoic (for the monitors) means as close to zero reverberation as possible, and zero is the only number which can be multiplied by any variation factor yet remain the same. The problem with reverberant approaches, be they reflective or diffusive, is that any given 'reverberation' time, produced by a variety of means, will produce noticeable variations in the perception of the subjective response of those rooms, and hence will disturb the room-to-room consistency of the monitoring conditions between rooms using different techniques of control.

In the 'non-environment', 'monitor-dead' rooms, one very strongly noticeable trend is towards recording and mixing on the large, in-built monitor systems, with the small close-field systems returning to their original position in the hierarchy as secondary references to a domestic world. Users report few surprises when taking the mixes away to play at home, in the car, or on headphones, which is hardly unexpected as the rooms remove one of the very large variables from the mixing environment, but by also removing the bias towards one form of reproduction (the domestic reverberation time) other listening environments are more fairly considered at the mixing stage.

17.7.7 Doubts

Two questions which frequently arise from people who have not experienced such rooms are 'Is there not a tendency to use too much reverberation in the mix, because the room sound is so dry?' and 'In the absence of any low frequency reinforcement from the room itself, do the monitors not sound bass light, and do mixes not subsequently leave the studio bass heavy?' In practice, the answer to both questions is 'No'. The RT that does exist in many control rooms is much shorter than that which would tend to be artificially applied for musical purposes, hence the two are not very closely related. The reverberation which *is* applied, however, is in relation to a lower background ambience, so can be heard much more clearly. In fact, one becomes very aware of the individual spaces in which the instruments were recorded. Differences in the choice of type and position of the recording microphones become much more clearly apparent, as do the aspects and defects of any artificial signal processing. Problems and distortions are heard where they should be heard, in the control room, which again can be used for quality control at the times of recording and mixing.

With respect to the relative level of low frequencies, as I stated earlier, one of the most noticeable aspects of these rooms is the tendency to rely much more on the large monitor systems, which if well designed allow proper relative monitoring of the whole low frequency band, so 'guessing' what is happening at 40 Hz becomes less of a lottery. Psychological adjustment to the short low frequency reverberation times seems to be accomplished quite quickly, and after one or two false starts, mixing personnel soon lock-on to the required techniques. In fact, some of the false starts are caused by the engineers and producers still applying subconscious 'correction factors' which experience has taught them to be necessary in more conventional rooms.

17.7.8 Noise

What is becoming more and more apparent, as the use of these rooms spreads, is that control room reverberation should be considered to be nothing other than unwanted noise. Like tape noise or non-linear distortions, it masks detail by introducing a noise floor below which it becomes very difficult to hear other noises, distortions, or low level signals. Many important aspects of a sound may exist at times in a region of, say, 65 dB below peak. A tape noise floor of -60 dB will make perception of these sounds very difficult, and may even mask distortions. The reverberant hang-over of a sound can in a similar way easily mask the low level detail of a sound existing two or three hundred milliseconds later, and it is many of these low level signals which provide clues to the spaces in which the instruments were recorded. What makes control room reverberation even more of a potential problem to monitoring than is tape noise, is that the tape noise is part of the finished item. Undesirable though it may be, the same tape noise which masks the problem in the control room will still be present to mask the selfsame problems in the homes of the purchasers of the recordings. The same cannot be said, however, for any masking due to control room reverberation. Although it may cover problems at the recording and mixing stages, the unheard problems may become only too apparent to critical listeners in the outside world – and, as stated previously, quality control is not their job, it is ours! They have paid for a finished product and they are justified in expecting the best value for their money.

Control room reverberation has long been a problem for inter-room consistency, and all that any of the conventional approaches have achieved is little more than pushing the response bumps around. Little consensus has been reached in precisely how RT control should best be achieved, but now that such a large percentage of end listeners will not be listening in conditions anything like the statistical norm, it has put the whole concept of control room reverberation standards in doubt. Given the additional clarity of perception being experienced in the 'non-environment' type of rooms, and the subjective removal of about 15 dB of masking noise from the mixing environment, surely the concept of any monitoring RT, other than as close to zero as possible, has no future relevance.

Chapter 18

Active control rooms

The new potential of active room design is not to be confused with the 'Assisted Reverberation' as used since the 1950s in the Royal Festival Hall, and later in Limehouse Studios. These were systems using tunable resonators, and multi-channel amplifiers to distribute the natural resonances to desired parts of the room. They were reverberation *enhancers*, not capable of reducing reverberation time, and only to a very limited degree were they capable of dealing with undesirable aspects of the room's standing wave acoustics. The new systems are digitally controlled, and are capable of 'learning' aspects of the room's performance, in order to be able to predict likely response errors and apply correction to the active control radiators.

Four quite distinct areas of control have become possible, all stemming from the same concept of system 'brain'. Firstly, active trapping, using a multi-point detection and control system, supplying anti-phase energy to negate the effect of standing waves and room resonances. These are effectively 'killed' at the room boundary, thus preventing reflexion back into the room. With these systems, loudspeakers are used as active absorbers, with the absorbed acoustic energy turning to heat in the voice coils of the control loudspeakers. The spacing of loudspeakers over an absorbent wall would be approximately such that the distance between loudspeaker centres would be half a wavelength of the highest frequency to be absorbed.

Secondly, an extension of this system would make possible a much more precise control over the acoustic characteristics of rooms in general. Almost every studio booking manager must be familiar with the problem of one producer who likes a rather 'dead' room and another producer who prefers a more 'live' room. Despite the studio having two, acoustically different, rooms, in 85% of cases the inevitable happens, and whenever the first producer tries to book time it is only the 'live' room which is available – more unbooked time! With the new, active systems, the rooms could be rendered acceptably similar, with programmable acoustics adaptable within limits to individual choice from the cavernous to being almost anechoic.

Some of these systems take a reference from the electrical drive to the monitor system, enabling them to cross-reference, and run a transfer function between the actual input signal and the perceived sound in the room. To the comparators would be added any pre-required room characteristics, and also the cancellation signals for the undesired room resonances, reflexions and standing waves. The systems become self-analysing, compensating for changes. A further extension of this principle would introduce an error signal into the monitoring system itself, thus enabling a third function for the system. The high level, largely minimum-phase monitoring systems which have been developing are capable of electrical phase and amplitude correction. This allows error monitoring of the sound within the desired listening area, which through a feed-forward system can provide a remarkable degree of linearity of the impulse response within the room itself. This technique has been possible for some time at a 'single point' in the room, but wider range coverage is only now becoming a practicable possibility.

Typically, a 2 ms delay on the monitor system allows the compensation to be realised via the feed-forward correction systems. To put 2 ms into perspective, if you move 2 ft further back from the monitor system, you will experience an additional delay of around 2 ms. The effect is inconsequential – it will not cause engineers to fail to press mute switches at the appropriate time. Aspects of this 'response linearisation system' have become available independently of the room control systems.

The fourth application is to the acoustics of the studio areas themselves. Modification of the technology can produce variable acoustics without the requirement for carrying in heavy drapes, corrugated iron sheets, wooden panels, or any other cumbersome devices. Although certain 'mean acoustics' may be desirable as starting points, the range of possible control is enormous, and the separation of the decorative aesthetics from the required acoustics is now realisable. Parallel walls are no longer a taboo; they may even be desirable, giving more orderly and hence more easily controllable patterns of reflexions. Layouts within the rooms themselves will also enjoy a new degree of flexibility.

Certain doubts have been raised about potentially 'disturbing' or 'stressful' aspects of the anti-phase systems, but all evidence from applications in aircraft and industry suggest that, when functioning correctly, the systems are very effective in *reducing* stress. There *are* implications for the sensory confusion of being in a visually 'dead' room with an aurally 'live' acoustic, but I am sure that this is by no means insurmountable.

18.1 Impulse reconstitution

The impulse reconstitution system would take a reference from the electrical output from a mixing console or control unit, and make comparisons with the acoustic impulse received in the proximity of the listener. An adaptive, digital, signal processing system would then compare the two signals, adjusting the drive signal to the monitor system in order to achieve the least error between the two signals. This error cancellation would take into account the irregularities produced in crossovers, amplifiers, and the loudspeaker drivers themselves. As the signal processing system must apply correction signals to the monitor chain, the greater the linearity inherent in that chain before correction, the lower would be the level of compensation. This would allow a greater achievable sound pressure level, all other things being equal, as a lower level of correction signal would need to be applied. The overload point of the monitor system would be a function of the sum of the desired output *and* the correction signal; hence the lower the level of correction, the greater the available power for the music signal. For this reason, the pursuit of highly linear monitor systems should be continued. The power of the new response linearisation systems should not be used as an excuse to worry any less about using inherently highly accurate monitors.

Figure 18.1 shows the results obtained by Clarkson et al.[1] from a very simple loudspeaker system using a version of this technique. Obviously, these results were encouraging, especially as the loudspeaker in question was of somewhat dubious quality. Developments of these systems are capable of learning the effects of the monitor system and the room. Whilst continuous monitoring of the error may be desirable, it would be possible to lower an array of sensing microphones from the ceiling to programme the system, retracting them once the system had adapted. Any changes in the room could be reprogrammed as required, simply by once more lowering the sensing array and resetting the system. It has also been considered possible to locate microphones in strategic, out of the way places, which could 'know' what they needed to hear when things were correct, which may in turn allow for continuous error monitoring to be achieved.

Figure 18.2 shows a typical room response at four different locations (the solid lines). The dashed lines show the corrected response after optimisation of the response by correction for position 0. From this it can be seen that the responses at positions 2 and 3 are quite clearly degraded from their original performance. Such a system would most likely be of the 'fix your head in a clamp' genre, it being unlikely that two or more people could benefit from any such system in one room. Figure 18.3 shows a similar application to an impulse response, once again with positions 2 and 3 suffering badly. Figure 18.4 shows a multiple least mean square approach, used to find a 'best fit' 'average' over a desired area, with which at all points measured, the dotted line of the corrected response shows that clear benefits have been achieved from the application of the above technique. At all points within this area, the impulse response would be significantly improved when compared to the response without processing. Assuming acceptably low non-linear distortion performance, then as all other factors of phase and 'frequency' response must be correct before an accurate impulse can be achieved, when the impulse response is correct, all other factors must be correct.

Should the processing system fail, the monitor system would *not* be rendered inoperative. As already stated, the greater the linearity of the unprocessed monitor systems and rooms, the easier the job for the processor, so if the processing system fits within the existing monitor chain, nothing else in the chain would be compromised or adapted to accommodate the system. Switching the processing system into bypass, or removing it completely, would return the monitor system to its original state. It is not as though one would have to limp on in a crippled condition until the processing system was either repaired of replaced. As the system would be dedicated to the main monitor system, there would be no interaction when switching to secondary, close-field monitors. No signal input into the main system would mean no signal output. It would not try to compensate for errors from a secondary loudspeaker system. Obviously, a second processing system could be fitted to the secondary monitoring system, but it would probably need only to be a less complex unit, more suited to application to domestic hi-fi systems.

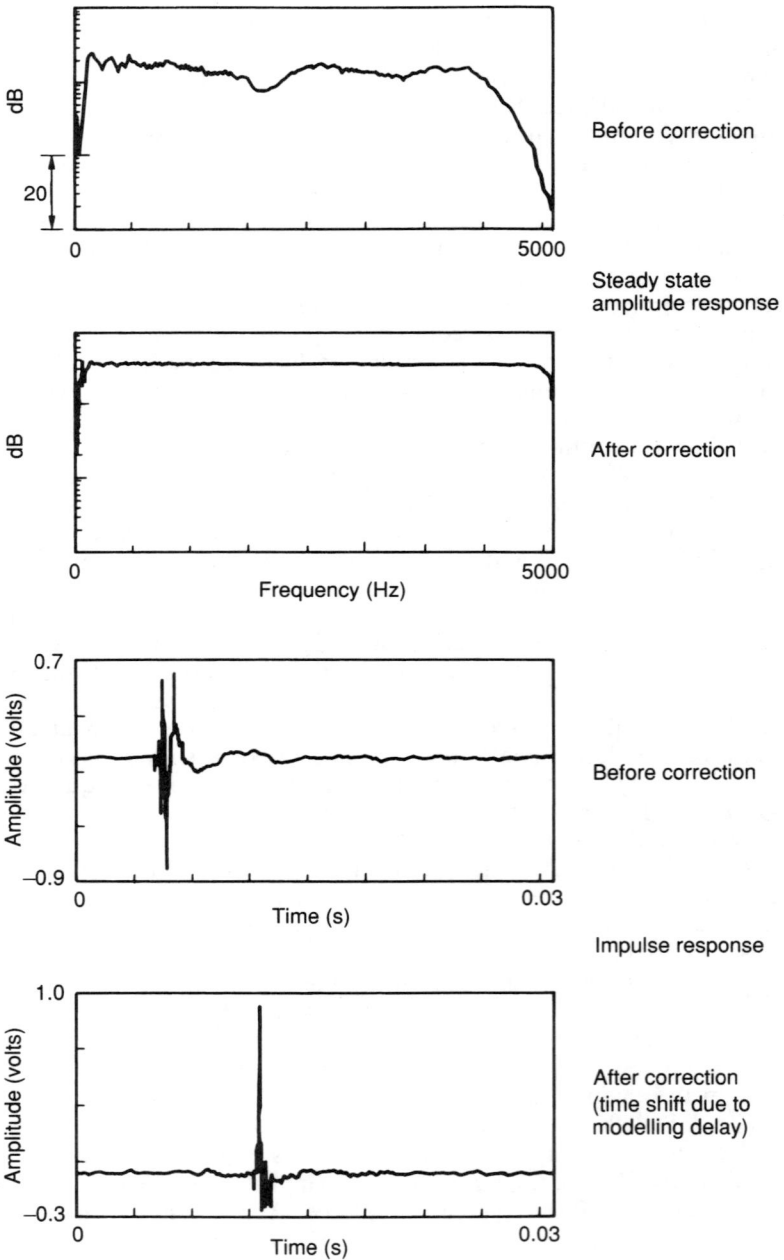

Figure 18.1 Spectral, phase and transient equalisation. (After Clarkson, Mourjopoulos and Hammond, 1985, *Journal of the Audio Engineering Society*, **33** 127–132 'Spectral, phase and transient equalisation for audio systems')

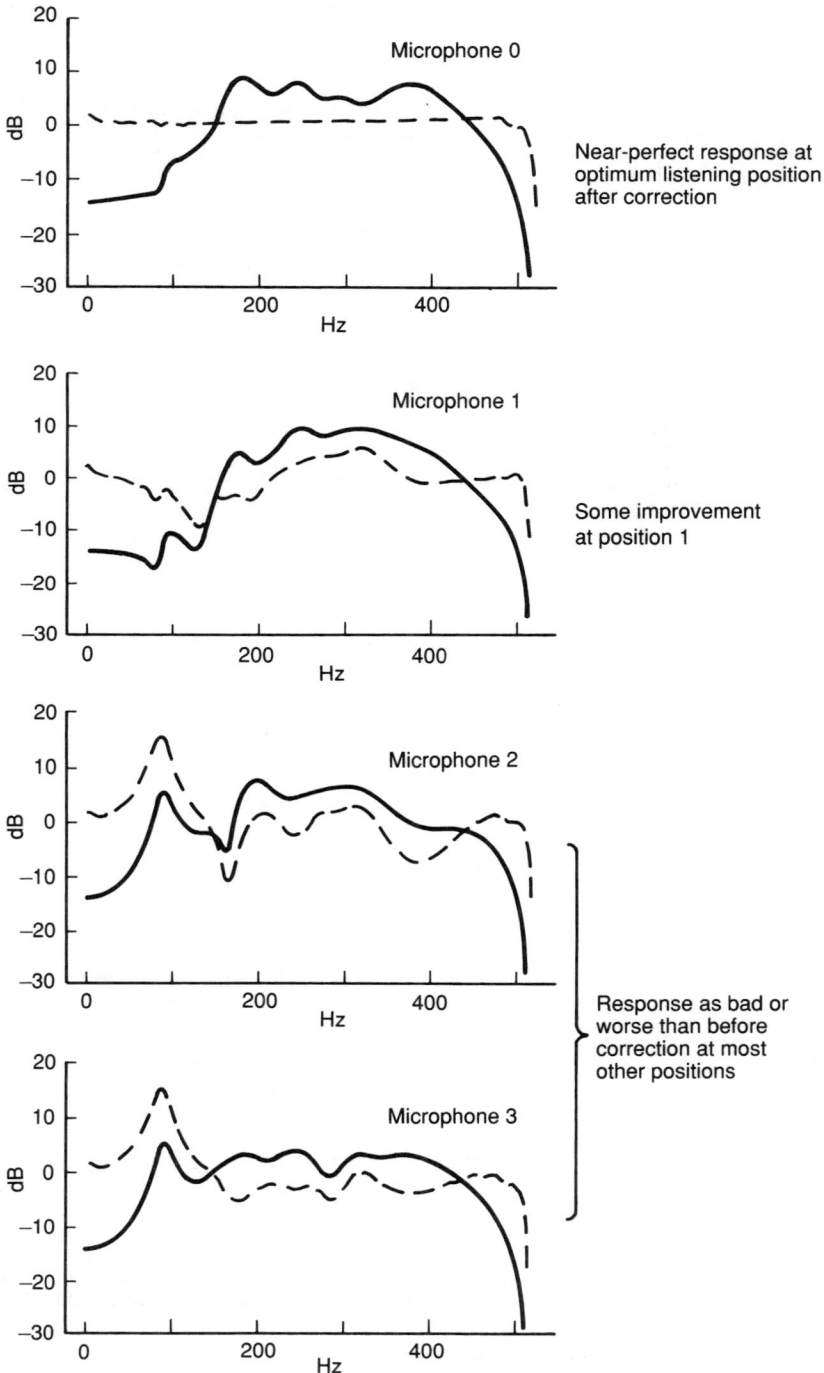

Figure 18.2 Single point equalisation of steady state response. Solid line, original response; dashed line, after correction for position 0

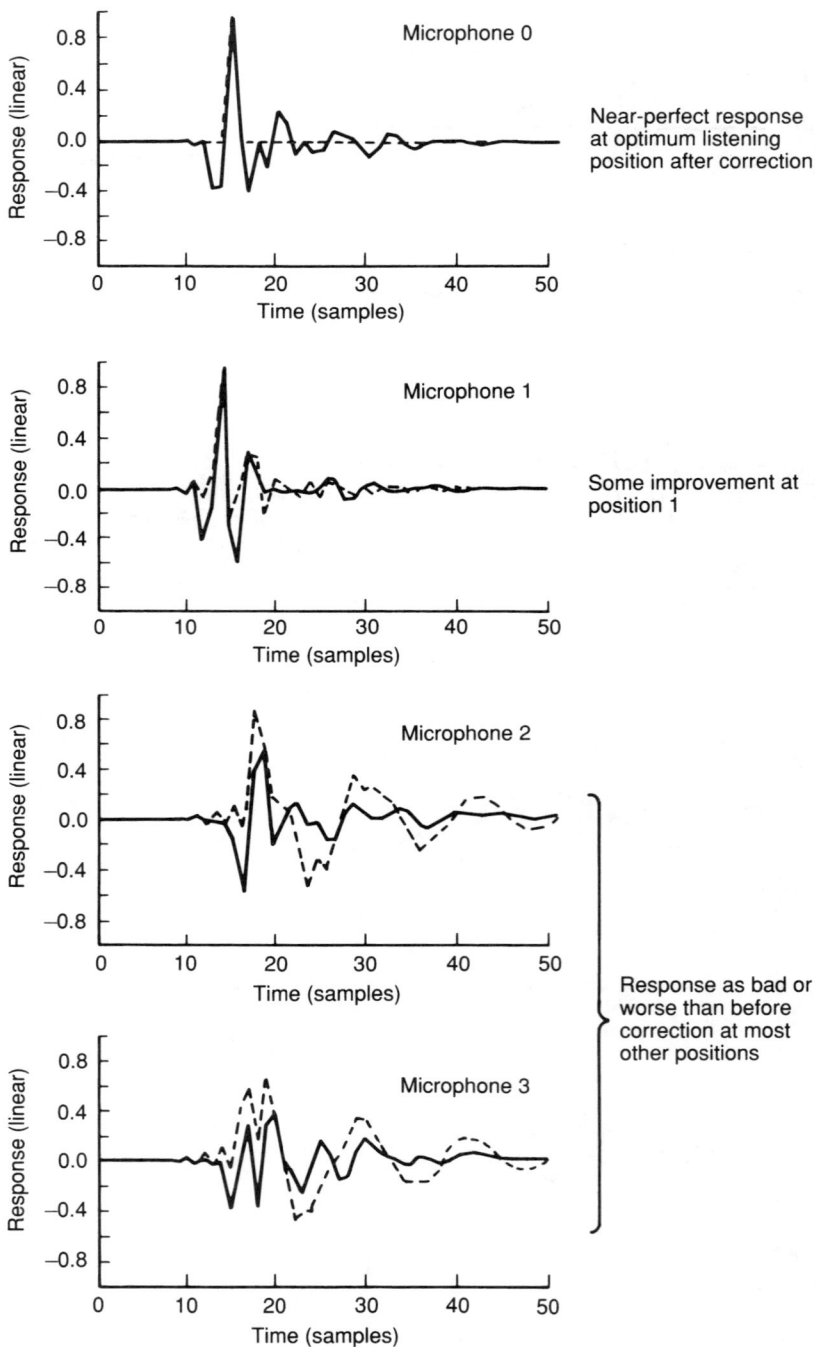

Figure 18.3 Single point equalisation of impulse response

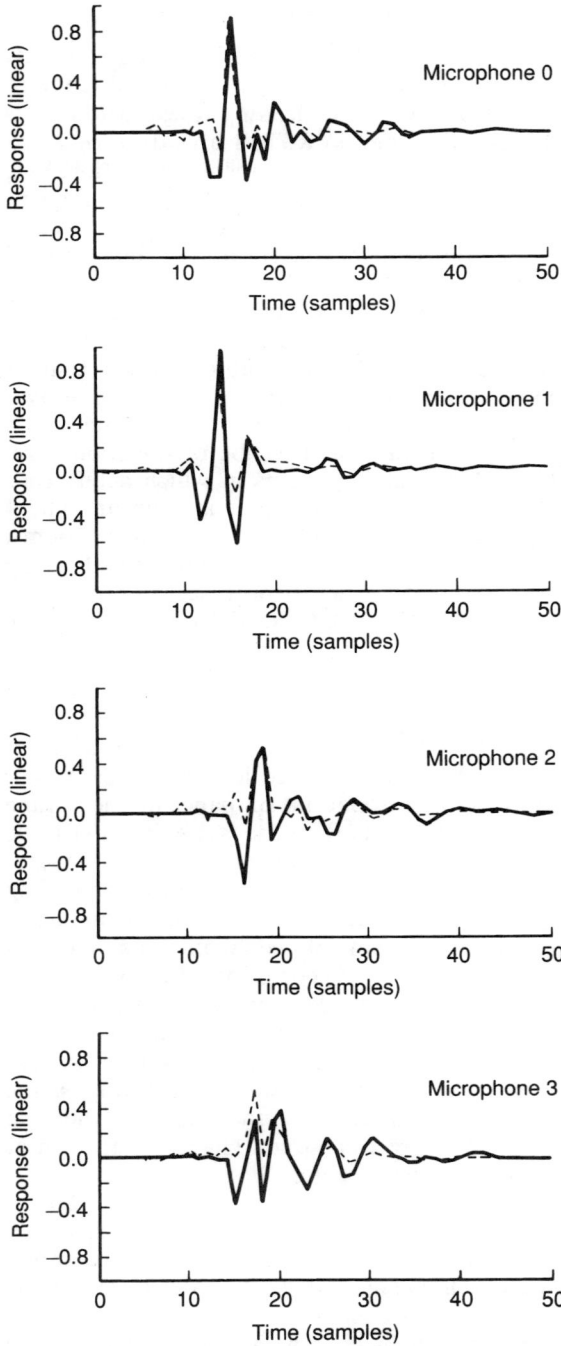

Though response at
position 0 is less
perfect than in
Fig. 18.3, *all* positions
show a great
improvement over
their uncorrected
responses

Steady state responses
show similar order of
improvement, nowhere
perfect but everywhere
significantly improved

Figure 18.4 Multipoint equalisation of impulse response

18.2 Active room control

Systems such as those described above could make enormous inroads into the age-old problems of response standardisation and consistency of performance. Even in the late 1980s, two of the leading researchers in this field, Steve Elliott and Phil Nelson at the ISVR, had already progressed some way down the road towards the in-car entertainment application; but, as can be seen, the applications are relevant wherever there is a requirement for accurate, high quality sound reproduction. Along with the second branch of the research, active room control, an enormous step forward could be achieved in so many areas.

That second branch of research, the active room control system, is somewhat more complex than the impulse reconstitution system, and, in turn, sub-divides into two further areas. The first sub-division is that of the suppression of unwanted room effects, the second is total control over the desired acoustic of any given room. As a rough rule of thumb, the complexity would appear to be a function of the cube of the room size. The first sub-division would apply mainly to rooms in which a reference signal was obtainable directly from an electrical input into an amplifier/loudspeaker system. Such applications would include studio control rooms, domestic hi-fi, listening and quality control rooms, and concert halls using public address systems – as opposed to acoustic performances of orchestras, operas, or similar. The second sub-division relates to those purely acoustic rooms.

18.3 Rooms with an electrical reference

The time-honoured method of controlling unwanted low frequency irregularities in rooms has been the use of resonators, or panel or membrane absorbers. One very great governing factor in their use is that in order to be effective they tend to need to be large, sometimes very large indeed, occupying 30% to 70% of the total air volume of a room. In recording studio control rooms in particular, space is often at something of a premium. More equipment and more people are being squeezed into many rooms which are no larger than those of a previous generation of studio control rooms. The subsequent compromise of acoustic control systems, together with the variability of the internal acoustics due to the changing quantity of personnel and equipment, is one reason for the current emphasis on the use of close-field monitors. By listening close to the small loudspeakers, the room variability has less effect on the subjective sound balance, rendering better day-to-day consistency of the perceived sound character, though at the expense of overall frequency and dynamic ranges.

Considering resonances due to the air space in a given room, the influx of large quantities of personnel and equipment will change that air space by ing air volume and changing the reflective paths within the room. des and resonances may well change appreciably, but to all intents oses the resonators, tuned to the empty or typical condition, will t at their pre-determined frequencies. The inevitable consequence overall, audible character of the room can change appreciably. systems are digitally adaptive control systems, which can absorb energy at the room boundaries, and cancel the effects of unwanted standing waves and resonant modes. Once again, an array of sensing microphones would compare the response in the room with the electrical input to the monitor system. If people did not wish to listen in anechoic conditions, then not all of the reflexions would be absorbed. On the other hand, response errors considered to be outside of acceptable limits would precipitate the feeding of an appropriate cancellation frequency to the room control radiators. The degree of energy absorption would be programmable, thus opening up great opportunities for the control of the room's general acoustic character. A degree of personal taste could be introduced into the system at this point, allowing for compensation to the individual likes, dislikes and preferences of the producers, engineers or musicians. The room could change from a 'recording' to a 'mixing' environment, to suit the moods or requirements of the personnel involved.

Within the concept of the above, there is a great deal of scope for imaginative development. Into the control signals can be injected the characteristics for tailor-made reverberation fields. The room control systems can tie in with the monitor impulse reconstitution equipment, allowing closer correlation to the initial development in anechoic rooms. Room characteristic which may be difficult to control via the impulse system can be subdued within the room itself, considerably easing the task for the monitor control system. Systems such as those described above could enable radical improvements to existing rooms, without the need for demolition and rebuild. The active system could fit within any existing acoustic trapping frames, with only a minimum downtime, loss of commercial revenue, and disturbance to the room.

18.4 Practical application

A typical example of the application of this technology would be to the problematical acoustics of a typical mobile recording truck. The constraints of the UK Road Traffic Act render limitations on the width, height and length of the vehicle. Although the dimensions of a typical truck, 30 ft × 8 ft × 8 ft, are still probably an optimal compromise, acoustically they are a nightmare. Symmetrical width and height compound the problems of the cross modes, whilst the parallel internal surfaces do little to break up the reflexions and standing waves. For reasons of ergonomics, and the minimisation of internal cross modes, the monitor loudspeakers often fire down the length of the vehicle.

Reflexions from the rear wall, twenty to twenty-five feet from the front of the loudspeakers, may give rise to standing waves back and forth along the vehicle. The resulting nodes and anti-nodes cause troughs and peaks in the low frequency response which cannot be equalised by conventional means. Once again due to ergonomic restrictions, only very limited changes can be made to the position of either the loudspeakers or the engineer's position with any view to minimising these standing wave effects. A study was undertaken many years ago to address the problem acoustically, but it was concluded that a bass absorber of four or five feet front to back depth would be required to effectively begin to control the low frequencies. Given the space premium in vehicles, this was considered to be out of the question.

An array of control loudspeakers in an installed depth of no more than six inches could effectively remove the back wall from the vehicle, below around 200 Hz. If the energy is absorbed at this boundary, the subsequent transmission through the structure, and hence the sound leakage to the outside world, would also be reduced. The finite distance from the loudspeakers to the back wall could be programmed into the system such that the arrival time of a sound wave could be anticipated by the control system, and nullified upon arrival, 'sucking in' the unwanted energy. Small wedge-shaped loudspeakers could be sited along the wall/ceiling junction to further control many of the other problems within such an environment. The prospects for control systems of this nature seem very favourable. It must now be quite clear that the potential for the development of these technologies is very far reaching indeed.

I have often stated my belief that monitor problems should be addressed within the monitor system whilst room problems should be addressed within the room. I have also stated that frequency problems should be addressed in the frequency domain whilst time problems should be addressed within the time domain. To a limited degree, active control does allow a certain amount of cross-domain corrections. Adaptive digital control, via the feed-forward around a delay, can address time, phase, and amplitude convolutions in such a way that conventional analogue means cannot.

As with almost anything else, there are problem areas inherent in the advances of the technology. Pre-echoes are produced as a function of many of these control processes, but their subjective audibility is time/intensity dependent, the sonically acceptable envelopes of which are as yet not fully assessed or documented. Whilst all digital storage and processing systems are still less than perfect, and are not universally considered preferable for all applications, undoubtedly they are here to stay, especially to deal with difficult conditions in which acoustic treatment would be either inappropriate or impracticable. Active control is still in its infancy, but as time and technology progress, many problems will be ironed out and enormous benefits will result.

The preceding brief discussion has obviously only scratched the surface of what is a highly complex subject. Commercial systems are now in professional use, providing response equalisation over a limited listening area in places such as post-production rooms, where a great deal of equipment is installed in a relatively small room, and where adequate purely acoustic control *would* be totally impractical. All that I have sought to do in this chapter is to highlight the areas in which digitally adaptive active control will almost certainly see expanding use in the future.

However, where fine detail monitoring is concerned, future generations of any such active correction processors will have to ensure that they themselves possess adequate bandwidth and sonic neutrality to prevent them from imposing their own limits on sonic resolution. I would suggest that DC to 60 kHz would be the minimum acceptable bandwidth for such units, to allow for not only the fact that even modern analogue tape recorders can have usable responses at 30 kHz, but also that future digital recording media will themselves soon be operating up to 60 kHz or even 100 kHz. Any components associated with the monitor system electronics should always have the widest bandwidth of any part of the recording chain. One cannot monitor the temperature of boiling water with a thermometer which only reads from 0° to 50°C.

I urge caution before rushing excitedly headlong into the absolute adoption of the new technology. The two words 'digital' and 'computerised' will be irresistible to many, but reputable manufacturers are aware of the exceptionally high degree of sonic neutrality the electronic systems must exhibit if some of these devices are to be no more than toys. I initially feared that over-zealous marketing could discredit the steady progress in this area, but happily, to date, the general promotion of the products using these techniques has been carried out in a responsible manner.

Active control devices are tools for application by acousticians. They are not to replace acousticians. One of the reasons why so much 'damage' was done by the use of graphic equalisers on monitors was that in an absurdly large number of cases it was seen as a piece of electronic equipment, hence its 'setting up' and general application was left in the hands of maintenance engineers, who often knew pitifully little about acoustics.

Somewhat ironically, as mentioned previously, the more that a room is suited for active control, the less it needs it: like an error correction system, it is better when it does not need to work. Active control is *not* a universal panacea for all acoustic problems, but it does already have its place in the acoustician's tool box, and as developments proceed, that place will no doubt become more important. Acoustic fixes to acoustic problems deal at source with the troubles. They therefore remove the problem. Active control compensates, very very cleverly, for the problem, but the acoustic problem still exists. Correction artifacts are therefore unlikely to be totally removed. On the other hand, for a troubled room in which acoustic control is for whatever reason out of the question, it is beyond doubt that active control can make significant improvements.

It is worth noting here that when current room response linearisation systems are employed to remove resonances and reflexions from the monitor response, they do so only from the monitor signal, and not for the sounds of speech and general activity in the room. They are thus very close in concept to the acoustical 'non-environment' rooms discussed in previous chapters. Whether the room characteristics are removed at source, acoustically, or whether a digitally adaptive filter is employed to deconvolute the room from the monitor signal, the end result is, in either case, relatively anechoic monitoring conditions in a room with its own general ambience. Philosophically, therefore, the two approaches are in relatively close harmony.

Where active control can be very useful is in providing the ability to switch between different desired equalisation curves. If, for example, a studio is to be used for multimedia work – say a film remixing studio which must also record pure music – any different compensation curves which may be required on the monitoring can be pre-programmed and chosen at will. Things may be beginning to change, but historically the optimum control room acoustics for mixing for film or for CD have not sat too comfortably together. A growing number of studios are now being built where multi-functional use will be the only way that they can survive in the future recording market.

Things are changing, and that may be a great boon for the proponents of active control. Progress is likely to be both swift and interesting, and I am sure that the necessary investigations which will be required in both audiology and psycho-acoustics will benefit all those involved with general acoustics. One thing which the major electronic companies do tend to have, which recording acousticians have traditionally lacked, is adequate research funding. In the past, much of the funding that was available for audible perception research was funded either by medical institutes seeking to cure hearing disorders, or communications companies (or Government War Departments) seeking to understand more about intelligibility. Unfortunately, the part of the brain which deals with the perception of the spoken word is quite distinct from the part that perceives music. This is one reason why so many 'laws' and 'rules' about hearing music are based upon total misconceptions. At last, we are beginning to see significant research funding aimed at the niceties of the perception of music. Much of this has already come from studies of digital data compression techniques and the audibility of its artifacts. With active control research now on line, this new source of audibility research will surely help us all.

Reference

1 P. M. Clarkson, J. Mourjopoulos and J. K. Hammond, 'Spectral, Phase and Transient Equalisation for Audio Systems'. *Journal of the Audio Engineering Society*, Vol 33, 127–132 (1985)

Bibliography

Ronald Genereux, 'Adaptive Filters for Loudspeakers and Rooms'. Audio Engineering Society Preprint, 93rd Convention, San Francisco (1992)

Chapter 19

Monitor systems for non-environment rooms

19.1 Requirements

The 'universal' loudspeaker system for all locations and all circumstances would involve technologies about which we can only dream, so given the world that we live in, we work for most of our lives seeking 'best compromise' solutions to specific problems. The advent of the 'monitor dead', 'non-environment' rooms has brought with it new demands as to where those best compromise points should lie for the designs of loudspeakers to be used in those rooms. As the rooms themselves seek to provide no environment for the loudspeakers, which are effectively driven into an anechoic termination, the *perception* of those loudspeakers is virtually the same as would be heard *in* an anechoic chamber. The directional, secondary room acoustic is there only to provide a pleasing ambience for the general activities of persons in the room.

What we always listen to, even in an anechoic chamber (which is never perfectly anechoic at all frequencies), is the room/loudspeaker combination; but as the reverberant and/or reflective nature of the room tends towards zero, then by far the overriding factor in what we hear becomes the loudspeaker system itself. This is indeed the whole purpose of the concept, for when we remove the room-to-room variability from our monitoring equation, we stand a much better chance of achieving mixing compatibility from one studio to another.

Sat in the engineer's chair, behind the mixing console, we are to all intents and purposes monitoring the axial response of the loudspeaker system, which by far the majority of reputable monitor manufacturers should have got together by now. So much of the variability in monitor systems has been in their off-axis responses, an aspect of design to which monitor manufacturers have paid various degrees of attention, from being a fundamental aspect of design, down to almost no consideration at all. There has then been an element of pot-luck when introducing monitors into the greatly varying architectures of control rooms, such that 'a great system' as perceived in one room can become a great disappointment when heard in another.

When rooms are to any significant degree reflective or reverberant, they will reflect back energy into the listening area which is spectrally characteristic of the surfaces from which they reflect, and of the on- and off-axis responses of the loudspeakers which excite the different reverberant and reflective fields. The aim in good control room design has been to achieve a uniform and smoothly spacially distributed total power response; total power being the direct energy from the loudspeaker plus the reflected energy from the room. The interaction of these responses is discussed in detail in Chapter 9, and no easy electronic rebalancing of this response is possible if the two are not linear in themselves. Only by active, digitally adaptive signal processing methods, such as described in Chapter 18, can this type of imbalance be corrected, but obviously, where possible, it is better to get the balance correct at the design stage.

In short, well designed reflective rooms of uniform reverberation time will show up very clearly any off-axis response aberrations of a loudspeaker system. A wide directivity loudspeaker system with a relatively uniform on- and off-axis response will highlight any deficiencies in the reverberation characteristics of a room. In between these extremes, for purposes of practical expediency, designers have variously selectively treated 'difficult' rooms using acoustic techniques, or have opted to control the directivity of the loudspeakers so as not to drive the acoustical problem areas of the rooms.

Given the sensitivity of our aural perception systems, however, it is hardly surprising that as there is a considerable non-minimum phase content in much of this variability, all the approaches tend to sound different; both from each other, and within their own groupings. Indeed, in all approaches except that of the 'non-environment', there is the further problem that any reverberant or reflective energy will have the tendency to mask much of the low level detail in the direct signal. The masking is often of a greater effect than the overall level of the reverberant energy would suggest, as the reverberant energy, unlike a noise type masking, is concentrated in bunches around the frequencies in the direct signal, as obviously it is the same direct signal which drives it.

So, in the 'non-environment' environment, we have few reflective problems to consider, therefore the obvious first choice for a monitor system would appear to be of the type with a wide and uniform pressure response. Certainly such a system would probably be heard in such a room to give a good account of itself, but in larger rooms of this type, it is difficult if not impossible to construct systems which can deliver enough high frequency energy before running out of steam. As the rooms give no help to the loudspeakers in terms of added loudness, then the loudness in each part of the room must be that from the direct drive from the loudspeakers. Consequently, in a large room, if the whole room is driven, then anywhere up to 12 or 15 dB of monitor power (up to 90%) can be wasted by driving entirely unused areas of the room. Effectively, a designer must chose the designated working area of the control room, then cover that area with as smooth a response as possible. If the achievement of this objective creates a directivity anomaly at, say, 50° off-axis, then it is usually of no great significance, as nobody is likely to be doing any serious listening from that position. The room will then absorb the unwanted sound: it will not be reflected back into the listening area to colour the response. The important thing is that when a listener does move towards the extremes of the usable area, the response should only change gradually as he or she moves progressively further off-axis.

In the highly absorbent 'non-environment' rooms, any erratic response changes would be unacceptable, as every different listening position in the room would be subject to a different frequency balance, which would lead to a great deal of confusion. Only in the 'hot seat' would the spectral response be as desired, and though this is accepted by many hi-fi enthusiasts in their homes, it would be unusable in the working environment of a control room, where personnel must inevitably move around whilst discussing the same musical balance. These rooms offer no smoothing of responses from any uniform reverberation. It is once past the ear, and once only!

19.1.1 Phase response

What does become very noticeable in rooms which are free of reflexions is that the differences in transient and phase responses of loudspeakers become much more obvious. An accurate phase response is necessary both because of its effect on the attack of notes, which carries much of the character of the timbre, and of the effect on low level signal detail, which carries much of the spacial information. I have referred in detail elsewhere to Manfred Schroeder's 'phase organ' experiment in which his 31 frequency pulse train with zero phase relationship could be made to produce 'notes' above the buzz of the train when any one of the 31 pulse frequencies was moved in relative phase, but which could only be heard under anechoic conditions. Any reflexions in the auditioning room would change his 'organ' from being able to play 'tunes' to rendering the notes completely inaudible. This point is worth repeating because it is highly important in relation to our perception of much low level detail.

A fast and accurate step function/impulse response has long now been a fundamental goal in my design approaches, and experience now in many non-environment rooms has only served to reinforce my belief that phase response, straightened out to the greatest practicable degree, is one of the main keys to neutral monitoring.

19.2 Practical layouts

Certainly in the larger versions of these rooms, the higher sound pressure levels required from the monitors, typically two to four times that of conventional rooms, has tended to dictate horn loaded drivers for the mid-range and high frequency sections of the in-built monitors. Such a design philosophy also allows for a minimalist approach, allowing the construction of systems capable of high output levels and low distortion, yet utilising only two-way systems. The single crossover point allows for fewer phase anomalies created either by relative loudspeaker positioning or crossover artifacts.

The tendency so far in such rooms has been to utilise drivers in a single vertical line, either with one bass driver mounted below a horn, or, in larger rooms, two bass drivers mounted with the horn in between them, with all units in the same vertical line. The latter system is depicted in Fig. 19.1, which was the system layout published by Shozo Kinoshita in the 1980s. This layout ensures that as listeners walk around the room, the relative

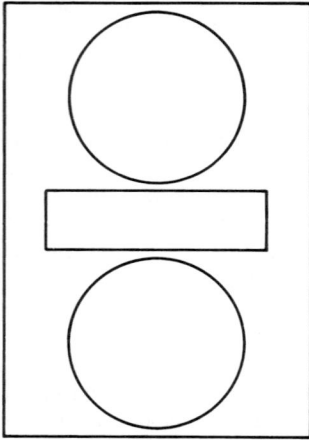

Figure 19.1 Kinoshita layout – 16″ LF drivers above and below the mid/high horn

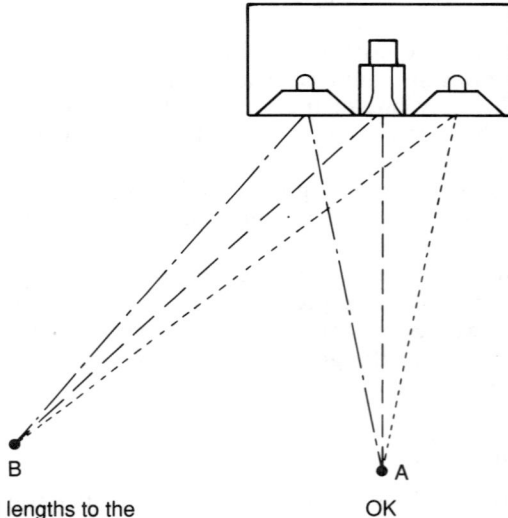

B

A

OK

Differing path lengths to the
three drivers will result in
time-smearing at crossover
frequency and poor LF summing

Figure 19.2 Horizontal mounting of monitor system shown in Fig. 19.1. At position A all is
well, with symmetrical positioning allowing equal arrival times from the bass drivers to the
ear. For position B, however, the path length, and hence the arrival time of the signal from
the left-hand bass driver, is shorter than that from the horn, whereas that from the right-hand
driver is longer

distance from each of the three drivers remains the same. Fig. 19.2, on
the other hand, shows how the path lengths to each drive unit would vary
(as would their relative phase response, and hence colouration of the signal)
if the cabinet was turned on to its side. In the horizontal layout, any
movement about the room would produce constantly changing colouration

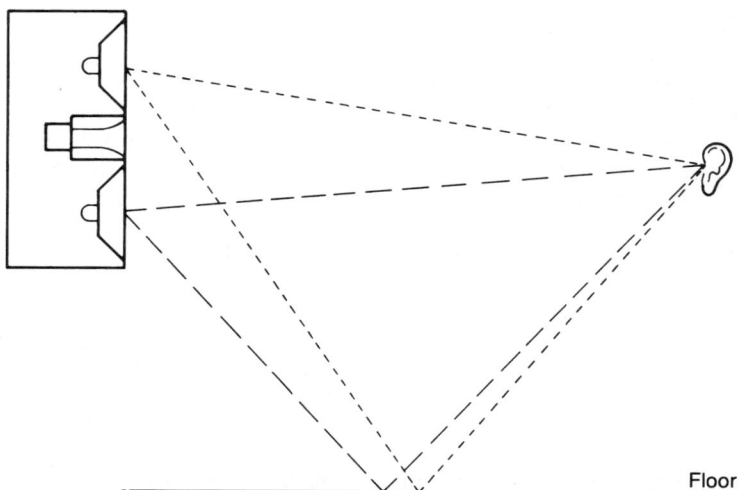

Figure 19.3 In vertical format, the direct paths to the ear are similar, but the vertical reflexion distances are considerably different. Were these reflexions in the horizontal plane, then image smearing would almost certainly occur, but thankfully, for most people, reflexions in the vertical plane are relatively innocuous in terms of image stability and definition

of the sound, plus imprecise and ambiguous stereo imaging caused by the multiple path lengths and differing arrival times from the different drivers. This would be incorrect for all positions except at a normal to the centre of the cabinet (position 'A' in Fig. 19.2).

The cabinets are mounted flush in the front wall, with the centre of the horn approximately 4' 10" above the floor. This height is a good mean between the height of the ears of a short person sitting down and a tall person standing up, so is a best compromise for normal daily movement of listeners.

19.2.1 Psycho-acoustic considerations

Although, when standing up and sitting down, movement relative to the three drivers (as in Fig. 19.2) does take place, it does so at 90° to the effect shown therein. Movement in the vertical plane, however, has little or no effect on stereo imaging, and, what is more, by far the majority of people are relatively insensitive to colouration-caused vertical anomalies. Figure 19.3 shows the typical reflexions which would be expected from a hard floor. When I originally read Kinoshita's paper on this layout, I initially had many reservations. Indeed, when I now discuss the ideas with other designers, many of them immediately point to the problem of floor reflexion irregularities in the response at the listening position, but in practice the effects are hard to detect by listening. In some instances, floor absorbers are used in the area where the direct wave would return to the ear from the floor. At worst, the effect produces a sharp notch from the bass drivers, usually in the 100–200 Hz range, but this has proved to be of little audible consequence. The directivity of the mid-range unit would tend to be such that the vertical directivity would ensure that the first direct wave could not strike the floor until such a point that it would, on reflexion, bounce up below the ear, and be absorbed by the trapping systems.

To recap on what was said in Chapter 17, were we to install a loudspeaker and an omni-directional microphone in a large anechoic chamber, then place a large reflective surface some feet below, we could measure the effect of the reflective surface on the overall response. If we were then to turn the whole anechoic chamber through 90°, with the reflective surface now in a vertical position (effectively the same as tipping the whole chamber over on its side), then nothing in the measurements would change. Despite the fact that no response change could be measured, if we were to stand at the microphone position and remain vertical, then our perceived sensations would be radically different, dependent upon whether the reflective surface was below us or to one side of us. When compared to the direct sound alone, the reflector to the side of the listener would be far more disturbing than when the reflective surface was below the listener.

There are two mechanisms at work here which do not always receive equal importance from different listeners. Firstly, and more obviously, with the reflective surface below the loudspeaker, the reflexion will arrive at the listener from the same horizontal direction as the direct signal, so it is not surprising that the horizontal stereo imaging of the sound will be little disturbed. A reflexion from the side, however, will come from a point which is laterally offset from the direct sound, so it is only to be expected that spacial smearing will result. The time smearing produced by either the vertical or lateral reflexion will be equal in both time and intensity from either condition; however, there is evidence to show that the subsequent colouration is not perceived to have equal significance for all listeners. By far the majority of people show a greater dislike for the laterally produced colouration than for the vertically produced colouration.

Again, human inconsistencies do not make the life of a designer any easier. In general, however, the floor reflexions are usually considered relatively innocuous. Based on this concept that floor reflexions are largely ignored by most humans, despite their disturbance to any measured response, the Kinoshita concept provided the basis for my considerations for the proposed monitors for my own versions of the new-style rooms.

19.3 Choice of components

Once such a layout of drive units has been chosen as appropriate, anything other than a two-way system is difficult to realise in practice. To help to outline the actual decisions which face a designer, I shall use the Reflexion Arts 234 as an example of the thought processes which led to one approach to finding a practical solution to the problems faced.

The two-way design tends to dictate a crossover point somewhere in the octave between 600 and 1200 Hz when dealing with this scale of loudspeaker system. Anything smaller than 15″ for the bass drivers would have no chance of supplying the required low frequency sound pressure levels. There is a very limited choice of 15″ loudspeakers, however, which can respond at 20 Hz yet still continue to produce a smooth and linear response above 600 Hz or thereabouts. The JBL 2235H is one of the few which can, and indeed its axial response is still respectable at 2 kHz; however, the physics of cone loudspeakers predicts serious trouble once the diameter of the driving area reaches one wavelength of the highest frequency to be used. This point is reached for a 15" loudspeaker at around 1 kHz.

The precise choice of which 15" loudspeakers to use can only be ultimately assessed by listening, but I knew from extensive listening tests conducted for my earlier loudspeaker system designs that there was an 'openness' and smoothness about the 2235H which I found difficult to find in other units. I knew of 15" drivers which sounded subjectively slightly deeper than the JBL, but I knew of none with the ability to match its overall performance from 20 Hz to 1 kHz. In earlier designs, I had used the 2235H in a three-way system, crossing over at 800 Hz into a mid-range unit, but in some early installations with adjustable crossovers, several people had quite independently chosen a 1 kHz crossover point as being most neutral in that design. I thus had well corroborated first-hand experience that I could use that driver at least to 1 kHz. In order to assist the low frequencies, two 15" bass units were used in parallel, and enclosed in a 21 cubic foot cabinet, tuned to resonate at 17 Hz. It would have been possible to support the response around 30 Hz by a higher cabinet tuning, but 17 Hz was chosen to give the smoothest amplitude/phase compromise, by keeping the cabinet resonance clear of most natural musical frequencies. Perhaps two 12" loudspeakers and a sub-woofer below 60 or 70 Hz could have been an alternative 3-way option, but simplicity was one of the key design aims in the search for neutrality.

The mid-range/high frequency unit needed to be capable of producing over 120 dB at one metre. It also needed to have the desired axial frequency response so that it could be used without equalisation, and should have an off-axis response which met with the criteria set down in earlier paragraphs for use in these rooms. The development of such a drive system was discussed at length in Chapters 10 to 12. The combination of the TAD TD2001 beryllium diaphragmed compression driver and the AX2 axisymmetric horn seemed ideally suited for the requirements. Certainly its axial pressure amplitude response was well inside the design criteria for monitoring purposes, and its directivity characteristics had been the subject of a paper presented to the Institute of Acoustics conference in 1990: 'Axisymmetric Horns for Studio Monitor Systems'.

The use of this horn fixed the crossover frequency, as the horn itself possessed a 12 dB/octave roll-off below its cut-off frequency of around 1 kHz. The concept of this combination appealed to me both in the way that it naturally fell together, and also because of its simplicity. A total system could be produced in two-way form, with drive units which could respond within the design specification limits from 20 Hz to 20 kHz, without the aid of any electrical contouring of the response. It was also based on units which had shown themselves to be capable of producing very natural and transparent sounds when assessed subjectively.

The amplification system followed my long-time tried and tested concepts, as described in Chapters 7, 20 and 21, of using electronic low-level crossovers and multiple amplifiers; with the whole electronic system responding from DC to 100 kHz to keep the transient accuracy, and the phase shifts well out of the audible ranges. The only interesting difference here, however, was that although a 24 dB/octave modified Linkwitz–Riley crossover had been chosen, the inherent 12 dB/octave roll-off of the horn below 1 kHz was incorporated into the overall design, so that whilst the electronic crossover had a 24 dB/octave low-pass section, the high-pass section was only 12 dB/octave, allowing the horn itself to contribute the 'missing' 12 dB/octave of the slope.

The enclosures are approximately 2 ft × 3 ft × 4 ft, with the front panels being 4 ft high and 3 ft wide. The boxes themselves were designed to be constructed of two layers of $\frac{3}{4}''$ chipboard, with a layer of 5 kg/m^2 deadsheet sandwiched between the two adhesively bonded layers. Three-axis bracing helped further to reduce panel flexing, and the whole was lined with $\frac{1}{2}''$ foam, glued to the side walls. There then followed another layer of deadsheet, and a final internal lining of either felt or Dacron, all layers being bonded with appropriate adhesives. The cabinets were designed for mounting in the heavy, dense, hard front walls of the non-environment rooms, which act as large baffle extensions against which the low frequencies can push.

19.4 Subjective and objective assessment

Figures 19.4 and 19.5 show the throat impedance, pressure amplitude and phase responses of the AX2 axisymmetric horn, compared to the corresponding plots of a good, professional rectangular horn. The improved smoothness of the AX2 is very clearly obvious. The plots of a complete system are shown in Figs 17.7(a) and (b) though these plots were taken with a symmetrical 24 dB/octave electronic crossover. The asymmetrical 12/24 dB/octave crossover as described above smoothes the response through the 1 kHz crossover region, as shown in Fig. 12.15. Figure 19.6 shows the AX2 directivity, and clearly demonstrates the gradual off-axis roll-off.

What was evident from the moment that the systems were first tested was the ease with which they could be adjusted to optimum high/low balance by the use of a known compact disc. Subsequent measurement showed that the initial set-up on music was within a dB of the measured optimum. A whole stream of people passed through the control room in the couple of months after completion of the first system, and the comments on smoothness, seamlessness, imaging clarity, general clarity, and spectral extension were all but unanimous. Certainly in the horizontal sense, the systems were behaving as coincident sources, with the 'phantom' location of the bass driver pairs being directly over the centre of the mid/high horn. Therefore, irrespective of the location of the listeners within the room, the sources remain subjectively phase coherent, thanks to our relative immunity from positional instability due to vertical (floor) reflexions.

To all intents and purposes, the monitors in the non-environment rooms behave like loudspeakers which are mounted in the outside wall of a building, pointing into a field. As such, the fall-off in sound pressure level is much more rapid as one moves away from the loudspeakers than one would experience in a conventional room. The monitor systems therefore require the ability to generate higher SPLs than normal usage would demand.

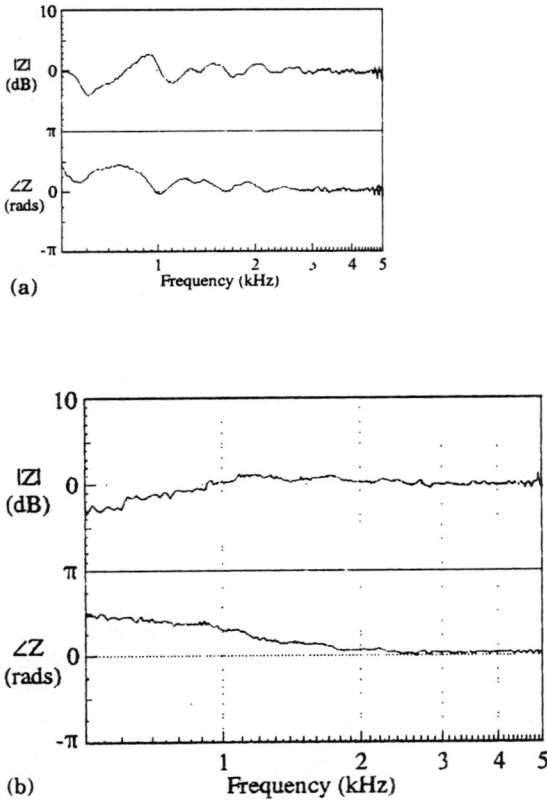

Figure 19.4 (a) Throat impedance plot of a typical, commonly used rectangular horn; (b) throat impedance plot of AX2 Axisymmetric horn

Another function of this 'free-field' listening environment is that whilst the axial response is all-important at the listening position, as no room reflexions exist to upset this response, any off-axis problems are certainly heard off-axis to a potentially greater degree than would be the case in a more conventional room, where some ambient smoothing would take place. In other words, whereas in conventional rooms the *off-axis* irregularities can produce *on-axis* disturbances via the reverberant field, in non-environment rooms such disturbances to the *on-axis* response are clearly impossible as there are no reverberant fields. But, where any *off-axis* problems which may be heard *off-axis* in conventional rooms are rendered less significant by being bolstered by the reverberant field from the linear, *on-axis* response, no such amelioration can be expected in non-environment rooms. If a 'hot seat only' priority exists in a non-environment room, then axial response accuracy alone will suffice. If, however, a 'whole room' coverage is required, then whilst the on-axis response requirements remain, the off-axis response becomes even more demanding than in many conventional rooms.

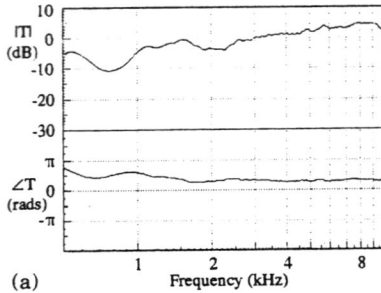

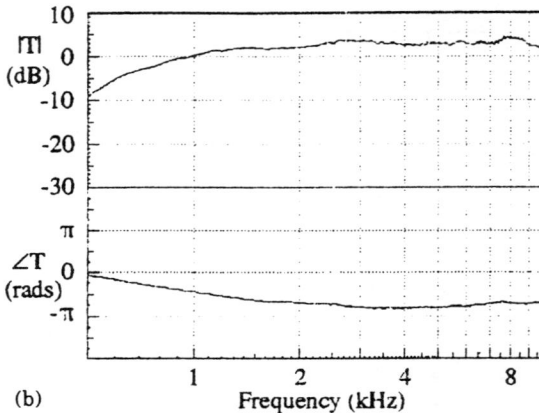

Figure 19.5 (a) Amplitude and phase response, in a room, for a typical rectangular horn and driver; (b) amplitude and phase response, in a room, for the AX2/TD2001 combination

What all the non-environment rooms have in common is that the monitoring is essentially the loudspeaker system. If steps are taken to position equipment such that reflexions are deflected away from the listeners and into the traps, then rooms of different shapes and sizes are capable of achieving a high degree of room-to-room compatibility. Different types of good quality monitor systems, providing that they meet the special requirements, are not perceived as being so different as they are often considered to be. Bad loudspeaker systems, however, are quickly revealed as such. For use in smaller rooms of this type, a variant of the monitor system described in the previous paragraphs was produced, using only one 15″ bass driver and a horn. Mixes travel between the rooms without any subjective changes in balance.

Once so much of the reverberant and reflective 'clutter' has gone from the listening environment, people rapidly begin to come to terms with aspects of their own audiology which had to them, until then, previously only been for the text books. It is essential that the monitor systems used in such rooms should have a good impulse response, or so many of the potential benefits will be lost.

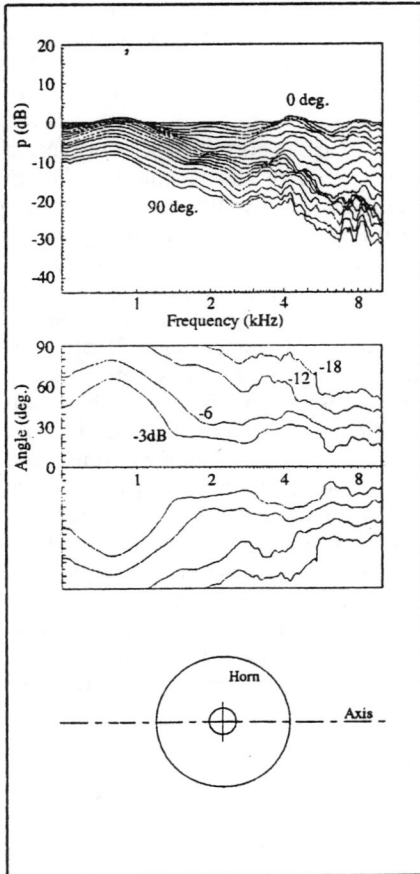

Figure 19.6 Directivity plots of AX2 axisymmetric horn

Sometimes, also, conventional taboos become irrelevant. For example, when designing the system described in this chapter, there were concerns at first about the fact that there were three signal sources at the crossover frequency of 1 kHz: two 15″ loudspeakers and one horn. The possibility was considered that the response of one of the bass drivers should be curtailed above 100 Hz or so, in order to prevent inter-unit interference above mutual coupling frequencies, and where the drivers would act as independent sources. In practice, however, when one is listening several metres away in a large room, the path length differences to the three drivers become an insignificant part of the wavelength. If one listens at reasonably close ranges, the narrowing directivity of the bass drivers, as the response is taken over by the horn, ensures that it is almost impossible to be in the direct field of all three drivers at the same time. After exhaustive listening tests, it was decided to dispense with the planned roll-off. On the other hand, in a more conventionally reflective room, the reflective surfaces would almost certainly in some areas be driven by very different path lengths to the individual drivers, and this would particularly be the case for reflexions originating from the floor or ceiling (see again Fig. 19.2). One should be careful not to take as gospel, truths which relate only to more conventional circumstances.

On the audiological aspects of such monitoring conditions, however, a few points are worth noting. Such a point is that of stereo imaging. When compared with normal rooms, the differences in stability between stereo pictures built up either from pan-potted images or recordings using stereo microphone technique are very clearly noticeable. The stereo microphone images, which contain positional data largely from time and phase discrepancies, support themselves well. Pan-potted images, relying solely on amplitude differentials, are frequently heard to collapse when listening well off-axis. This is exactly what *should* happen, but the comment is frequently heard from those who do not realise this that the stereo can be less stable than under more conventional monitoring conditions. On the other hand, from within the designated listening area, the imaging is very clearly defined and highly stable. It cannot be said that such rooms do not support a wide stereo listening area, but only that the width of the area is dependent upon recording technique, which is precisely what audiology would demand. In general, as soon as people become accustomed to what is happening, everything rapidly becomes 'normal' to them. Once the subconscious takes control, the rooms become very quick and easy to work in, as so many uncertainties disappear. A further example of how clearer monitoring can help us to understand more about our own hearing mechanisms is demonstrated in the following anecdote.

In the mid-1980s, a well known studio designer began telling the world that his new rooms possessed monitoring which not only had accurate stereo width perception, and a front to back depth which was clearly discernible, but also that the monitoring had 'height'. Personally, I had never experienced this, but at the end of the 1980s, I had a phone call from the owner of a studio which I had just completed. He asked me if I could call in when I was next in the area, as he was experiencing some strange effects from the new monitoring. About a week later I called in, and he asked me to listen whilst he panned a hi-hat from left to right, then asked me what I thought caused the effect. (He was an intelligent man, so I presumed that he was not referring to the fact that the sound went from left to right as he turned the pan pot.) I could hear nothing unusual, so asked him what, precisely, he was referring to. He then asked why it went up in the middle. I explained that the whole concept of pan pot law had been a bone of contention for many years, with different opinions on whether the centre position should show a rise of 3 dB, 6 dB, or several points in between. The studio owner immediately countered, 'No! Not up in level – up in height!'

Somewhat taken aback, I asked to listen again, but try as I might, I could perceive no such effect. We concluded that the 'problem' was no real disaster, but that I would look into it. The next time that I was at Southampton University, I asked an eminent audiologist about the problem, and was told that the way that the pinnae (the outer ears) collect the 8 kHz region has a great bearing on the perception of the height from which a sound is appearing to emanate. Clearly, I was constructed differently to both the above studio owner and the aforementioned studio designer.

This 'problem' occurred at a time when I was beginning to shift my design philosophies towards producing rooms with an absolute practical minimum of monitor reflexions, and shortly afterwards a very odd (but fortuitous for me) event took place. I was in the process of completing a studio when another prospective client sent his most respected engineer to look and listen to my latest studio. Unbeknown to me at the time, he reported back that, in this studio, you could hear height in the image, and he insisted that I should be the designer chosen to build the new studio complex. I only recently discovered this when chance caused me to find myself co-engineering a project with the very same engineer, and when he remarked to me, 'I really love these rooms because of the sensation of height in the image.' He then related to me the whole story.

Once again, I sat in a room of my own design, with a monitor system of my own design, and heard absolutely nothing of this phenomenon which had just brought me so much work. The truth is that no conventional microphone technique, when recorded onto a two-track recorder, entirely unprocessed, can carry height information *per se*. What is more, any perceived 'height' in the image will, upon investigation, usually not relate to the actual height of the instruments as they were recorded.

Less acoustic confusion in the rooms leads to less clutter entering the ears. This in turn highlights once again just how different human perception can be from one person to another, and is just one more reason why what is 'right' in monitoring remains such a subjective, individual decision.

19.5 Conclusions

In principle, all good quality loudspeakers should perform at their optimum in non-environment rooms, as their responses will be less coloured than when auditioned in conventional rooms. However, when one is freed from the constraints of having to take into consideration a reflective or reverberant environment, life for the designer can be much simplified. The only really significant additional demands are the need for a greater output capability due to the lower subjective loudness in a room of very low reverberation time, plus the great emphasis that must be placed on a need for excellent transient/phase response, as anechoic listening environments can highlight time/phase artifacts which would go largely unnoticed in more conventional surroundings.

The following two chapters also deal with monitor system design; in the first case, large monitors for more conventional control rooms, and in the second case, monitors for close-field listening. What should be evident from taking the three chapters together is that whilst many of the design philosophies and targets hold true in all cases, there are others which must be adapted, modified, or at least relocated in the order of design priorities in each individual case.

What experience has shown is that, in non-environment rooms, the majority of multi-band, multi-drive unit systems, i.e. 3-, 4- and 5-way designs, often have difficulties in 'knitting together', and these difficulties are exposed quite easily when moving around the room. These are, however, systems which are entirely acceptable in many other environments. Once again, as with so many aspects of studio design, specificity to circumstances continues to be a major factor, and many 'gospel truths' must be viewed with caution.

Chapter 20

An integrated approach to the design of a studio monitoring system

During the summer of 1984, to meet the new demands of the then current recording studios, I began considering the requirements for a new range of studio monitor systems. It was my opinion that a studio monitor system should be designed in a manner which took into account not only the components of the system itself, but also the environment and music with which it would be used; and most importantly, the human aspects of the people with whom it would be used. The culmination of this overall design philosophy was the Reflexion Arts Model 235 monitor system. The first system went into commercial operation in March 1985 and the number of systems soon multiplied.

20.1 Background

The previous few years had seen not only the advent of digital recording, but also the proliferation of computer-generated keyboard instruments, with a considerable shift of musical personnel from the studio to the control room. The many hours of preparation and programming of these instruments, together with the necessity for close communication between the musicians, programmers, engineers and producers, had inevitably led to larger control rooms. Keyboard rigs needed to be set up with sufficient space for everybody involved to move freely about.

In order to achieve similar acoustic levels as would be produced in smaller rooms, the larger control rooms required greater output capability from the monitors. The advent of digital recording preserved transients which would have been lost on analogue tape, so, in addition, the dynamic range of the monitoring and its ability to handle repeated and higher transient signals were also required to be correspondingly greater. Keyboards produced two further problems. Firstly, the ability of computer and digitally generated sounds to produce signals of a very unnatural nature, often with extraordinarily high intensity signals concentrated in very narrow frequency bands. Secondly, with the control rooms now becoming the studios in which the musicians were playing, the monitoring systems needed to be able to produce, when required, 'live' volume levels. As many musicians 'play off the volume' for inspiration, the monitors must be able to re-create the levels of a concert stage in order to generate the necessary excitement to give the performer and producer the atmosphere required to create the appropriate feel for the music. Turning it up afterwards will not help! It is the frame of mind of the musician at the time when the music is recorded that dictates the feel of the track.

20.2 Basic requirements of a system

The above requirements dictate a system with a high output capability, fast response to large transients, relative indestructibility to cope with keyboard 'accidents' (they do not always put out the level that you were expecting), flat acoustic output to the extremes of the audio spectrum, low distortion, and a well balanced tonal character, independent of level. Let us look at these things in more detail.

Low distortion and high power handling will largely be down to the choice of individual drivers. Flat acoustic output could be achieved by equalisation, or attenuation of the more efficient units. However, these factors could work against us in other ways. If we choose a system for use with a single amplifier and high level crossover, we would be required to match the efficiencies of the drivers. A mid-range driver with a 6 dB greater sensitivity than the bass driver would produce a peak in the middle unless attenuated. Any attenuation used would involve buffering the unit away from the low impedance drive of the amplifier. This would bring the risk of power level and frequency dependent variations in response, due to the driver impedance deviating from its nominal value at different frequencies.

If we drove each individual driver, or pair of drivers, from their own independent amplifiers, no constraints would then be placed upon our choice of drive units. The optimum units could be chosen purely for their desired sonic and directivity characteristics. Furthermore, any attenuators or crossover components, which may come between the loudspeaker and the amplifier, serve to reduce the control which the amplifier's damping factor may exert upon any resonance or momentum-induced cone or diaphragm excursions. Tight control over the cone movements, especially of the bass drivers, may well be severely impaired by passive, high level crossovers.

By separating the amplifiers, absolute maximum use is made of their headroom and transient handling ability. A further advantage is somewhat less apparent. Consider a sudden, low frequency peak, when fed into a system driven by one amplifier. Should the peak exceed the amplifier's output ability, harmonic distortion will be produced. The harmonics, of a higher frequency than the fundamental, will pass through the crossover into the high frequency drivers. This will not only produce unpleasant audible distortion, but will introduce high level 'spiky' overload signals which the HF drivers may have difficulty in handling. Such repeated overloads can cause listening fatigue, and also may cause premature failure of the HF drivers.

Should a similar overload occur in a multi-amplifier system, such low frequency overload distortions cannot enter the HF drivers, as they are not coupled to the same amplifier output. This results in three beneficial effects. One, the highs continue to be heard as clearly as ever, untainted by the LF distortion. Two, no unnecessary strain is put on the HF drivers, which helps towards long and consistent life, and also reduces diaphragm fatigue. Three, a bonus, the LF drivers, having a response severely limited at the higher frequencies, cannot reproduce the majority of the LF amplifier distortion. In effect, the sudden LF overload passes through the system almost imperceptibly, and without the risk of straining or damaging the MF and HF drivers. Together with the appropriate choice of units with suitable power handling, this goes a long way towards the desired indestructibility and consistency of sound quality, negating any need for adjusting equalisation to help compensate for 'tired' or fatigued drivers. It also reduces fatigue on the ears.

Another aspect of performance which helps to reduce fatigue is a 'flat' acoustic output. Chapter 9 discussed at considerable length the pros and cons of equalisation on monitor systems. The aim in the design described in this chapter was to choose drive units which performed effortlessly over their intended range of use, with the intention of avoiding the need for the application of any equalisation. There could be situations where extreme high or low frequency response needed to be tailored to the positioning of the loudspeakers in certain rooms, and here a certain degree of 'frequency response' modification could be required. However, with a system driven from a suitable electronic crossover, and with up to four individual amplifiers, smooth adjustments can be made to relatively small frequency bands by adjustment of gain controls only. To the ear, this seems to sound much more natural and lifelike, and much less fatiguing than correction by means of equalisers.

Limiting can also be dangerous on monitors. What is it that is limiting, your monitors or your mix? It may well be imprudent to mix at ridiculous levels, but occasionally, in practice, that may be what the circumstances demand. Even if it is only on peaks, monitor limiting will suppress transients, and you may find yourself putting too much top on tape to help compensate for the lost peaks. What is more, clipped peaks produce clipped reflexions in the room, which affects any ambience in control rooms which possess it. Given the flexibility of the split amplifier system, the careful choice of drivers and amplifiers should alleviate the need for limiters, while still not putting the drivers at risk.

Obviously, with excessive amplifier output capability, some damage *can* be done, but with the appropriate choice, this risk can be greatly reduced while still allowing for transient headroom. To achieve a power bandwidth down to 20 Hz and for good bass transient ability, DC amplifiers were my first choice. Output power selection would depend upon the power handling and efficiency of the drivers, so this choice would be left till last. The response of the crossover should also exhibit 20 dB of headroom over loud working levels, whilst responding down to at least 5 Hz as the 3 dB down point, or possibly even to DC.

20.3 Choice of drivers

Driver choice is affected by considerations of type of music, so I suppose that I should state here that this monitor was intended for use in studios recording mainly electric and electronic music. At low frequencies, I felt that the best compromise between tightness and the extension of the bottom end tended to be that produced by a bass reflex cabinet of suitable design, loaded of course with the appropriate driver(s). The choice of size

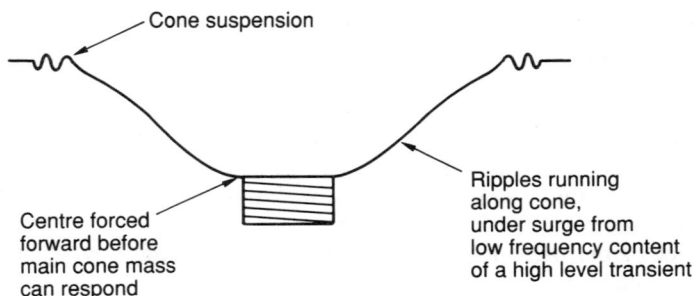

Figure 20.1 Large cones can tend to produce a 'boom' rather than a 'thud' from the bass end

of bass drivers, within normal limitations, provides the options of 10″, 12″, 15″ or 18″ units, either used alone, in multiples of one size, or even mixed. The apparently obvious choice would be to use 18″ units, with their ability to produce great low frequency outputs. One drawback frequently found in 18″ units, however, is the difficulty in preventing such large cones from flexing under high level transient inputs. This can serve to reduce the 'punch' from the loudspeaker, and together with the resulting harmonics from the flexing cone, tends to produce a 'boom' rather than a 'thud' from the bass end (Fig. 20.1).

By contrast, 12″ drivers require far greater cone excursions to move the same volume of air, and there are drawbacks to having a small surface area for the generation of the low frequency sound. As the suspension systems for 12″ and 15″ cones tend to be similar, it can be appreciated that the suspension of the 12″ driver receives far greater long term punishment than that of the 15″ driver. Furthermore, in order to achieve a very low resonance, the 12″ cone must be either more heavily built than its 15″ counterpart, or its suspension system must be even more compliant. The first option reduces efficiency and requires greater output capability from the amplifier, which produces a lower total acoustic output per watt of power handling capacity. The second option reduces the driver's ability to cope with transient overloads without damage or strain.

Musical instruments with large, low-frequency contents, tend to *be* large. Natural sounding low frequencies usually come from sources which produce a relatively low air pressure from a large source area. Even though it may produce a similar reading on a spectrum analyser, moving air in this way undoubtedly sounds different from the high pressure, small source area approach of 12″ or smaller cones. Once again, this refers back to the frequently misleading results where subjective audibility and simple measurement do not coincide.

The root of these problems lies in the fourth power law in the relationship between cone area and acoustic output. If cone A has twice the area of cone B, which would be typical of the relationship between a 15″ and a 10″ driver, the smaller driver would have to move not double, but four times the distance travelled by the larger cone in order to produce the same SPL. To travel four times the distance at any given frequency requires the small cone to travel at four times the speed of the larger cone. Four times the speed means four times the effects of any propensity towards Doppler distortion, especially at the high excursion distances associated with small cones and high output levels.

In order to achieve the desired low frequencies of resonance for the smaller cones, they tend to have to be made heavier than would be strictly necessary purely for structural integrity, the main effects of which are two-fold. Firstly, the increased mass may produce a smaller cone of similar overall weight to a comparable, larger 15" cone; but even half the weight, travelling at four times the speed, would produce double the momentum when the cone was in motion. Increased momentum will produce a greater tendency towards overshooting when reproducing transient signals, demanding a greater damping or braking effect from the amplifier in order to control the cone movement. More momentum means more of a tendency to overshoot, which means less accurate reproduction of transients, which means more 'woolly' bass.

Secondly, the greater mass to area ratio reduces the efficiency of the motor system. Reduced efficiency means more drive power is required to achieve the desired output. More drive power means more heat in the voice coil. As the voice coil temperature rises, inevitably the voice coil resistance will rise. In turn, the increased resistance means that less power will be available from the amplifier, therefore power compression can take place. Once again, undesired compression can distort the output signal and reduce the overall faithfulness of the reproduction. In general, larger cones require lower cone speeds for any given output, which in turn reduces Doppler distortion tendencies, and the unpleasant shearing effects as high speed, large cone excursions tear through the air, instead of gently pushing and pulling on it. Larger cones, when travelling slower for any given output, usually possess lower momentum, hence have a reduced tendency to over-shoot, and consequently an improved transient response. They are usually more efficient, and therefore suffer from a much lower degree of power compression. The 'effortless' sound produced at low frequencies by large drivers is not only perceived, it has a very real base in the fundamental electro-mechanics of the system.

In practice, I believe that 15" drivers tend to achieve the best compromise for low frequency use. The use of a 2 × 15" system gives a further increase in output due to the mutual coupling achieved when the units are positioned relatively close together. When the cones of two drive units are separated by less than one wavelength, the forward movement of the air displaced by one driver gives the other one more to push against. More work is therefore done, and an extra 3 dB of output can be expected from two drivers at less than one wavelength's separation, as compared to the same two, parallel driven drive units, at separation distances of greater than one wavelength. The 2 × 15" choice gives us a large sound source area, a relatively rigid cone, reasonable efficiency, high power handling, a good transient response, and a long, stable, life expectation.

To achieve a clear response, down to sub-audibility, a low resonance, say 15 Hz to 20 Hz, would seem optimum. This usually demands the use of a roll surround type of cone assembly, which allows relatively free movement for good low frequency capability, especially in reflex type enclosures.

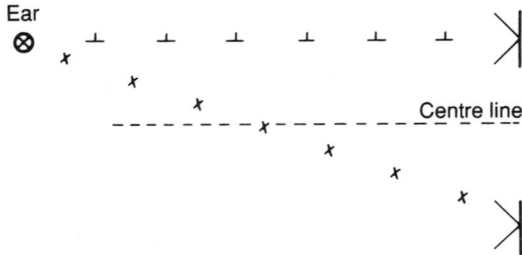

Figure 20.2 Pressure wavefronts do not reach the ear simultaneously if the ear is not precisely on the centre line

20.4 Mid-range drivers

The mid-range is the most critical and most contentious area of monitor design. The choice of drivers mainly falls between cones, domes and compression horns. Cones and domes have been more generally considered to produce a softer, less harsh sound than compression horns, which, strictly speaking, are two separate components: the driver and the flare. When very high output levels are required, the more efficient compression horns are capable of much greater acoustic output. Thus, at high levels, they may actually be sweeter than a hard pushed cone or dome driver.

Whilst initially hoping to use cones for all of the frequency ranges, I soon began running into problems above 1 kHz. To achieve good high frequency and transient response whilst maintaining high efficiency, the moving mass of a mid-range driver must be rigid yet very light. As increase in size means increase in weight for any given material – cone and dome mid-range units cannot just be made larger in order to be louder. They also certainly cannot be unduly increased in diameter, as one would begin to run into directivity problems once the cone or dome diameter exceeded one wavelength of the highest frequency to be reproduced. Their efficiency would tend to decrease as their power handling increased, and would thus face a situation of diminishing returns. It is also difficult for a small mass to dissipate the heat which is generated in the unit at high output levels. For typical efficiency of, say, 2.5% for a cone unit, for every 20 watts supplied by the amplifier, 19.5 watts are turned into heat within the driver, only around 0.5 watt being transmitted to the room as sound energy.

Doubling up the number of units obviously doubles the power handling, but each doubling of output power only provides an additional 3 dB of headroom. Another problem is that, unlike at low frequencies, at higher frequencies a small source is desirable as the wavelengths involved become shorter with rising frequency. Unless the listener is equidistant from the twin drive units, some phase cancellation will occur, blurring the sound and impairing the stereo imaging (Fig. 20.2).

For high output levels, therefore, the compression horn seemed to be most suitable, subject to the reduction of any perceived 'harshness'. Several options are open to help to reduce undesirable effects. Firstly, use cone drivers as far up the frequency range as possible, whilst still remaining on the good side of the efficiency/frequency response curve. Secondly, do not attempt to drive any horn system beyond the point where the response begins to tail off, or where the directional characteristics of the flare begin to 'beam' beyond acceptable limits, and lose their even distribution over the required listening area. Thirdly, in order to maintain the output within the lowest distortion range of the compression horn, provide adequate power handling and efficiency for the desired acoustic output. Fourthly, the design of the flare itself can contribute significantly to the 'timbre' of the sound, although frequently this cannot be easily determined by instrumentation. Subjective acceptance tests of the combination are therefore a practical necessity. (Chapters 10, 11, 12 and 13 were written some years after this system was designed. The research discussed therein was a result of the problems in the design of the system described here.)

If the output of the cone drivers can be maintained up to at least 1200 Hz, this allows the use of a 1″ instead of a 2″ compression driver. In general, 1″ drivers have improved high frequency output and a generally smoother response, with less tendency to the 'barking' character of some of the larger compression drivers when used to lower frequencies. The smaller horn throat also improves high frequency integrity, with far less of a tendency to 'beam' than many of the larger diaphragm/horn-throat combinations.

The flare itself can be contoured to cover the desired listening area, giving a sufficiently wide directivity to allow all personnel concerned to perceive the same sound balance. Furthermore, the flare can control the directivity to prevent unwanted reflexions from points in the room where sound need not be directed, though certain audibility/directivity compromises may be involved here. The general requirement to achieve this would be in the order of 100 degrees horizontal directivity with, say, 40 degrees vertical directivity, presuming that most ears in the appointed listening area will be between 3 feet and 7 feet from the ground.

The highest frequencies which can be smoothly and comfortably reached by such a combination of 1″ driver and flare, whilst maintaining adequate directivity, would still seem to be around 8 kHz, though certain recent systems have improved upon this. An optimum overall format from my personal design criteria was now beginning to unravel itself: cones to 1200 kHz, low distortion, low colouration, high output, 1″ compression horn to 7 kHz and a separate unit for the top octave or so (7 kHz to 20 kHz).

20.5 Specific choice of drive units

At the very bottom end of the frequency spectrum, two 15″ Gauss 4583F drivers were chosen. A 4 × 12″ Gauss 2831 option was considered, giving a roughly similar overall source area, but, with 8 ohm coils, a parallel arrangement would produce 2 ohms, which is unacceptably low for most conventional power amplifiers to produce their best. Individual amplifiers could be provided for each pair, but this was considered cumbersome. A series parallel arrangement was considered unsuitable, for then not one of the drivers would be directly connected across the amplifier terminals, as each would be in series with at least 4 ohms from the other drivers in the group (Fig. 20.3). This would reduce the ability of the high damping factor

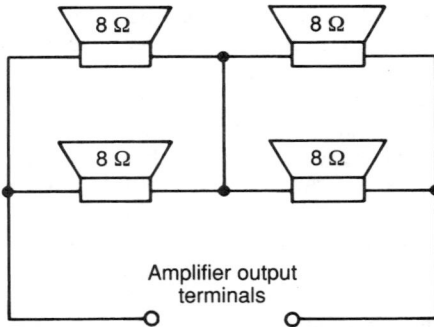

Figure 20.3 Series/parallel connection of four 8 Ω drivers

of the amplifiers to control cone excursions, and, once again, the 'tightness' of the bass response could be unacceptably compromised.

A pair of 4583Fs in parallel would give 4 ohms, ideal for most power amplifiers to produce full potential output power, and not making excessive demands from the current rating of cable, connectors or power supplies. The voice coils are rated at 400 watts each, though with the roll surround of the 19 Hz units the RMS power handling of each driver is rated at 300 watts. Six hundred watts RMS was duly considered adequate power handling for the bottom end of each speaker system, especially given their individual efficiency of 97 dB for 1 watt at 1 metre – 1200 watts RMS down to 20 Hz for a stereo pair. By 600 Hz, the 4583Fs are losing their natural character of reproduction, so 300 Hz was chosen for the upper crossover point for the bass drivers, leaving a full octave of reasonable response above the crossover point.

Whilst still delivering full output at 800 or 1000 Hz, many 15″ drivers have a distinct lack of life in their subjective reproduction above 500 Hz, but this is not easily measured. Quite probably, many maligned drivers have been unfairly judged when being used beyond their ideal ranges. Some of this can be down to the driver manufacturers themselves, publishing measured responses within specified limits, then quoting 'usable frequency ranges' even beyond these limits. This 'usable frequency range' can probably be interpreted as that range in which there is still some audible output within 10 dB of the optimum range. In reality, the 'usable' range usually ought to be less than the published frequency response, as frequently, towards the upper end of the range, not only does the response become ragged, but more importantly, the tonal characteristics are no longer desirable.

Once again, the provision of monitor equalisation has allowed drivers to be pushed beyond audibly desirable operating envelopes, and well outside the range in which they can produce a natural timbre. Many monitor systems employing 15″ drivers and 2″ mid-range horn drivers, crossing over at around 800 Hz, suffer the most in this area. The bass driver performance is compromised by the necessity to choose a unit capable of performing reasonably well in the bass *and* lower mid-range areas, which is then asked to meet a mid-range horn, itself frequently operating below a frequency range which would be considered optimum for studio purposes. The physical layout of the drivers in the cabinet is also often inappropriate when three drivers (2 × 15″ units side-by-side, and a horn above) are producing 800 Hz at the crossover point. Phase disturbance is inevitable when moving, or in reflexions.

It was for the above reasons that 300 Hz was chosen for the first crossover point, leaving a gap of two octaves before the mid-range horn takes over. Consistent with the overall design philosophy, a cone unit was required to bridge the gap. The JBL 2122 was considered to be probably the most suitable, available unit for this purpose, with a very smooth frequency response, high efficiency, low distortion and 75 watts RMS power handling. The 2122 10″ unit was also considered to give an improved attack when compared to similar 12″ drivers, the natural yet startling response to a snare drum appearing to confirm this assessment. The relatively light weight cone, roll surround and prodigious magnetic flux all contributed to the excellent transient response and high acoustic output. The unit has since been superseded by the 2123 driver, the latter having an efficiency at 101 dB and a 250 watt power handling, but, subjectively, I still prefer the 2122, which is now only manufactured in special, limited runs. It has also been used by Westlake Audio in their BBSM 15s.

JBL considered the market for the 2122 too restricted, in the light of the higher power devices of similar size being made available to the sound reinforcement market by other suppliers. The 2122 was therefore superseded not on grounds of improved sonic performance, but of increased total output. Such is the free market economy!

So, at 1200 Hz, we arrive at the decisions about the compression horn. The flare chosen was a modification of a design by Malcolm Hill, manufactured by ASS from a glass-fibre/resin mix with a heavy loading of powdered slate. The cavities in the moulding were filled with a resin/silica-sand compound of high density and the whole of the rear surface was then coated with a rubbery application to further damp any potential resonances. This flare was chosen for its relatively uniform axial frequency response over the range of its intended use, and good prior reports of its general sonic properties. Although constant directivity designs were considered, offering a more uniform polar pattern control with respect to frequency, by their nature they do not have a flat axial frequency response. By spreading the output of the driver uniformly with respect to frequency, the constant directivity devices, by definition, must have an axial response closely approximating to the total power response, which must in turn correspond to the power output response of the compression driver – most of which have outputs which tend to fall off above the lower to middle mid-range. Their subsequent reliance upon equalisation precluded the use of constant directivity horns with this particular design philosophy. Further repercussions on the sonic properties of constant directivity horns are discussed at greater length in Chapter 12, from the findings of the cepstral analysis.

20.5.1 Compression drivers

Compression drivers themselves can become the source of endless, quite unfruitful argument on the subject of what is considered to be correct, as the mid-range is probably the area which produces the most intense conflicts of likes and dislikes. The options initially offered for this design were the Emilar EK175, and a combination of Coral driver with JBL titanium diaphragm. Straight JBLs had also been used. All of the above drivers can produce a substantially flat, axial response from 1200 Hz to 8000 Hz, and also have similar sensitivities and around 100 watts power handling capacity. The choice was entirely down to subjective, personal, audible preference. The Emilar produces a slightly softer sound, whilst the Coral and JBL units exhibited somewhat brighter characteristics. A fourth possible option was the TAD TD2001. This driver has a beryllium diaphragm, with tonal characteristics somewhere between the aluminium diaphragm of the Emilar and the titanium diaphragm of the JBL. This is the driver now used so successfully on the AX2 horn, but with the RA1 described here the axial response was too bright. This highlights the degree of specificity in driver/ horn 'best' combinations. The 'best' driver for one horn is not necessarily the 'best' for another.

Studio monitor loudspeakers are a means to an end – tools to achieve the optimum, overall, final mix. They are sometimes tailored to take into account the human aspects of life in a control room. Although other drivers could, no doubt, be used, and truly dozens were tried, the four mentioned above appeared to give the smoothest transfer and closest match to the units chosen for the frequencies immediately below and above the mid horn. Ultimately, the Emilar won out on overall system neutrality when combined with the RA1 horn, though it must be added that, for different purposes and usages, other drivers may have been preferred. The Emilar was deemed to be the most neutral sounding choice for *this* system, across the widest range of programme material. Some of the more exciting sounding drivers which were tried were ultimately deemed to be 'too good to be true'. None of this, however, implies the general superiority of any one driver, *per se*.

By 8 kHz, the natural response of these mid horns is beginning to tail off, so once again, according to the overall design philosophy, no attempt would be made to equalise the falling response; it would be allowed to dovetail neatly into the response of a suitable high frequency driver.

20.5.2 High frequency driver

In order to match the rest of the system, the HF driver must have a high acoustic output capability, high reliability, and a smooth response from 7 kHz to 20 kHz. It must also have good horizontal directivity and be as close as possible to a point source. Of all the units available, the JBL 2405 'slot' was considered to be the one which was best suited for this purpose. With its very lightweight diaphragm assembly, it provides exceptional highs of crystal clarity, without the requirement for any additional equalisation. With a 40 watt power handling capacity, a 105 dB sensitivity, and a huge magnet system, the acoustic output capability in the top octave is awesome. The directivity angles of the 2405 also closely match the polar pattern of the chosen mid-range horn. Even on loud music, the slots would typically only be running at 1% of their rated output.

20.5.3 Total system choice

To recap, the final driver choice was as follows:

20 Hz to 300 Hz: 2 × GAUSS 4583F 15″ bass drivers, each of 300 watts RMS power handling, and 97 dB sensitivity.

300 Hz to 1200 Hz: JBL 2122 10″ cone, mid-range unit of 75 watts RMS power handling. Or 250 watts for the 2123, and 101 dB sensitivity (which were used in some very large rooms.)

1200 Hz to 7 kHz: EMILAR EK175 with horn as described, capable of handling 100 watts with a 108 dB sensitivity.

7 kHz to 20 kHz: JBL 2405 of 40 watts power handling, and 105 dB sensitivity.

All the above sensitivity ratings are for 1 watt at 1 metre.

Each unit is essentially flat over the frequency band for which it is being used, eliminating the need for equalisation circuits and producing a very clean, natural, 'unequalised' sound. Small adjustments may be made to the overall frequency response of the system by adjusting the gain of the individual amplifiers, thus providing a 'shelving' response adjustment over each frequency band, without introducing any unnecessary circuitry.

20.6 Cabinet requirements

Once the drive units had been chosen, they required housing in some form of suitable enclosure. The cabinet used was chosen to have a final volume of air, after loading and internal treatment, of around 12 cu. ft. A tuned port was used, with a relatively large surface area to preclude any tendency towards 'breathing' noises, which sometimes accompany the higher velocity of air moving through smaller apertures.

The material chosen for the cabinet was 'Multicapa', a type of 1″ ply, developed in Galicia, North-West Spain, and comprising of alternating layers of softwood laminates and compressed eucalyptus fibre. This material is very dense, exhibits excellent resistance to warping, and is also very easy to work with. Its construction gives good suppression of panel resonances, while maintaining great strength and damage tolerance. The enclosures were heavily cross-braced, side to side, front to back, and top to bottom. Each internal surface was also braced, then treated with a series of heavy and compliant counter-layers to further damp any spurious resonances.

The initial treatment consisted of covering all internal surfaces with a mixture of underseal, heavily loaded with silica-sand. When this was thoroughly dry (an important note as residual fumes given off can affect some loudspeaker cone adhesives) the surfaces were treated with a PVA adhesive and covered with ½″ high density, reconstituted polyurethane foam. Upon this was added a further layer of PVA adhesive, then a layer of mineral-covered bituminous felt, additionally supported by a 'quilting' of large-headed galvanised nails. Later versions replaced the bituminous felt with Noisetec LA5, a plasticised deadsheet with heavy mineral loading. The above combination gives remarkable resistance to the onset of structural resonances in the enclosure itself, and also provides an appropriate internal absorption to effect the desired tonal qualities from the bass end of the spectrum. For further control of reflexions, especially with respect to any harmonic chatter from the front baffle, one sheet of BAF wadding or a dense cotton-waste felt was applied to the rear surface, and hung as a curtain inside the cabinet volume.

At very low frequencies, and, consequently, long wavelengths, the absorbent property of padding is minimal, hence the use of the highly lossy composite lining to reduce the amount of LF energy impinging upon and subsequently vibrating the cabinet walls. All that we are intending to do is to reasonably control any resonant modes which may be excited within the enclosure, preventing them from striking the acoustically partially transparent loudspeaker cone, and subsequently passing through into the listening area. It is in the lower mid region where the 'timbre' of the sound is most readily affected by internal damping, and where more careful consideration must be given to any such damping materials.

One advantage of crossing over at 300 Hz is that without the compromises which are usually required, the two ends of the bass spectrum can be given their optimum enclosure treatments. The 10″ lower mid-driver could then be placed in a separate chamber of around half a cubic foot. This chamber was mounted on the front baffle and consisted of a roughly 10″ cube. The primary function of this sub-enclosure was to prevent the 10″ cone from going into orbit when the two bass drivers punched inwards. The separate acoustical treatment facility was a further advantage. Although a cube may at first sight not be the ideal shape from a standing wave point of view, by the time that the cone and magnet assembly had been introduced, the box became far from cubic. The break-up was further enhanced by the addition of a diagonal half-width sub-divider. The small enclosure was then lined with polyurethane foam, further suppressing undesirable resonances, and a further treatment of cotton waste felt. The outside of this sub-enclosure was treated with the same underseal/sand mixture as the main enclosure, in order to reduce any resonant tendency in the wooden walls. Lead-in wires to the drive unit were sealed with silicone rubber.

The compression horn and 'slot' were mounted in the same vertical plane as the 10″ driver (Fig. 20.4). They were placed as close together as practically and aesthetically possible, in order to maintain the closest approximation to a co-incident sound source. Aesthetics may initially seem to be a somewhat peripheral subject, but its importance is frequently underrated. Sight is the sharpest of our senses: it has the ability to distort our other perceptions by overriding them. When watching a TV whilst listening through a less than perfectly co-located hi-fi system, in many instances we soon forget that the sound and vision sources are not from the same place. A very great number of the hours worked in a control room are spent staring towards a pair of loudspeakers. A symmetry of the loudspeaker's physical layout can condition the brain to expect symmetrical sound sources. It is a point of psychology rather than pure acoustics, but, once again, the loudspeakers are designed to be used by human beings, with all their peculiar idiosyncrasies. These aspects cannot be ignored just because they have no bearing on the results of measuring instruments. All in all, though, I am still a firm adherent to the philosophy of vertically symmetrical driver alignment, and the concept, happily, seems to be becoming the norm. Chapter 19 discussed the acoustic reasons at greater length.

Figure 20.4 Completed 1985 version of 235 as described in this chapter. Shown with optional cone mid unit on top, which directly interchanges with the horn

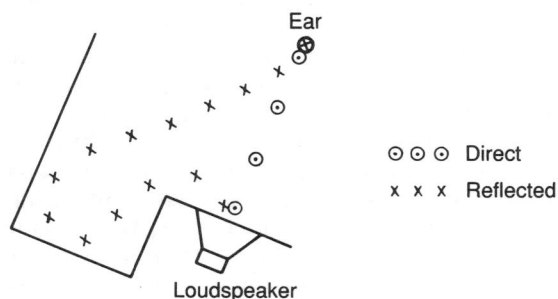

Figure 20.5 Differing lengths of sound paths with corresponding phase/time delays between loudspeaker and ear

20.6.1 Wiring and mounting regimes

The boxes were wired up with cable of sufficient gauge (6 and 10 mm²) to easily pass the high transient currents associated with such systems. The systems were primarily designed for mounting such that the front baffles were flush with the front wall of the room. This allows no areas around the cabinet sides for the bass to take any but the direct path to the ears (Fig. 20.5). Two versions of the 235 were eventually produced. Enclosures designed for flush mounting having front baffles ½″ proud of the edges of the cabinet sides, to allow wall finishes to butt up neatly. Free-standing versions were produced having their front baffles recessed one inch. The one-inch lip is of purely visual origins, and whilst the purist may maintain that cabinet edges should chamfer backward from the baffle, extensive listening tests on cabinet designs of this size could determine no audible difference whatsoever! On purely aesthetic grounds, the lips remained.

20.7 Power amplifiers

The final choice of amplifiers were combinations of Crown DC 300As and D150As. The 300As deliver 350 watts into 4 ohms for the bass drivers, and 170 watts into 8 ohms for the low mids. The D150As are capable of driving 80 watts into the 8 ohm compression horns and 40 watts into the 16 ohm slots. This combination appears to achieve the optimum compromise between headroom, indestructibility, and providing sufficient acoustic output when put into an operational environment. Solid state protection circuits could have been incorporated as breakers at the loudspeaker end, but they have not proved necessary. Anyhow, most protection systems can introduce some potential for non-linearity, and thus, where possible, I believe are best avoided.

Many different amplifiers were tried with these systems, and, by general consensus, the decision of greatest suitability fell to a genre of amplifiers of which the Crowns were typical examples. From time to time, other suitable amplifiers were used. But, when producing something as a reference, then to vary each individual installation, even by as much as substituting a relatively similar item, one soon loses the absolute certainty gained from repeated installations of identical systems. Some people had reservations about the 'old' technology of the Crowns but repeated listening tests showed some ostensibly superior units to match the system less well. The subject is further discussed in Chapter 6, Section 6.4. I am loath to change, merely for the sake of change: I prefer to stay with what I know until an improved component part is proved superior in the actual circumstances of use.

To give some indication of the actual power distribution within the above system when responding to average musical programme, the following figures may prove interesting. With 300 watts into the bass drivers, the low mids draw around 30 watts, the high mids around 4 watts, and the slot tweeters rarely more than half a watt. The very low power consumption of the MF and HF horns is a testament to their efficiency. Domes or cones at comparable SPLs for the MF and HF would draw around 100 watts and 20 watts respectively. Dissipating the waste heat could be a problem. The mid-range and high frequency horns achieve close to 50% efficiency, every 1 watt of electrical input producing around $\frac{1}{2}$ watt of acoustical output. Five watts into the mid horn would produce around $2\frac{1}{2}$ watts of voice coil heating, whilst $\frac{1}{2}$ watt into the tweeter would produce around $\frac{1}{4}$ watt of voice coil heating.

In the case of the direct radiating units, which achieve typically only around a 2% efficiency, the 100 watts required to drive the mid-range unit to the same acoustic output as 5 watts into the horn would produce around 98 watts of voice coil heating. As we all know how hot a 100 watt light bulb can become, the implications for thermal compression and voice coil fatigue in the direct radiators should be clearly apparent. Similarly, a 1″ dome tweeter under these circumstances would be attempting to dissipate around 19½ watts of waste heat from its 20 watts input and 2% efficiency. At such high levels, direct radiators certainly feel the strain and can be prone to premature failure.

Some trial systems were built using MOSFET amplifiers on the higher frequency end of the system in response to the wishes of certain customers. The one major limitation of most MOSFET amplifiers is their watt-for-watt inability to deliver the immense current surges demanded by the low frequency components of high power transients. Although generally considered unwise, in this instance it was possible to mix amplifier types. The low frequency crossover point of 300 Hz is in a far less critical area of hearing than the low-mid to mid crossover point of 1200 Hz. Bi-polar amplifiers can be used on the low frequency end of the spectrum, with MOSFET amplifiers driving all other stages. Were bi-polar amplifiers used on the lower mid as well as the low end, the transfer to MOSFETs would take place coincident with the switch from cone drivers to compression horns.

This change in amplifier and loudspeaker types at the same frequency was not considered to be conducive to the smooth transfer of sound as the music moves up the spectrum. The MOSFET option therefore was chosen to maintain bi-polar amplifiers on the low end, with MOSFET amplifiers on all other stages. After long field trials, however, the all bi-polar option has generally been deemed preferable on overall balance. One note of caution here is that extreme care should be taken whenever mixing amplifier types to ensure that they are correctly phased. By no means all amplifiers give a positive-going pulse on their red output terminal in response to a positive-going input pulse. Always check manufacturers' data, or measure them.

20.8 Crossovers

Much of the basic crossover philosophy used for these systems was discussed in Chapters 7 and 13. The overall design philosophy of the system called for each drive unit to handle its own range effortlessly, and the crossover frequencies were chosen to cover the optimum ranges of the selected drivers. One octave beyond the crossover points, however, some of the drivers exhibit some irregularity as the response tails off. With a 12 dB/octave roll-off, a 4 dB peak in the drivers' response one octave above the 3 dB down crossover point is only 11 dBs below the system's smooth response at that frequency. This was definitely able to be detected as colouration in the sound, particularly on certain instruments. The problem was all but removed with the use of 24 dB/octave crossovers.

Unfortunately, though, despite the drive units having been chosen to have characteristics complementing and matching each other as closely as possible, the first choice of Linkwitz–Riley 24 dB/octave slope was somewhat abrupt in its transfer from one drive unit to another. This was noticeable especially at the change from the cardboard cone to the metal diaphragm at 1200 Hz. It was by no means distressing, but a slight change in timbre was noticeable, particularly on rising sections of strings. The crossovers with Butterworth 18 dB/octave slopes were initially adopted as, upon listening, they were generally considered to offer the smoothest and most pleasing overall performance. They also had the advantage over the 12 dB/octave units in their capacity to reduce the fatigue on the compression drivers by more rapidly reducing their input at frequencies below the crossover point. Indeed, the extra 6 dB increase of slope reduced by 75% the power delivered into the drivers one octave below the crossover point. Ultimately, with the development of Colin Clarke's hybrid slope design, a 24 dB/octave crossover became realisable having the sonic transfer smoothness of the 18 dB/octave designs (see Chapter 7).

20.9 Developments of the basic system

Accepting that not all control rooms could accommodate monitor loudspeakers of the size of the 235s, a more compact version, the single 15" model, 238, dispensed with the 10" loudspeaker and operates as a 3-way system. In this model, the crossover point from the 15" bass driver to the 1" mid horn was chosen as a compromise at 800 Hz, later to be changed to 1 kHz. This necessitated a change of bass driver, as the Gauss 4583F unit, whilst excellent in the lower register, had no subjective 'life' in its reproduction beyond about 500 Hz. The Gauss was substituted by a JBL 2235, the 'subjectively usable' response of which extended at least one octave higher than the Gauss, whilst compromising only slightly on the performance of the Gauss at the low end. Despite not having the extreme ruggedness of the Gauss, in the smaller rooms and less punishing situations envisaged for the 238s the JBL has proved to be an excellent unit. At the time of writing (1995), to the best of my knowledge, none of the original dozen pairs of 238s supplied in the mid-1980s have since been replaced. They have had a lot of happy purchasers.

As an option, all models could have the slots crossed over passively from the mid-range horns with a 6 dB/octave internal crossover. The crossover components were incorporated in the cabinets and could be connected, if required, on the rear terminal blocks. Although not providing quite the same degree of flexibility as the actively crossed over units, the 7 kHz passive option has been used with entirely acceptable results.

As a further option, a cone mid-range unit could be fitted to any of the systems by means of a quick and simple conversion. The cone option box is fitted to a front plate, identical in size and fixing holes, to the mid-range horn. The cone driver is less efficient than the compression horn, but by means of pre-set amplifier gain control levels, the systems can be set up for rapid interchange. For some orchestral music, for closer range monitoring, or for situations where the full output potential of the system is not required, this option may be considered desirable. Where cone mid-range units are used, the HF slot is usually changed for a bullet, in order to match more closely the directivity pattern of the cones. At close range, the narrow vertical directivity of the horns and slots can be problematical, as the act of standing up or sitting down can involve a vertical movement of the ear, beyond the ± 20° vertical coverage of these units (Fig. 20.6).

20.10 Summing up

As with several other chapters of this book, the intention has not been to imply any absolutes, but once again to examine the many branch lines down which a thought train must travel, and to show the mental gymnastics

Head of standing engineer

Narrow directivity
angle of
conventional horn
in the vertical plane

on axis

Head of sitting
· engineer

(a)

(b)

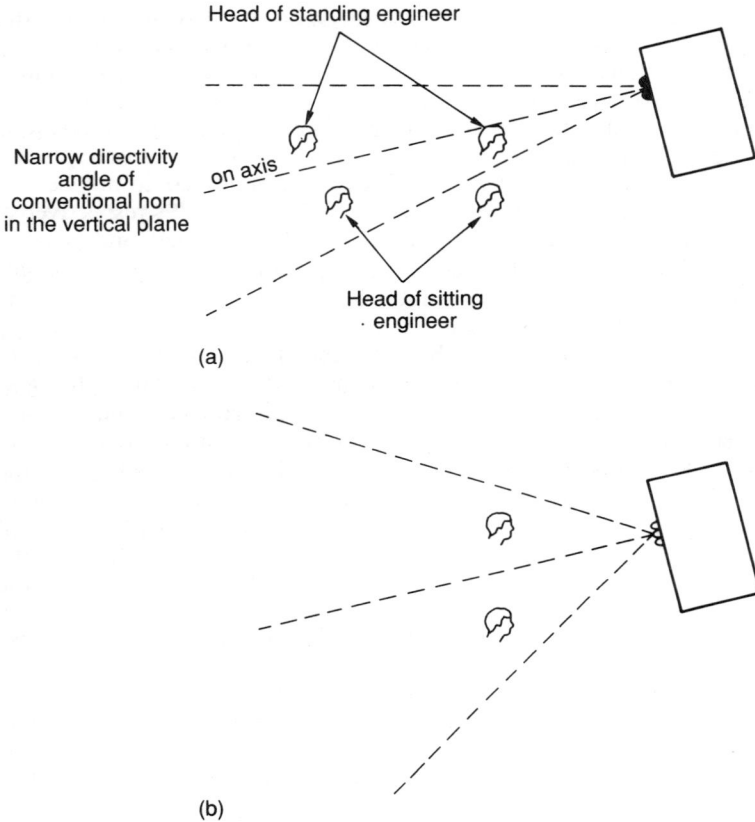

Figure 20.6 (a) At close listening distances, with a narrow vertical directivity remaining within the directivity angle can be difficult when moving from a standing to a sitting position; **(b)** with a wide-directivity drive unit, much greater freedom of movement is acceptable at close quarters

which may be required to reach the overall compromise which a designer may deem to be optimum for the uses intended.

Whilst perfection is by no means claimed, the principles discussed are obviously viable, as the systems in use designed around these concepts are standing the tests of time. The only field failures have been a slight loss of 'sparkle' from the aluminium mid-range diaphragms, but only in the larger systems in the larger rooms after 18 to 24 months of very heavy, high level use. Diaphragm technology is a subject of continual research, but no users have yet baulked at the prospect of a biennial diaphragm replacement programme, which works out at around £1 per week in maintenance. Titanium diaphragms do seem to remain consistent over longer periods of time. Notwithstanding my own preference for the tonality of the aluminium diaphragmed Emilar compression drivers, some users requiring consistently very high SPLs and being quite happy with the tonality of the JBL 2425s and 2426s have chosen to use these devices. Remember, there are no absolutes. As Chapter 12 shows, different people may have very different perceptions of what is 'right', even under highly controlled conditions. Not only subjective preferences, but also perceived similarities are very much still in the ears of the beholders!

Sadly, more recently, the demise of the Emilar Corporation created supply problems for the mid-range drivers. A number of suggested replacements were tried and tested, but the previous sonic match could not be re-established. The 235s and 238s were updated to 236s and 239s respectively, with the TD2001/AX2 combination being 'lifted' from the loudspeaker design described in Chapter 19. These replaced, in one unit, the mid-range and high frequency horns of earlier models, and thus the 4-way system became 3-way, and the 3-way became 2-way. However, whilst the overall axial performance remains compatible, the inevitable differences in the directivity between the rectangular and axisymmetric horns renders the two systems dissimilar in overall performance when used in conventional control rooms. Consequently, the new versions are not direct retro-fits for the older ones. They must be considered to be different loudspeakers. Such is reality!

Chapter 21

Close-field monitor design – a practical approach

21.1 Background

My first encounter with anything resembling close-field monitors was over twenty-five years ago. I used to work for Pye Studios, which was then one of the leading recording complexes in the UK, with two eight track studios, four disc cutting rooms, a handful of tape copying rooms, and two reduction rooms – remix suites in current parlance. The consoles in the main control rooms were Telefunken in studio one, and a Ken Attwood/Derek Sticklen construction in studio two. Equalisation was largely outboard, and patched in when required. These valve (vacuum tube) consoles were subsequently replaced by germanium, then silicon transistored Neves. The installation of early Dolby noise reduction systems sent us searching for hums of which we had previously been unaware, masked in the 60 dB or so signal-to-noise ratio of the 8-track, 1″ Scullys.

Monitoring was largely carried out on Lockwoods. These were fitted with 15″ Tannoy 'Silvers' and 'Reds' and driven variously by Radford, Pamphonic, Pye and Leak valve amplifiers. Despite the seemingly rudimentary nature of the equipment, we achieved some outstanding recordings – even by today's standards. The ability of engineers such as Bob Auger, Ray Prickett and Howard Barrow had to be heard and seen to be believed. They were grand masters of their art form – the big session. Everything had to go down more or less at once, as eight tracks did not leave too much scope for overdubbing.

Around this time, the sale of singles reached an all-time peak, and their importance to the record companies' finances became enormous. How a single would sound when played via a transistor radio seemed to become paramount, so into the studio's list of equipment came the Pye 'Black Box' record player, complete with turnover crystal cartridge for 33/45 rpm or 78 rpm discs. Why they were sold in the shops as the 'Black Box' I don't know; to the best of my recollection they were polished, veneered wood! Anyhow, before a mix was finished, an acetate would be cut in one of the disc cutting rooms, then played on a record player for appraisal. The curtailing of frequency response would often lead to a readjustment of the levels of vocals, and guitar solos in particular, which were often left exposed by the rapid roll-off at the top and bottom end of the spectrum. A suitable compromise level was usually achieved, all parties leaving satisfied.

It should be remembered that, in those days, there really was a chasm between 'studio' and 'domestic' equipment. There were a few people with systems which could be deemed to be 'hi-fi' by today's standards, but these people were only fractions of one per cent of the record buying public, and probably even fractions of point one per cent of the singles buying public. As we earned our livings courtesy of this record buying public, they did, at least, deserve some consideration in the making of the music. Even to this day *their* enjoyment in listening to *their* record collections should be a high priority to the recording industry, but nonetheless it is still incumbent upon the professional echelons to try to maintain, and wherever possible to advance, the overall achievable standards.

21.1.1 Domestic limitations

The question which used to be asked of any recording prior to the cutting of the disc was, 'Will it track?' With CDs and cassettes now taking the lion's share of the market, though the 'tracking' problem of vinyl discs still exists, it is no longer the priority that it used to be. What you could put on tape, and even what you could put on to disc, bore no relationship to what could be reproduced by the record player; engineers were very conscious of what could ultimately be reproduced in people's homes. If an engineer achieved an 'amazing' bass sound on the recording tape, it was of no consequence if it merely caused people's record player needles to jump out of the groove. When thousands of singles were returned to the shops because they 'stuck', or 'the needle jumped', the engineer would be most unpopular. His prima donna attitude to achieving 'amazing' sounds in the studio would soon leave his reputation in tatters as being unrealistic and unprofessional.

To some degree, a new version of this old theme has emerged. The extended frequency range of digital systems, such as compact disc and digital audio tapes, now exceeds the performance capability of most modern domestic hi-fi. This time, however, it is the loudspeakers which are the weakest links. Much domestic hi-fi still does not like high levels of 20 Hz, but thankfully the days of the 'Black Box' record player are a thing of the past. Albeit in a more up-to-date form, a domestic reference is still a necessary requirement.

The old divide between 'domestic' and 'studio' quality has now greatly shrunk. Indeed, much domestic equipment will surpass the performance of some studio equipment. The main difference now is in the realms of consistency and longevity in continual use; hence the price difference. It is far more expensive to build maintainable equipment than to build disposable equipment.

21.1.2 Early close-field approaches

I suppose that the first accepted 'standard' for close-field monitors came with the influx of Auratones from the USA. These were more on a par, certainly in the mid-range, with the average 'music centres' of the mid 1970s. Tough and durable, the Auratones were a useful tool for producers and engineers alike in their search for an 'average' system. By the early to mid-1980s, however, the average domestic system was progressing beyond the Auratones, and more advanced close-field systems were being sought.

The domestic reference began to change to a specific, close-field monitoring requirement. Many engineers were finding that the Auratone monitoring had other advantages beyond the domestic reference. Vocals and guitars became easier to balance in the crucial mid-range area where a substantial part of the information is carried. Reverb was made easier to judge in terms of desired length and quantity. Stereo imaging appeared to be more simple to judge. Control of the overall mix seemed achievable with much less effort than before. Certainly during mixdown, the tables were turning.

By this time, the bulk of the mixing was frequently being done on the small monitors, with less frequent cross-reference to the main monitors. The large monitors were now mainly used to check for any lack, or excess, of extreme bass, of extreme top, to listen for noises, to solo instruments, and, ultimately, for the final 'playback' to check that all was well; and also to give everybody concerned a little well deserved excitement.

The strength and failing of the Auratones lay in that they were single driver systems – no crossovers, no tweeters, no mechanical displacement of high and low frequency drivers. To all intents and purposes, throughout the most critical mid-band frequencies there were no significant phase shifts! Asking this from a large studio monitoring system was a tall order. To achieve the extended frequency range and adequate acoustical output, multi-driver systems were virtually mandatory. Many such systems had inherent problems of time and phase accuracy, and even the so-called time and phase 'corrected' systems could cause as many problems as they solved. Larger, multi-driver systems must also be listened to at some distance, as the outputs of the various drivers do not 'gel' until the listener is six to eight feet away. The greater distance the listener is from the loudspeaker, the more the acoustics of the room begin to take effect. Reflexions and reverberation in the room become confused with the effects and reverb in the mix.

A room with twenty people and little equipment in it will sound much different from the same room with only a few people, but piled high with hard surfaced keyboards. The early domestic/close-field systems were more 'secure', more consistent, less upset by environmental changes. What is more, they were small enough to take from studio to studio as 'known' references.

21.1.3 More advanced requirements

I think that one reason why such good recordings could be made on such antiquated equipment as was available in the 1950s and 1960s was that each person in the recording chain had some understanding of the other people's problems. Recording engineers understood the principles of disc cutting; indeed, most could cut a disc if they had to! They also understood and accepted the limitations of their equipment. Just as racing drivers must understand the performance limits of their cars, it is the people who achieve the best compromise within their limitations who achieve the best overall results. I believe that it was this breadth of knowledge which enabled such excellent recordings to be realised all those years ago. The very term 'recording engineer' was a description of a very skilful job, usually only acquired after many years of comprehensive training. By the late 1970s, the term had become more related with the specialised job of a balance/mixing engineer, many of whom had little in-depth knowledge of the other aspects of the processes involved between the musicians playing, and the records appearing in the shops.

It seemed that demands were growing for a close-field system which went as loud as the main monitors, had a full frequency response, were small enough to fit into a briefcase, had all the cohesiveness of the Auratones, were indestructible, and cost less than £50. OK, I am exaggerating – slightly! In the quest for this Holy Grail, just about every decent loudspeaker around was blasted to destruction then dismissed as rubbish. It was a very unfair situation, as many people were just not aware of exactly what they were demanding from the loudspeakers. We are all governed by the laws of physics. We can demand whatever we want – but if it is impossible, we cannot have it!

It was the turn of Yamaha next to provide the industry standard – the NS10. Precisely how and when the NS10 made its first entry into studio use, I am not sure, but its penetration was widespread and rapid. By late 1982, they seemed to be springing up like mushrooms. Not that many people claim to actually like them, and as a hi-fi loudspeaker, the purpose for which they were designed, they were a commercial failure. Despite this, they have now held much of their position in studios for almost fifteen years. Clearly they fulfilled a need, even though few people can explain exactly how they managed it. They are hard sounding, they lack transparency, they can go reasonably loud, and they are short on high top. In fact, they are characteristic of many of the large monitor systems which they were ousting on their first appearance. What they are, however, is relatively consistent, especially when used at close ranges.

They could be quite punchy and exciting on certain types of music, so it seems that in many ways they were more representative of many studio monitoring systems than they were of domestic hi-fi, yet people felt that work mixed on the NS10s in the studios travelled well into the domestic world once released. They also seemed to sit at a crossroads, somewhere between the Auratones and the large studio systems. Unfortunately in many respects, they sat there a little too comfortably, and many people began to use them as the reference, often working on the NS10s alone. The inherent limitations in such a practice were that the lower octaves went almost totally unmonitored, and the sonic hardness which they possessed masked any tendency toward hardness in the mixes, caused by the clashing time and phase characteristics of many of the ever increasing number of digital effects which were being used. As a reference which most people in the industry are familiar with, I am all in favour of their use, but as a sole reference, their use can leave too many questions of balance unanswered.

Certainly in their early versions, they suffered a very high rate of driver failure in studio use, but this is not a criticism that I would level at Yamaha. There was a label on the back of each pair of NS10s which quite clearly stated that they should not be used for reproducing the sounds of tapes fast-winding, sine waves, music with the tone control knobs set in the maximum position, signal generators – they just could not be more specific!

So, what do our 'specialist' engineers do with the NS10s? They leave them switched on when the tape is winding back, leave them on when the machines are being aligned with reference tones, solo instruments, then swing the equalisation controls on the console from minimum to maximum to find the sound that they are looking for – then criticise the loudspeakers when they finally expire.

Yamaha's response to this was to produce the studio version, uprated in power, and designed for horizontal mounting, though this placed the time/phase characteristics through the crossover point at a disadvantage by widely separating the drivers on the horizontal axis. Nonetheless, in this form, they are still in widespread use. When the studio versions first appeared, many people claimed to prefer the original version, though whether this was a real preference, or just a reaction to a slight difference from a 'comfortable' familiarity, is hard to know. With the studio versions, the problem of failures was somewhat reduced, but the general limitations of the design still existed. Yamaha's next development was the 3-way NS40, but it failed to emulate the success of the NS10, perhaps by neither being sufficiently neutral to provide a significant step forward in clarity, nor possessing that 'magic' formula which had caused the NS10 to be so widely accepted. Bearing all of these problems in mind, including the human aspects, it can be seen that the design of an ideal close-field monitor is no mean feat. In fact one thing that it is certainly *not* is cheap!

21.2 Practical design considerations

The design philosophy used in the development of the Reflexion Arts 250 close-field monitor system may help to outline the maze of dilemmas involved in effective, 'close-field' design. The aim was to produce a close-field monitor system, achieving the most transparent and natural reproduction, whilst being relatively compact, loud enough for all general purposes, and relatively abuse tolerant. It had been felt that many existing close-field systems were lacking in low frequency response, sometimes leading to poor subjective LF judgement where they had been used as the main mixing reference. Where the Auratone really scored was that it was a single driver system. The entire, highly critical mid-range of frequencies was unencumbered by crossovers or mechanical misalignments. If, in practice, the whole concept of close-field stereo monitoring had its roots in the era of the Auratone, then we should perhaps look there to find out just what we need. Essentially, we are looking for an Auratone with an extended frequency response, higher overall output capability, greater definition, and, generally, the best quality available from such a system!

21.2.1 Choice of crossover frequencies

The easiest way to maintain an accurate phase response over the most crucial range is to continue to use a single driver as far as possible. To fulfil our previously stated design criteria, however, a system using a single loudspeaker seems currently unable to meet our needs. Consistent with our specifications on size and output power, a two-way system would be the minimum required to extend the frequency range to our desired limits. If the low frequency drivers are required to go down much below 100 Hz, it becomes difficult to achieve the correct tonal characteristics from them much beyond 1.5 kHz, irrespective of what any measurements tell you. Drivers which sound good at low frequencies rarely have the appropriate timbre in the mid-range. Notwithstanding this, and even pushing these drivers to 2.5 kHz, this still requires the high frequency driver to take a great deal of upper mid-range punishment, with the attendant risk of premature failure. Such a combination also places the crossover frequency in a very sensitive area of our hearing.

5 kHz would seem to be the lowest desirable crossover frequency for reliable HF driver performance, given the overall size and SPL requirements for studio use. Almost inevitably this leads us into a three-way system, as 40 Hz to 5 kHz is too much to ask from a single driver. Again, I am not talking about measured responses here, I am talking about perceived sound quality. Concentrating on the mid-range, a response from around 600 Hz to 6 kHz would seem to be desirable. A 5″ driver would handle this range well, also maintaining the single driver philosophy over the most critical range. It should also be substantial enough to take a reasonable amount of punishment, yet small enough to be essentially a point source. Once we add a tweeter to handle the top couple of octaves, the first of our problems begin to reveal themselves.

As we approach 20 kHz, adequate, efficient response can only be maintained if the diaphragm is sufficiently light to follow such high frequency waveforms. With a wavelength of around one inch or less at 20 kHz, we must be looking for a very small source to avoid cancellations from waves emanating from opposite edges of the diaphragm. In practice, we are looking for something such as a 1″ dome, placed as close as physically possible to the mid-range driver. The moving mass of a 1″ dome tweeter is in almost all instances small and light. If it is not small, we have phase problems, if it is not light, we have high frequency efficiency problems. If it must be small and light, then how can it dissipate large amounts of voice coil heat? The heat dissipation in the tweeter must inevitably place a ceiling on the total power capacity of the entire system. Doubling up on drivers is no good, as our point source would be lost; so this is another reason for utilising a 3-way system. The HF unit can then come into play at a much higher frequency than in a 2-way system, thus handling a lower proportion of the overall system power.

21.2.2 Programme handling

In normal programme, the percentage of the total musical power above 5 kHz is well down in single figures. If the input to the HF driver is restricted to frequencies above 5 kHz, the total output ability of the system should still be adequate for close-field purposes. Given a typical sensitivity of around 90 dB for 1 watt at one metre, a power handling of 20 watts for a tweeter should allow a total system output of at least 110 dB at one metre. Such figures are realistic, but only achievable by keeping the upper crossover point as high as possible.

We must, here, understand the relevance of the warnings against the reproduction of sine waves. Say we have a system with bass drivers capable of handling a 100 watt sine wave signal. We cannot reasonably rate the system at 100 watt RMS as the power handling capacity is frequency-dependent. Should we feed in a tone of 100 Hz from the console oscillator or a test tape, we could set the monitor volume level to give a power input to the drivers of, say, 60 watts. All should be fine as long as we remain at 100 Hz. The amplifier system will hopefully have a flat frequency response; so, should the oscillator or test tape be switched to 10 kHz, the amplifier will continue to put out 60 watts. This time, however, courtesy of the crossover, all the power will go into the HF dome. Before the magnet assembly can begin to conduct heat away, the sudden surge of power in the voice coil of the dome will cause it to burn out.

Think about it. It is not designed to glow white hot like a light bulb; neither is it made of high melting point tungsten, nor surrounded by argon. Hence the warning about the application of sine waves and the fast winding of tapes. Sine waves concentrate power at spot frequencies, so the power is not being distributed evenly throughout the system. Fast-winding a tape concentrates the bass energy in the higher frequency bands, resulting in similar problems. This warning should also be heeded in respect of certain synthesisers, again capable of 'unnatural' high frequency outputs. With many acoustic instruments, their natural power output closely corresponds with the power handling distribution of most loudspeaker systems, therefore with conventional music few problems arise.

From this it can be seen that if it is compactness, wide range, high output, and quality that is wanted, the manufacture of totally idiot-proof systems is really not feasible. Like any other thoroughbreds, close-field systems must be treated with a reasonable measure of respect. If you really want to hammer the loudspeakers, then you will have to revert to the big monitor systems. Incidentally, we cannot utilise the more powerful HF drivers from the big systems, because they just will not 'gel' with the rest of the systems at close range. Their size, shape and directivity patterns are usually designed for 'longer throw' systems.

21.2.3 Durability

Suitable bass drivers may also suffer from a degree of fragility, but for different reasons. Many domestic loudspeakers have a restricted bass response – for some very practical reasons. Domestic systems are largely in the hands of amateurs. People who have music systems in their homes cannot all be expected to understand the subtleties of loudspeaker design. Children may frequently play with the music system, so many of these systems must be capable of withstanding a degree of domestic abuse. If we want to achieve 60, 50 or 40 Hz bass response at an acceptable level from a small box, we must reduce the resonant frequency of the driver low enough to be effective down to our design requirements. This can be achieved in two possible ways. We can increase the mass of the cone/coil assembly, and/or reduce the stiffness of the suspension.

The first option has two serious drawbacks. Increasing the mass will increase the momentum of the moving parts, so the efficient ability to accelerate and decelerate will be impaired, tending to produce a poor response in the upper ranges. The problem can be overcome to some degree by larger, more powerful magnet systems; but this is unwieldy, expensive, and does not always have the desired effect. The result may tend to be a less 'tight' bass response, and a requirement for inordinately powerful amplifiers. The second option is to reduce the resonance by increasing the compliance of the suspension – that is, by loosening it. The only effective drawback here is in terms of mechanical fragility. In professional circles, this should not be too problematical. In domestic use, however, a certain amount of abuse tolerance is essential – warped records, fridges turning on, people plugging things in with the volume turned up, and so forth.

On the grounds of cost, and therefore durability, these LF response options are not always open to the domestic manufacturer. If we are looking for a bass driver for a professional close-field system, and want deep bass from a small box, we will, of necessity, be looking at a relatively freely suspended driver. Again, due care and respect for the equipment will be called for, so it would be unwise to solo a bass drum with the monitor volume turned up high; the main monitoring system should be used for such purposes, then go back to the close-field system at an appropriate volume level. This is one of the reasons why the close-field systems are still to be used in conjunction with a large system. They do not entirely replace the large systems. If a small system must be totally fool-proof, it will not sound as good! The two are just not compatible, so that choice must be made.

21.3 Choice of drive units

It should be stressed that in the model 250, the drivers were chosen for sounding 'right' together. Specifications tell you nothing about the sound of a unit. Computers cannot design loudspeakers to sound a certain way. Such a choice is a very human thing, being judged on the performance on a very wide range of music, so the choice had to be made by listening. The units used in the production models were all JBL: the 044 dome tweeter, 2105 mid-range, and two 115H bass drivers.

In the crossover regions, transfer from one driver to another had to be smooth and natural. A distinct change in timbre as frequencies pass from one driver to the next was unacceptable. The fundamental of C above middle C will come predominantly from the bass drivers; the C one octave above will come predominantly from the mid-range unit. When heard consecutively from one loudspeaker system, the result ought to sound as though the notes came from the same piano – if indeed they did! Selecting units in isolation is useless. There are times when choosing a 'better' unit can be a retrograde step as far as the system is concerned.

21.4 System arrangement

Once we have selected the drivers, we have to think of how to arrange the system. In order to maintain their stereo imaging, smoothness, and general integrity of sound, one of the most crucial criteria for close-field monitors is

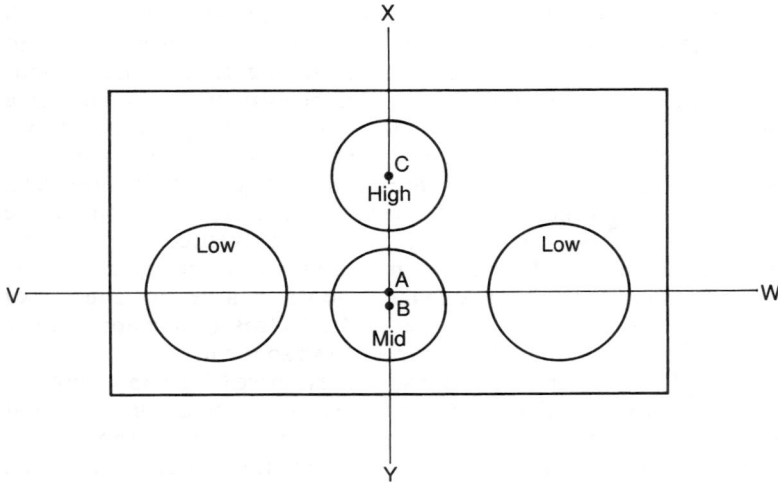

Figure 21.1 X–Y Common vertical axis. V–W, Horizontal axis of the two bass drivers producing a 'phantom' source for the two which falls at point A, very close to point B which is the source of the mid-range unit. A, B and C, the acoustic 'sources' of the low, mid and high frequencies, forming a very tight pack, necessary for an approximation to a co-axial, point source, for the whole system

phase/time response. We should still be aiming as closely as we can towards a point source consistent with wavelength. The backbone of our problem is broken by using one single, small, 5″ driver to cover the entire mid-range. Crossover points can thus be chosen to be sufficiently high that phasing begins to become more academic as the wavelengths shorten – say 5 to 7 kHz – and low enough that the wavelengths become longer than the cabinet itself – say 500 to 700 Hz. With such a system, as long as arrival times are kept in check, awkward phase interference effects can be minimised.

At the crossover points, two things are likely to affect the phase response: physical positioning, and the crossover itself. Let us look first at the effect of physical positioning on the high frequency crossover point. We were still aiming as closely as possible at a point source for the entire frequency range. One reason for choosing a one inch dome for the high frequencies was the very small source area. It also possessed a directivity pattern matching closely that of the chosen mid-range cone driver. The phenolic resin 1″ dome transferred the sound very smoothly from the paper cone of the mid-range driver. (It should be noted here that, along with various soft fabric domes, we also tried aluminium and titanium domes. They would, however, just not subjectively 'sit' properly in the system as a whole.) The dome was located on a vertical centre-line passing through the middle of the mid-range cone. The two drivers were then located as close together as physically and practicably possible (Fig. 21.1).

Besides the physical coupling, there is also the electrical coupling to be considered. In choosing the crossover parameters, the options available to us cover high level or low level, active or passive, and the frequency and slope rate of the roll-off. The frequency response of the chosen mid-range unit begins to die a natural death at a rate of around 6 dB/octave beyond 5 or 6 kHz. This mechanical roll-off has all the same properties as an electrical roll-off, phase shifts included, so it would be no use thinking that putting a 24 dB/octave electrical crossover at this frequency would bring about an in-phase response. It would not! Each 6 dB of roll-off produces a 90° shift at the crossover point; therefore, 24 dB/oct = 4 × 90° = 360° = full circle = 0° phase shift. The mechanical roll-off will add to the electrical roll-off in every way, hence we will end up with maybe a 30 dB/octave composite roll-off, and back to the original 90° phase shift.

The other disadvantage of a very high rate of roll-off at this point was the 'brick wall', sudden transfer of sound from the mid to high frequency drivers. Matched in character as they were, 24 or 30 dB/octave was still too abrupt for a natural sounding transfer from the paper cone to the phenolic dome. We could remove this mechanical roll-off effect either by lowering the crossover point out of the range of the roll-off, or by using a different driver with an extended HF response. The first option was rejected, as we had already concluded that we must keep the crossover point out of the most sensitive range of our hearing, hence above 5 kHz. The second option was rejected, as the driver had been specifically chosen to sound right in the system.

The conclusion was to use the smooth mechanical roll-off to full effect, bringing in the HF driver via a capacitor, giving a 6 dB/octave, electrical roll-off below 5 kHz. Audibly the transfer was smooth and natural. Under normal circumstances, a 6 dB/octave crossover is the only one which will sum to unity on-axis in steady state terms, whilst also being phase accurate in respect to its ability to reconstruct a square wave at its output. The effect is a somewhat fortuitous result of the ± 45° phase shift at the crossover point, counteracting what would be a lump in the perfectly in-phase summed output. 6 dB/octave is too shallow a slope for most practical purposes, but this instance is one of the few in professional monitoring where it can be used.

In the 250, the only addition to the HF crossover was an inductor, but strictly speaking this was not in the crossover. The dome had been found to have a resonance in the 1 kHz region, and with a 6 dB/octave crossover slope, the response at the 1 kHz resonance was only about 10 dB below the 6 kHz response level. The inductor was chosen to suppress the resonance by increasing the LF roll-off. As this was well below the main crossover point, other than removing the resonance its effect on phase or the general character of the sound was inconsequential.

So, to the low frequency drivers. The crossover frequency had been chosen to be 700 Hz – precisely why, we will discuss later. Once again, ideally, we were looking for a co-incident source for the entire system, and for once we were in luck. As the wavelengths below 700 Hz are around eighteen inches or more, we can close-couple two 6½" bass drivers, placing them tightly either side of the mid-range unit. Their 'phantom' source point will then be on the centre of the line passing between the centres of the mid and top units, being arranged to fall very close to the centre of the 5" mid-range cone. Such a coincident sound source is not possible with any single bass driver system (see again Fig. 21.1).

Of course, a fundamental component of the low frequency system is the cabinet. Much of the thinking for the 250 followed closely that for the 235 described in Chapter 20. Construction materials and technique were basically similar, but the 250s utilised a cabinet of just over one cubic foot after the insertion of treatments, drive units, and the one litre sub-enclosure for the mid-range driver. The cabinets were provided with two ports, giving the options of 45 Hz tuning with both ports open, 35 Hz with only one port open, or a sealed box response with both ports closed. Different circumstances of use, different musical programme, or personal choice dictated the 'best' option. Looking back, the 35 Hz tuning received the most widespread use.

21.4.1 Electrical considerations

The low frequency crossover was the next item to be considered. The mid-range driver was easily capable of a flat response down to around 300 Hz, with the bass drivers responding up to around 2 kHz. Choosing 700 Hz as the crossover point meant that there were smooth responses of at least one octave at either side of the crossover point. With this absence of mechanical roll-offs in the crossover range, there would be no attendant phase shift or slope-rate problems to complicate the electrical crossover. Consistent with the design philosophy of the larger monitor systems described in the previous chapters, it was decided not to place any components between the bass drivers and the power amplifiers. This inevitably entailed the use of bi-amplification, the advantages of which seemed ultimately to far outweigh the drawbacks. To have the full damping factor of the amplifier controlling the bass drivers really is the only way to consistently achieve the desired, low frequency transient response.

The option of bi-amplification left a decision on the choice of slope-rate for the electronic crossover. From the point of view of phase, 24 dB/octave would be the ideal choice. In the passive form, 24 dB/octave can have problems, especially in terms of component tolerance and power loss, but in the electronic form no such restrictions apply. In the case of the 250, the sonic characteristics of the bass and mid-range drivers were well matched, but even so, sweeping the crossover frequency up and down produce a change in timbre. The crossover frequency was ultimately chosen to be 700 Hz, being the subjectively audible optimum for the most smoothly consistent neutrality of sound on the widest range of music.

With the drivers having responses a good octave either side of the crossover point, mechanical roll-offs and phase shifts were well out of range of compounding, and hence upsetting, the parameters of the electronic crossover. This allowed more flexibility of choice over the crossover parameters, and, especially as mechanical roll-offs are likely to change, both with age and between batches, it restricted likely variations to out of band areas.

When buying a high quality loudspeaker system, nobody would think of a conventional, passive crossover as being a separate, add-on unit. In this case, the electronic crossover must equally be considered an integral part of the loudspeaker system. Any old crossover at the right frequency just will not suffice. The parameters of the crossover were designed to precisely match the requirements of the whole system. Crossovers should always be considered as part of a package with the loudspeakers.

Ideally, even the amplifiers should be specified in such a system. The constraints of the laws of physics preclude small diameter low frequency drivers from attaining the sort of efficiency/frequency range characteristics of larger diameter units. The drivers chosen for the 250s had sensitivities in the order of 89 dB for 1 watt at one metre. This necessitated 8 dB more drive (about 7 times the power) than that needed for the bass drivers of the large monitors described in Chapter 20. Despite being used at closer listening distances, the 250s nonetheless require all the amplifier power of the larger systems. As with the larger systems described in the previous two chapters, the 250s were designed around the use of direct coupled amplifiers. With certain other types of amplifier, the very solid bass can tend towards a more rounded sound. This is usually due to a function of the amplifier's inability to supply high, instantaneous peak currents, normally a function of circuit type rather than output power. Very large power supply reservoir capacitors help greatly in this respect. It is worth remembering again that one of the most common causes of the premature failure of HF units in conventional systems is the distortion from over-driven bass units passing through the passive crossover, and producing impossible to follow, spiky waveforms for the HF drivers. Bi-amplification precludes this possibility, and hence goes a long way towards consistent, long-term operation.

21.5 Two-monitor systems

It does appear that we are now reaching a stage where we have different monitor systems for the recording and mixing stages of operations. With musicians spending so much time playing their instruments in the control rooms, the large systems are still invaluable. They are used for soloing instruments, experimenting with sounds, listening for noises, and providing musicians with a level to which they may be accustomed on stage; any one of these factors on its own necessitates a large system. The large systems are far more accident tolerant during setting up, or when recording where there may be a risk of feedback. The current pattern seems to be to use the large systems mainly during recording, with frequent reference to the close-field systems. The converse seems to apply for much of the time during the more subtle stages of mixing, where operations should be under much greater control.

21.6 Power handling

On the subject of power handling and output capability, just how should we rate the power handling of such a system as the one described here – and is it relevant in any way? The answer to the first question is that I do not; and to the second question, rarely! The aim was to produce a system which was 'loud enough' for all intended purposes, and with enough headroom for clean transients and a measure of damage tolerance. The design specification looked for a broadband output, per cabinet, in excess of 105 dB per watt at 1 metre, and a frequency range of 40 Hz to 20 kHz. Such figures as these are really the only ones which count. Power handling figures, especially in multi-amplifier, multi-driver systems, are usually very misleading and are of little relevance, as will be seen from the following figures.

In the 250, there are two bass drivers handling an approximate frequency range of 40 Hz to 700 Hz. Each unit is rated by its manufacturer at 60 watts of pink noise with 12 dB/octave roll-off below 40 Hz, and peak-to-average ratio of 6 dB. This is endured for a period of 2 hours to meet I.E.C. 268–5. Their sensitivity is 89 dB, averaged from 100 Hz to 500 Hz. No mention of RMS, and no mention of sine waves. They are simply not built for listening to high level sine waves, and who wants to anyway? It should be pointed out here that a similar sized driver with a heavier coil and double the power handling (3 dB up), but with an 86 dB sensitivity (3 dB lower), would produce only the same sound pressure level at the overload point, yet would take twice the amplifier power to achieve this. In this instance, doubling the power handling capacity may well look good on paper, but in practice could well be a backward step as thermal compression could be more of a problem. Everything must be taken in balance.

The mid-range units used in the 250s are rated by their manufacturers as handling 50 watts continuous programme. Continuous programme is defined as 3 dB greater than a continuous sine wave input. The sensitivity is rated at 94 dB SPL for a 1 watt input, swept between 500 Hz and 2.5 kHz, and measured at 1 metre distance. The high frequency driver is rated to handle the top range of 60 watts of pink noise, filtered 12 dB/octave below 2 kHz and above 5 kHz. These last figures are not very useful in our case, as we are only using the HF driver from around 5 kHz to 20 kHz.

So many times, engineers have asked me of a loudspeaker system, 'Oh, how much power will they take?' As explained in Chapter 13, without reference to sensitivity and frequency, the answer 'seven bananas' would carry as much relevance as '50 watts'. There still appears to be a macho value to power handling, but it is absolute nonsense.

What they really want to know is, 'How loud will they go?' This can probably only be realistically specified in terms of pink noise, band limited to the frequency range of the system, and measured at one metre. The rated output level should be, say, 3 dB (half the power) below the point where the first part of the system begins to show any signs of stress. This should give us a conservative, long term rating, and act as a reliable indication of how loud the system will go when used on music. Pink noise is a good approximation to a normal musical signal, which, after all, is what they were designed to produce. I use the word 'normal' there with some trepidation. If somebody so chooses, they could incorporate into their music a high level swept synthesiser tone which, I think, must be concluded as being 'abnormal' programme. Such a swept tone would be akin to tones from the console's oscillator or from a test tape, and, as we have already discussed, this can lead to problems. The rule of thumb should be: 'Whatever the programme material, if the system begins to show signs of stress – turn it down!'

Before setting about the design of the 250s, many engineers and producers were questioned in an attempt to define the acceptable 'normal' level for the use of close-field monitors. The first prototype 250s were easily capable of producing a 10 dB higher level than had been requested, which should have provided an adequate margin to ensure a long and consistent life. But no. In every case, when the systems were put out on trial, they were being used at maximum. Everybody thought that it was great that they went so loud, but our safety margin had instantly withered to zero. Such are the conflicts of physics and art!

21.6.1 Overloads in practical use

It seems that whatever common sense may dictate, small monitors will be used to breaking point. The general rule of thumb seems to be: 'The level at which a monitor system will be used is 3 dB beyond its point of destruction.' Very similar to the old motor mechanics' adage: 'Should a torque wrench not be readily available, tighten the nut till the thread strips, then back it off a quarter of a turn.'

A loudspeaker system which carries protection systems can also suffer constraints on sonic performance under certain drive conditions. I am reluctant to compromise audible performance for the sake of over-protecting a system, and *was* so especially in the case of the 250s, as they had so much in-built headroom to protect them from all normal risks when being used for the purposes for which they were designed. They are close-field monitors, not small sound reinforcement systems. They were designed for professional use, under the assumption that it would be professionals who would be using them.

21.7 Conclusion

To date, the subjective characteristics of loudspeakers cannot be effectively quantified in terms of figures, but people hearing the 250s have told us that they like them and that they can use them. 'Use' is the operative word, as, above all, monitors are tools. They are designed to help the producers and engineers achieve a desired objective. Carpenters are not in universal agreement about the finest saws and chisels; tennis players do not all prefer the same type of racquet. There is no reason why recording engineers and producers should all decide to use the same monitors.

I tried to design a loudspeaker system with the greatest degree of accuracy and neutrality that I could currently achieve. I cannot categorically state that it is better or worse than any other system. That decision is in the hands of the users. I believe that the system is accurate and natural, and I am very pleased with the users' responses. Being a designer and not a marketing person, I had no wish to produce yet another competing loudspeaker system in a field already saturated. I had felt that there was an unfulfilled requirement which was at that time not being addressed by the industry, at least not without compromising on grounds of price, and therefore quality. Beyond a certain point, one becomes involved in the application of processes of diminishing returns. Every 10% improvement doubles the price, then each 5%, then each 2%; but if the improvements are desired, then these costs are reality. That is one major difference between designing for a professional market and designing for a consumer market. The cost-effective criteria of the latter cannot always be directly applied to the former if the ultimate achievable performance is demanded.

We begin with a somewhat vague brief of what is required, progress through a veritable jungle of paradoxes, dilemmas and constraints, then conclude with an arbitrary result. As can now clearly be seen, all is not quite plain sailing in the design of small loudspeaker systems. After reading this, at least a few people may now realise just how much must be considered in the design of a close-field system. Small does not mean easy.

Taking the technical and human considerations fully into account shows that there is much more to it than selecting a couple of loudspeakers and putting them in a box.

One other point which may be worthy of note is the similarity of physical layout of many of the systems which are currently on the market. I am sure that many of the thought processes which have been described in the preceding pages were the same for many designers and manufacturers. They are logical steps, entirely dictated by the laws of acoustics and psycho-acoustics. There are many advantages of a dual bass driver system which cannot be achieved by other means. The layout shown in Fig. 21.1 is the one which comes closest to the co-axial, point source concept, whilst retaining the flexibility of driver choice and the many other advantages of the discrete driver option. The relationship of structure to function is well known throughout science. In biology, it applies very powerfully, from the levels of molecular biology right through to the macro systems of the organisms. Probably, as powerfully as anywhere, it relates to the very ears through which our loudspeaker systems will be judged. Some shapes just have to be!

Chapter 22

Human factors and general observations

22.1 Problems in the comparison of loudspeakers

Loudspeaker manufacturers frequently comment that their small monitor systems do not receive the quantity of editorial reviews in the professional recording press as do many other products. Magazine editors and many users would doubtless love to see truly meaningful comparative tests, but such tests as are often envisaged are usually both flawed and controversial. Consequently, many editors avoid them as they often provoke a huge backlash. In general, different loudspeakers sound more different than different amplifiers, and whilst much can be gleaned from the specification and general review of an amplifier by an experienced individual, such is not the case for loudspeakers, where conventional written specifications relate only very poorly with perceived sonic performance. The only seemingly viable means of comparison would be listening tests, but if results are inconclusive, then what practical purpose would they serve?

No one loudspeaker can be all things to all people, and as with musical instruments, different manufacturers largely judge the success of a design by the degree of acceptance by professional users. However, the 'rightness' of a musical instrument is generally accepted as the entirely personal, subjective choice of the individuals concerned, whereas the 'rightness' of monitor loudspeakers is usually related to a more objective, definitive 'rightness': 'The closest approach to the original sound' as one manufacturer so aptly put it. Initially it would seem that 'the best' loudspeaker would suit all purposes, but this is not necessarily so. To use an analogy, one could decide that one wanted the best aeroplane wing. It would be reasonable to assume that British Aerospace were likely to have the know-how for such a requirement, but it would be futile asking them simply to produce 'the ultimate wing'. Which ultimate wing? For what flying speed? For what altitude? To carry what weight? For what minimum landing speed?

22.1.1 Location

Before attempting any comparative assessment of small monitors in particular, one must ascertain with great accuracy the precise conditions of location, mounting and driving for which the manufacturers have designed their systems. Loudspeakers designed for mounting in a relatively free space will have very different low frequency characteristics if mounted next to a wall, as was discussed in detail in Chapter 4. Indeed, those characteristics will be different again dependent upon whether the aforementioned wall is either behind or to one side of the loudspeaker. If the low frequency directivity is 360° or thereabouts, then in free space the low frequency power output is balanced such that a more or less uniform overall response exists on-axis. With a wall behind the loudspeaker, the mid and high frequency output will still continue to project in a largely forward direction as before, but the low frequency energy cannot propagate through the wall, or at least not to any significant degree, so it is reflected back into the room and reinforces the forward radiating part of the low frequency output. The on-axis response will thus be raised by something in the order of 3 dB.

If the wall in close proximity to the loudspeaker is to one side, then the same sort of low frequency augmentation will take place, this time with the energy which would have travelled to one side of the loudspeaker being reflected back to reinforce the axial response. In this case, however, some mid and high frequency energy may also be reflected off the side-wall, dependent upon its nature, and whilst the overall on-axis energy may remain reasonably uniform, the reflexions at the shorter wavelengths of the mid and high frequencies will be time delayed, due to the additional length of their reflected paths being a greater proportion of their wavelength than of the low frequency reflexions. The resultant phase discrepancies will cause comb filtering, and hence both colouration and time smearing of the signal (Fig. 22.1).

An extension of this location problem occurs when the choice is made arbitrarily as to whether to mount the small monitors on top of the mixing console, or on stands behind the mixing console. When mounted on top of the mixing console, typically on the meter housing, the flat surface of the console will act something like the side wall in the last paragraph, largely reflecting at all frequencies. If the loudspeakers are mounted on stands behind the console, there will be a space between the loudspeaker and the console which will allow the low frequencies to 'breathe' and escape below the console, approximating more to the free space mounting, but here the middle and high frequencies may still encounter a reflective surface in the top of the mixing console. Different consoles will reflect to different degrees at different frequencies.

Whilst the side wall or the console reflexions may measure quite similarly, the reflexions from the side wall will most definitely be perceived differently to the vertical reflexions coming back up from the console, both in terms of perceived colouration, and the effect in stereo imaging. Given all of these variables, and even just scratching the surface to add a few more, such as the proximity of corners or other low frequency reinforcing structures, or even the mid/high frequency reflexions of equipment racks, it must become very evident that any small monitor system will probably rarely encounter similar listening conditions for any two pairs of units sold. This is one reason why I have avoided manufacturing or selling loudspeaker systems unless I have had some control over the acoustic design of the rooms in which they would be used. I sympathise very much with loudspeaker manufacturers who have faced totally uninformed criticism of their products, after having been auditioned in entirely inappropriate circumstances.

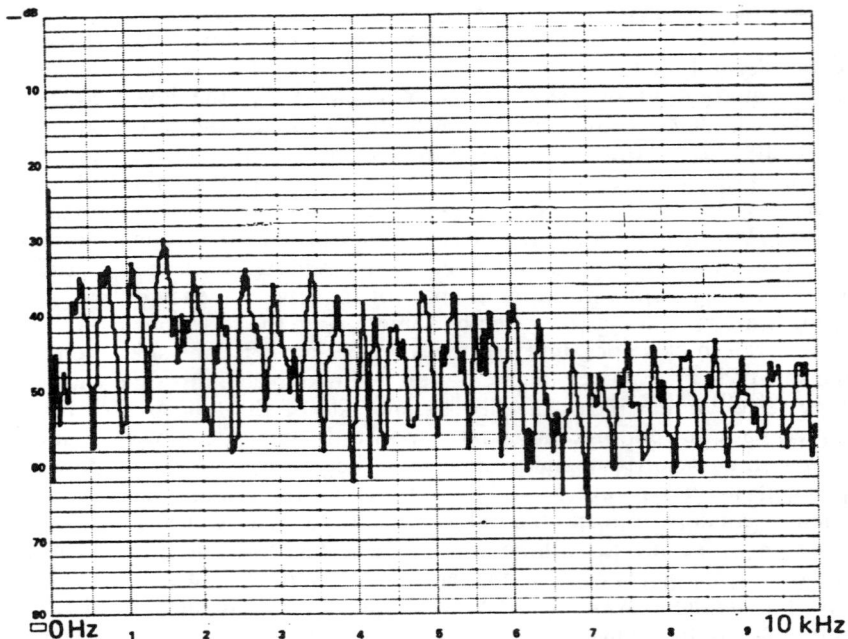

Figure 22.1 The averaged power spectrum of a signal with one discrete reflexion. Comb filtering is revealed clearly in a linear (as opposed to a log) plot, where the regular nature of the reflexion-produced disturbances is plainly evident.

In the instance shown here, the additional path length of the reflected signal over the direct signal was just under one metre, producing comb filtering with dips at a constant frequency spacing of just under 400 Hz

The first problem in any meaningful test set up for auditioning small loudspeakers is therefore to locate them in a position preferred by their manufacturers – the location in which they were *designed* to perform best. If one then attempted to set up, say, six pairs for A/B/C/D/E/F testing, then what would happen if, say, two or more loudspeakers demanded the same location for optimum performance, or the preferred location for some of them obscured the direct signal path for the others?

22.1.2 Interaction

The second problem which arises with any sets of loudspeakers in close proximity to one another is that the drive units, or their tuned boxes, or both, may resonate in sympathy with the driven units; hence they can cause, like mixing console resonances, either absorption or resonant over-hang. If the amplifiers are left switched on and connected to all of the loudspeakers, then some electrical damping will be effective in vastly reducing the vibration of the non-driven loudspeaker cones, but there is little that can be done to prevent a tuned box from resonating as long as its tuning port is left open. For all of the above reasons, the setting up of more than a couple of pairs for comparison in optimised locations begins to appear as something of an intractable problem.

22.1.3 Drive amplifiers

As discussed in Chapter 6, amplifiers can be influential on subjective loudspeaker performance, so in any attempts at comparative trials of loudspeakers, very great care should be taken to ensure that only sonically appropriate amplifiers are used with each loudspeaker to be tested. If listening tests are performed using just one type of amplifier, then this may be unfair on some of the loudspeakers in the tests. Loudspeaker manufacturers should always be consulted on preferred amplifier types or topologies which they consider to be the most suitable for their products. The use of inappropriate amplifiers would invalidate the findings of any comparative listening tests.

22.2 Design priorities

Chapters 2 and 3 discussed at great length the different order of priorities which face loudspeaker designers, dependent upon the type of music, be it rock, classical, electronic or whatever; and the type of recording technique such as close mic'd, stereo pairs or direct injection. Furthermore, would the loudspeakers be the original source of sound, as would be the case for computer-generated music, in which case the monitor loudspeaker during recording would be an extension of the music production system; or would the monitors always be in the reproduction chain, as when recording an actual acoustic event? In the first instance a 'buzz factor' for the musicians would be a possible requirement when the loudspeakers were part of a music production chain, whereas when recording an actual acoustic event, the loudspeakers would not be in the same room as the musicians, so no 'buzz factor' would be relevant. Given these different priorities, the characteristics would be likely to be different for each personal concept of what is achievably 'right'.

The fact is, different individuals simply do not necessarily perceive anything, by whatever senses, in the same way; and from this it follows that even from the same set of unanimously agreed outline specifications, there will be differences in the end product.

22.2.1 Human sensory inconsistency

Maybe the following parable will help to explain, in terms not of sound, but of all of our other senses. Let us now imagine that three people go to dinner at a fine restaurant, and all eat similar meals. During their after-dinner conversations, they concur that the meal was excellent, and that they all thoroughly enjoyed it. There was agreement that the sauces were fine, the main ingredients were fine and the presentation was fine; yet known only to themselves, each individual probably placed a subtly different emphasis on each aspect, which because of the limitations of the spoken word, was not easily communicable to the other diners. Nonetheless, so impressed were they all by what they had experienced, that they decided to open a restaurant each, and give instructions to their first class chefs on exactly what they wanted in replication of this original meal. When each had honed his or her preparation to 'perfection' the three restaurants opened simultaneously, only to serve what were effectively three recognisably similar, yet equally recognisably different, meals.

Clearly, despite their seemingly outwardly uniform opinion of the common meal, their inner perceptions of the points of what were possibly only subtly differing priorities led them, when creating the meal for themselves, and in their attempts to do their best, to highlight and magnify the small aspects which had given them their individually most pleasurable sensations. From sampling the meals of the three new restaurateurs, it may become evident that one had realised the greatest satisfaction from the sauces, another from the main ingredients of the meal, and the other from the presentation. When the meals were prepared separately, their different perceived preferences, which had not been apparent at the after-dinner discussion, had obviously existed all along. Furthermore, it is impossible for any other person to know what any of the three diners actually perceived in relation to any of their senses, or to know any hierarchical order in which their senses internally organised themselves.

Which of the three meals was 'best' was probably not a realistic question to ask, and despite all being different, in their own way each was excellent. No doubt either that, to each restaurateur, *their* offering was the best from their point of view, and most accurate in terms of reproducing the original meal; but what of the views of the clientele? In all probability customers would try each restaurant, maybe several times, and many may settle on one particular favourite, yet by the end of the year, each restaurant may still have the same number of customers. Indeed, some customers may choose to still visit all three, possibly leaning towards one or another for different occasions or circumstances. The opinions of the customers may therefore be entirely inconclusive in finding any outright winner, the choices being influenced by personal opinions, moods or tastes.

Above is a good analogy of the choices made between different high quality loudspeakers. In reality, most designers tilt their compromises to their own beliefs and preferences. Their customers are frequently people with similar sets of preferences and priorities to themselves, and who make many of their choices based on this empathy. The refinement which a designer seeks for future models perhaps follows this path, so the users and designers reinforce each other's opinions such that the refinement follows an almost deterministic course. From the point of view of some people, the refinements of different designers should eventually converge, but for a number of factors, this is not necessarily the case.

As I have stated elsewhere at other times, loudspeakers as we know them are too far from any true reality for convergence to be realisable in current practice. Individual designers, clients and manufacturers will tend to go in their own directions; and even so, this is still notwithstanding the aforementioned effect of the three restaurateurs, and the individual tendencies to make for ourselves, slightly larger than life, those aspects of a sound which give us pleasure. It is of course this fact which creates such a diversity of choice of pianos, guitars, drum kits, hi-fi systems and many other things.

Two additional points relating to individual choice may help to further develop this point. I remember one well known and successful studio in London where two of the equally respected and successful engineers swore by their own desired settings of the high frequency drivers above 6 kHz, but their individual settings were 3 dB apart. As they both considered their individual settings to sound most accurate when compared to an original sound, then given that the same signal path was used, each of their perceptions of the original, then via loudspeaker sound fields, must have been very different, with their own heads and pinnae masking in different ways the different sound fields. Remember, no loudspeaker even vaguely represents an original sound field unless that original sound was itself produced 'synthetically', electronically generated, and first heard via a loudspeaker. Hence one great dilemma: should a monitor system attempt to mimic an original, acoustic sound, or should it be as representative as possible of music when reproduced via a majority of end user loudspeakers? This question has a very great influence on monitor compromises, often unwittingly, and even by experienced personnel.

22.2.2 Unrealistic comparisons

The hi-fi fraternity are often also well off the beaten track when trying to assess monitors. One technical engineer to whom I recently spoke in a studio was trying to compare his hi-fi with a type of studio monitoring system which he had not previously heard. His first comment was that his hi-fi 'specials' at home went deeper. Whether they did or not I am not sure – most likely he perceived them to go deeper in his room. When I related this story to a musician friend of his, the friend said to me that the technician really believed in his home system, and on one occasion had brought his loudspeakers to a studio to demonstrate. The musician, intrigued by the expectations of some wonderful sound, asked a recording engineer to start a DAT machine, then went into the studio to begin playing the drums so that he could return to the control room to listen to the recorded drums, via the wonderful loudspeaker system. Once the technician realised what was about to happen, he shouted 'NO! NO! You will destroy them!' Clearly he was trying to make an unqualified and misleading comparison by criticising a monitor system not by others of its kind, but by judgement alongside something which was entirely unsuitable for monitoring purposes. That which can be achieved on a small scale at low levels cannot always be achieved on a large scale, and what is achievable from a small hi-fi loudspeaker is not always realisable from a large monitor system.

For example, a 24 ft high human being would be four times normal height, but $4 \times 4 \times 4$ times (height $\times$ width $\times$ depth) or 64 times normal weight. Human bodily systems would not function with such disproportions; surface area of skin and lungs would not increase in proportion to weight and energy requirements. Such a human could never breathe properly; in fact the ratios are too wrong to even *be* human. In most physical systems, scaling is not always feasible if functions are to be maintained. Stereo, for example, in a large auditorium, will rarely function properly off the centre line. If a person was to move 20 ft away from the centre line, 100 ft back from the loudspeakers, then although the angular displacement may be within the range of movement from one side of a mixing desk to another in a control room, the increased time differentials between the arrival of the sounds from the left and right loudspeakers would be so much greater than those perceived in the control room that the whole stereo perception system of the listener would be upset by the 'out of range' time delays. This is one reason why so many sound reinforcement systems in large auditoria use central clusters: once above a certain size, stereo will simply not work except for the people listening very close to the centre line.

It is little wonder that monitoring is one of the most contentious issues in the recording world, because despite being expected to be the major point of reference for any recording process, systems are inherently 'inaccurate' in absolute terms, and disagreement exists widely on precise specifications. If such conditions exist for the 'experts' then little wonder that the studio personnel, who have *not* spent a lifetime dealing with the problems, feel a little insecure at times with their 'elastic tape measures'.

22.2.3 Realistic, but inconclusive, comparisons

Given all of the similarities in the problems which monitor designers face, one of the main contentions between the various camps has centred around the question of 'to use, or not to use' horns in the monitor systems at frequencies not below 500 Hz. I think that almost everybody in the studio world agrees that horns below 500 Hz are not suitable for monitoring, though I do know of some people in the hi-fi world who still think that full range horns are 'the answer'.

Until the mid-1970s I had spent ten years firmly believing that I was set for life against the use of horns. For the following ten years, given the requirements of then current studios and the monitor systems available, I tolerated horns as a 'necessary evil'. Nonetheless, I did some very good work, engineering and production-wise, using such systems as Eastlakes, Westlakes, UREIs, JBLs, Altecs, Tannoys, and more besides. When I restarted monitor design in 1984, I spent a great deal of time examining my memory banks to help me make a difficult decision on whether or not to use horns at all. The only fact which kept the debate at all alive in my mind was that the old Tannoy Dual Concentrics, despite being for my requirements short on output, woolly in the bass, and idiosyncratic at the top, had stood the test of time and were rarely, if ever, included in any debate on the pros and cons of horns. By many they were, are, and undoubtedly still will be loved, and certainly the clarity of their mid-range horns was something which had never caused me any problems; unlike the old Altec 604E which had caused me much grief. Although the two were of similar size, and both were co-axial systems for a 'point source', their means of achieving their goals were radically different. In the early 1970s, if I could have had the middle of the Tannoy with what there was of the top and bottom of the Altec, I would probably have been much happier.

As described in Chapter 20, after settling on the drive units for the bass, low-mids, and top of that monitor design, I was left with a gap of around $2\frac{1}{2}$ octaves from 1.2 kHz to 6 kHz in my decisions as to which drive system to choose for the upper mid-range. As control rooms were becoming ever larger and digital demands put greater strains on monitor systems, output level and reliability were quite high on my list of priorities. Roughly around the same time, Roger Quested began producing monitor systems, and in the early days, like myself, used both Gauss and JBL units at low frequencies.

He was looking for a mid-range device, and was directed towards the ATC dome by the much experienced Mike Cotter, a man with great knowledge of the entrails of drive units. At the time, Roger was looking for a mid-range unit of smooth performance and high output capacity, with a generally 'large hi-fi' performance. Although Mike Cotter had pointed Roger Quested in the direction of the ATC dome mid-range, the early failure rate, albeit at very high levels, caused me some concern.

Mike understood my reservations, and said that he could only think of one horn/driver combination from which he had had excellent sonic performance reports in various uses, and that in his opinion it was not archetypically horn sounding. The combination mated an Emilar EK175 compression driver with a Malcolm Hill/ASS horn. I asked ASS to make modifications to the construction material, and was very happy with the results. The model 235, as the 4-way system became, has been in production since 1985, and has proved very reliable indeed, even in circumstances of foolish abuse. The indestructibility allowed me tremendous freedom in crossover design, and the system requires no protection or limiting of any sort. Nevertheless, I still had hopes of achieving the best of both worlds, and it was in the light of that that I initiated the research work culminating in the AX2 as described in Chapters 11 and 12.

Make no mistake, I am not claiming any superiority here over any other approach, but merely wish to point out how different designers can take different approaches. In my opinion, Roger Quested is a very good designer who produces very fine monitor systems. I would never turn my nose up if asked to engineer or produce using Quested monitors: I should be more than happy to use his systems. Furthermore, it is interesting to note that in the last ten years, whilst I have taken Eastlakes, Westlakes, UREIs, JBLs and Tannoys in part exchange when replacing such systems, I have never replaced a Quested system. Nor for that matter have I replaced systems of Neil Grant or Andy Munro, or other dome oriented designers. On the other hand, whereas all of those latter three have also replaced many other systems of different types, to the best of my knowledge none of them have ever replaced one of my systems. It would seem that the users of certain generic system types can be very partisan. Back to the three restaurateurs, I think!

As time progressed, Quested's use of bass drivers has changed to units with much lower sensitivity than those used originally in the early models, so from that point of view our systems have drifted further apart. In a magazine article, Roger Quested explained the philosophy of his approach towards efficiency, drawing the conclusion that he could achieve what he felt was a sonically more desirable end result from systems which may require enormous input powers. I felt that I could achieve what *I* felt was sonically most desirable from the additional headroom capability of high efficiency drivers. A spin-off from higher efficiency is that less watts per dB SPL are necessary from the amplification systems, anywhere between $\frac{1}{4}$ and $\frac{1}{10}$ the power. I know of many places, certainly in Lisbon where I am as I write this, where the 5 KVA demanded by some of the new Quested systems would consume the maximum power available from one phase of the power supplies to many of the premises. Low efficiency systems also produce more heat, which in 40°C summers would require more air conditioning, so more power from that phase and so on. Sometimes, that sort of power consumption is simply impractical. Obviously it could be argued that without adequate power reserves, a studio should not be built in certain premises; but in reality, things grow, they develop, and often many other factors begin to dominate choices. No rights or wrongs here, just examples of further pros and cons to different approaches.

Referring back to a couple of paragraphs ago, there are no doubt many people who will note that the majority of monitor systems which those of my design have replaced have been systems with a horn loaded mid-range. There are people who would claim that, in general, horn loaded systems would not be likely to be chosen by people who had become accustomed to the 'low distortion' of soft domes. Equally, there are other people who would claim that there is an effortlessness to the transients of good horn systems which they miss on some direct radiator units. If such large bodies of people exist on each side of the fence, then it must surely suggest that each camp is receiving something which they need from the loudspeaker systems of their preference. In the past years, I have been working long and hard to try to bridge the gap, and though much progress has now been made, I am fully aware that no matter how confident I could become that the gap had almost closed, others would equally, under the influences of their own choices and sensory perceptions, no doubt decide quite differently.

Well designed systems are not really in any competition, as the users will almost automatically follow many of their instincts, and tend towards the ones with compromise points more in line with their own wants, needs and likes. If everything was too uniform, life could become a little tedious. Hopefully from the above description, it will be seen how difficult it can be to gain any overall consensus, even when people are in a position of having great control over the use of their products. When loudspeakers are sold 'over the counter' to whoever wishes to purchase them, the problems can be a hundredfold greater.

22.3 The need for a range of choice

Some years ago, I was telephoned by a man in Yeovil, England, who asked if I could visit his studio because something was wrong with the monitoring. He had bought a pair of medium sized Genelec monitors which he had liked very much when using them in a studio in London. I went to see his studio, and he explained that he had paid a great deal of money for these loudspeakers and was most disappointed in their performance. How could they be so different from the pair in London? The room had little acoustic control and was greatly colouring the sound. The sort of acoustic control which the room required was out of the question because the owner was in the middle of a long-running project. The room also had little space to spare, and after the purchase of the Genelecs, money was tight. However, the owner was very, very lucky! I moved the loudspeakers about 4 feet and he suddenly exclaimed 'That is it! That is the sound I heard in London!' Had it been a room of different shape, size or construction, the same solution may not have been possible. Obviously, the position of the loudspeakers was such that they were driving some problem room modes; but in a different room, it may merely have been a case of the move driving other, different, problem modes, so out of the frying pan and into the fire.

Somewhat similarly, I refurbished a small control room in Miraflores, Portugal. The room had been too live to produce any real clarity, particularly in the mid and high frequency regions, and was of such a shape and size that the best solution seemed to be to reduce the overall reverberation time to very low figures. There was a large window in the room, out of which went much of the low frequencies, so the low frequency build-up had not been too great. When the work was completed, the owners were delighted by the new clarity and imaging, but now they considered their monitors to be bass heavy. The loudspeakers themselves were not bass heavy. What had happened was that the mid and high frequency reverberation of the room was now much reduced; it no longer reinforced the overall energy in those frequency bands. The large window still 'lost' much of the low frequency energy, so the subjective result was an overall reduction in the perceived mid and high frequency levels, with the overall level of perceived bass remaining as before, though what *was* perceived was heard with much greater definition.

The loudspeakers which they were using were Audix, with a bass port on the rear of the cabinet. Such a loudspeaker was no doubt designed for mounting in a relatively free space, but given the very narrow front to back nature of the room described here, less than 3 metres, inevitably the loudspeakers were forced back, very close to the front wall. As mentioned at the beginning of this discussion, when a loudspeaker is balanced by the manufacturer for a relatively uniform on-axis response when mounted in a relatively free space, it is assumed that some bass energy will be 'lost' behind the loudspeaker. When a wall is placed close behind, this rear radiated energy will be forced forward to produce an excess of low frequencies on-axis. Once again, these people in Portugal were also lucky, but in this instance moving the loudspeakers was out of the question.

The purpose of the port on the cabinet was of course to augment the low frequency response of the loudspeaker, as compared to a sealed box of the same size. It turned out that this augmentation was not dissimilar to that produced by the location of the loudspeakers too close to a wall. Given the 'live' state of the room prior to refurbishment, especially in the mid and high frequency ranges, the 'double boost' provided by the port and the positioning had helped to provide a reasonably uniform though 'muddy' frequency balance, but when the mid and top were controlled in the refurbishment, the bass became predominant.

By sealing the port, the overall balance of the axial energy was restored to something more akin to a lifelike response. Certainly the owners of the studio and the engineers using it considered that they now had an overall sound vastly superior to that before the refurbishing. Had the loudspeakers not been of the tuned port type, but of similar overall frequency balance to the open port response, such a rectification would not have been possible, and a change in loudspeakers would have been required.

In the first of the two cases mentioned above, Genelec could easily have been accused by the uninformed of a lack of consistency in their production batches, implying that two nominally similar pairs could sound very different. This would patently not be true, and I have personal experience of the

excellent consistency of Genelec products. In the second case, the uninformed could equally have made statements to their colleagues to the effect
that 'This room is now very smooth, but the Audix loudspeakers are bass
heavy', which would also not be true. So many loudspeaker manufacturers
must receive enormous amounts of entirely unwarranted, uninformed,
negative criticism as a result of such misapplication of their products, when
in reality no blame could be attached to the loudspeakers themselves. To
the credit of both the above-mentioned studio owners, they had the wit to
consult an appropriate person first, before making rash judgements.

The audio industry is moving apace into the realms of more 'home'
recording in untreated and frequently unsuitable rooms. This situation
cannot be controlled, so each and every room will be different. Half a
dozen such rooms may perform as best they can by choosing different
loudspeakers for each room. The smoother rooms may suit smooth, wide
directivity loudspeakers; rooms with problematical side wall reflexions may
suit loudspeakers with a response more concentrated on-axis. A room
where the loudspeakers must be placed in corners may suit loudspeakers
which are nominally bass light, to offset the 'room gain' caused by the
augmentation of the response by the corners at low frequencies. A larger
room, on the other hand, may suit a loudspeaker with a nominally more
'bass heavy' response, as much of the low frequency energy will spread far
and wide. No one loudspeaker design can suit all of these rooms, and whilst
acoustic control of the rooms may be the 'correct' answer, it is unlikely that
we will see a situation arising whereby all control rooms are specifically
designed as such.

22.4 Psycho-acoustic realities and artistry

It is absolutely unreasonable to take an untreated room, pile in a load of
equipment, all sited so that the knobs are in easy reach, put a pair of
inexpensive loudspeakers in some physically convenient location, and then
expect to achieve adequate monitoring. Far too many people consider the
recording process merely in terms of equipment. This concept seems to be
being reinforced by the ever increasing use of entirely electronic or
computer-based sounds. Where no natural sound ever exists, the discrepancies in any recording chain are not quite so obvious as they would be if an
actual, physical, acoustic instrument was there as a reference. A rather
similar parallel exists in the 'virtual realities' of computer games, which
stimulate our senses in a very believable way, but can be so removed from
any true reality that, after extensive use, instantly readjusting to the real
world can be somewhat strange. Likewise, people spending many hours in
the 'world' of intense work in a studio can begin to believe that what they
hear is 'the real truth'. In fact as one puts more and more energy,
concentration, and emotion into a project, the project itself can become a
reality of great intensity. The ear and brain become conditioned to the
room and monitor responses, and any *other* representation, even on a
much more 'accurate' or wider range system, may be rejected as 'wrong'.

The emotional and psychological aspects of an artistic process, such as
music creation and recording, means that we must be very careful that we
look at our assessment of the processes involved in a highly objective way.

Quality control is not an artistic function – in fact it is one of the few purely objective features of the whole exercise. Any sort of recorded rubbish, no matter how foully distorted, is allowable as long as that is the way that the artiste intended it to sound. It is then down to the record buying public to choose whether or not to buy it. What is not artistically allowable, however, is for noises, distortions, poor frequency balances, harsh sounds, or other afflictions to affect the recording, just because the people recording them could not hear them on their monitoring systems. Because the recording process is now very much in the hands of people who have never worked on 'state of the art' equipment in superb studios, the knowledge of what is possible in terms of detail resolution is often lacking. Experience is often gained by trial and error on the simplest systems, rather than being passed down from highly experienced personnel in high quality studios. The results can be, as one British magazine put it: 'the duff leading the deaf'. If only more people appreciated the true importance of good acoustics and monitoring, then probably they would realise that the extra cost is an excellent investment, allowing them to squeeze every last ounce of performance out of the equipment which they have; often saving costly updating and upgrading of equipment, and giving a new pleasure, ease and confidence to the whole process.

22.5 Acclimatisation

Some of the above-mentioned psycho-acoustic tendencies to reject as wrong any sudden change, even to something more 'right', does of course have its visual parallels. There is the classic experiment of giving a person a pair of inverting spectacles, through which they see everything upside down. After about a week of wearing them all day, and every day, the person will begin to adjust to this new 'reality' until eventually all appears to be normal. Once full acclimatisation has taken place, removal of the spectacles will leave the person seeing the world upside down, without the spectacles. A further period of readjustment will be necessary before all, once again, returns to normality.

I remember speaking to Dave Hughes, then at Rebis Audio, who had been developing a range of sound effects with his partners. Some of their 'wonderful' effects, which they had demonstrated to many interested persons at the factory, had not gone into production, because in daily use the effects 'wore off'. Some of these involved selected harmonic distortion production, and phase shifting. On first use, many people liked the effects, but as the day progressed when using them in a mix, the amount used to create a desired effect would gradually need to be increased. By the end of a day's mixing, to some people the effect all but vanished, but on returning to the studio the next morning, the amount which had been used on the previous night's mix was clearly heard to be excessive. Rebis decided not to manufacture these products, as the effects were too transient.

I remember also speaking to Eric Radcliffe, the producer and 'Eric' of Yazoo's (Allison Moyet and Vince [Erasure] Clarke) classic 'Upstairs at Eric's' album. Eric is the owner of Blackwing Studios in London. He is a graduate in laser physics and a very perceptive individual. He was telling me of the potential similarity between a rhythmic backing track and the noise from an air conditioning unit. After a suitable period in a room, the brain tends to ignore the noise from air conditioners or tape machine fans. Only when the noise ceases is one suddenly aware of its prior existence. By ten in the evening, an engineer may begin a rough mix after having laid down a vocal an hour or so before. With the vocal still fresh in the mind, he or she completes a seemingly satisfactory mix, only to be horrified the following morning upon realising just how low the vocal really was in the overall mix. The engineer's brain gradually suppressed the repetitive, rhythmical backing track as a 'noise', just as it had suppressed the sensitivity to the repetitive, rhythmical beating of the cooling fans. On returning to the studio after a good night's sleep, the engineer is once more aware of the cooling fans – and the backing track.

The three examples referred to here may go a considerable way towards explaining just how easy it is for people to become acclimatised to unsatisfactory monitoring, but when this does happen it is usually the record buyers in the outside world who suffer the rude awakenings.

22.6 Implications for realistic listening tests

Loudspeaker manufacturers seek gaps in the market for which they may produce systems to fill. There are a vast array of different loudspeakers which are used for monitoring; and where room control by acoustic means is not feasible, this available choice of systems is an absolute necessity, even without the added complications introduced by differing personal concepts of what is 'right', and all the other factors discussed earlier both in this chapter and elsewhere. A listening test comprising numerous loudspeakers and several respected engineers and producers may well seem to be an attractive proposition to make good reading in a magazine, but it may achieve only very little in practice. Even if the problems of positioning, driving, and interference from other units could be overcome, the test would still serve little purpose other than to show which loudspeakers suited best not only the room in which they were auditioned, but the very position in that room. Audition in a different room may produce very different results, as could different music, a different set of producers, different power amplifiers, and a whole host of other variables. In Figures 4.1 to 4.4, we saw how the response of one loudspeaker in different positions, even in one room, could be more different than the responses of different loudspeakers sequentially placed in the same position in the same room, showing how the rooms can dominate to an enormous degree.

One apparent light at the end of this long, dark tunnel, is provided by the prospect of digital signal processing using adaptive digital filters and modelling delays, as discussed in Chapter 18. By such means, the loudspeakers can be driven with a signal containing the inverse of their response errors, and antiphase drivers can be superimposed to neutralise room time response problems. However, all of the correction inputs are superimposed on the musical signal, so the greater the degree of correction required, the greater needs to be the headroom of the monitor system in order to accommodate both the musical and the correction signals simultaneously. Slightly ironically here, the more linear the room/loudspeaker combination, the more suitable it is for correction, but the less it needs it!

However, neither digital nor acoustic control is likely to become a norm in the expanding number of 'home' facilities, so it would seem that for as long as the wide variability in rooms exists, then there is a requirement for an equally wide variability in choice of small loudspeakers, in order to find the most suitable match. As all loudspeakers and all rooms are 'wrong', so it is down to individuals to choose which ones are least wrong for them. This should not preclude or discourage progress in the search for a greater degree of 'rightness', but a little more widespread knowledge by the users could save a lot of grief both to creativity and pocket, and may also help to save many unfortunate manufacturers from undue, uninformed, negative criticisms of their probably quite worthy products. Even for these reasons alone, any listening tests intended to find a generally 'best' monitor would be 90% invalid.

22.7 Ouch!

In Chapters 5 and 12, I referred to an anechoic recording of an acoustic guitar chord, which when played through different loudspeakers was perceived by some people to change in its tonality, yet to other people its tonality remained constant but they perceived a different notational inversion of the chord. Dr Diana Deutsch of the University of San Diego, California, has recently published findings showing how even the individual perception of rising or falling pairs of notes can be perceived differently from person to person, dependent partially upon the place of birth of the individual concerned, and also upon the key in which the experiment is performed. It would appear that the median pitch of the language and accent in which we each learn to speak as a child pre-determines for all of our lives certain aspects of our musical perception. Within a given piece of music, people brought up in Southern California or Southern England may not even hear the same tune: but there again, in a different key, maybe they will!

22.8 Audio/visual analogies

Hopefully in the preceding discourse, I have been able to explain some of the reasons why different designers, different users, and different listeners have opted for different compromises in the search for what is 'right' in different circumstances. There is also a visual analogy to this, of which I am sure that many people are not aware, yet it serves well to illustrate and reinforce very strongly the concept that when 'technical' performance is less than perfect, compromises for 'best' overall performance will be circumstantially dependent. No photographic film is perfect, but Kodak, Fuji, Konica, Perutz, or whichever other companies, all produce films which either they think are the best compromises, or fit into their perceived niches in the market place. Many professional photographers know well the strengths and weaknesses of most commonly used films, so they may choose different films for different circumstances to achieve what they believe will be the best photograph. However, just as a loudspeaker fails to produce sound in exactly the same way as any acoustic instrument, the camera does not 'see' light in quite the same way as the naked eye. Hence no one combination of film and lens will be optimum for all occasions.

For example, in the morning, say three hours before midday, one would think that the light would be similar to that in the afternoon, three hours after midday; but not so. In the morning, there is a general blue shift in the light spectrum, compared to that at midday, and in the afternoon there is a corresponding red shift. We are not generally aware of this as we go about our daily routines, but when fixed on photographic paper, the effect is evident. The higher the quality of resolution and overall performance of the photographic system, the more readily the effects will be noticed, so, for the best professional uses, either compensating filters must be used, or a different film must be chosen. What is more, if we purchase a 'standard' film in London, then purchase a nominally similar film in Nairobi, the films may in reality be balanced differently. The reason for this lies in the subtleties of the balances of human skin tones. It is difficult to produce one film which optimally reproduces both Caucasian and Negroid skin tones in the most natural ways, so in countries where 'white' skins dominate the population, the films generally on sale may be optimised for 'white' skin tones, and where 'black' skins are predominant in number, the nominally similar film may well be optimised for the natural reproduction of 'black' skin tones.

The above problem could, like the morning/afternoon problem, be addressed by the use of filters, but the film manufacturing companies realise that the vast majority of people using their products will not be so equipped, nor are they even likely to be aware of the problem, so the compensation is made in the film itself. All of this is closely analogous to the classical/rock music design conflicts, and whilst some of these compromises could also be addressed in many instances by filters or other external means, the production of different loudspeakers, which the users can choose for their own tastes and needs, addresses the problem somewhat more practically on the larger scale, and with less chance of inappropriate use of any such add-on compensation.

During this somewhat lengthy chapter, we have looked at many of the underlying reasons for both the necessity and inevitability of having a wide range of available monitor loudspeakers. Rooms and drive signals create the necessities; human perceptive differences and individual opinions create the inevitability. Hopefully, our digressions into food and photography have made evident the individuality of all our sensory systems; but at the same time, whilst all of these differences create a range of different monitors, there is no justification whatsoever for the use of bad monitors or bad monitoring environments. Good monitors misapplied or mislocated can cause some of the inconsistency of tonal balance which is currently evident when listening to any random batch of CDs, but the current tendency towards unduly harsh sounding mixes, I feel, lies in bad monitor systems.

22.9 Bad monitoring – the root of many evils

Almost all of us have now come to expect that in the world of electronic signal processing, the march of time brings two almost certain benefits: more for the same money, or the same standards for less money. A digital delay in 1972 would cost around the same as a digital delay in 1992; but whereas that sum of money in 1992 would represent maybe two weeks' average wages, in 1972 it would have represented possibly three months' average wages. Not only that, the 1972 unit would offer a performance of such limited standards by comparison to the units of today that it would probably not even find acceptance in domestic musical use. 'Real' price reductions have also benefited mixing consoles, tape recorders, and many other parts of the signal path.

However, at the two electro-mechanical extremes of the recording chain, whilst developments have been made, no such evidence exists to show that we are now getting more performance for less money. High quality microphones remain expensive, and the general rush to buy up almost any 'quality' microphone from yesteryear would suggest quite strongly that sweeping performance improvements have not been as prevalent as in electronic equipment. There are some very inexpensive microphones which have appeared in recent years, with very respectable performances at much more affordable prices than in days gone by, but the 'high-end' remains high in most senses.

22.9.1 Loudspeaker evolution

There is no doubt that at the loudspeaker end of the chain, great developments have been made since 1970 or thereabouts. It is probably true to say that virtually no monitors remain in use from that era, except for the odd few Tannoys or old Quad Electrostatics in some listening rooms. This is in very stark contrast to the microphone situation: almost every microphone considered excellent in 1970 will still be in use today, unless broken beyond repair or where spares can no longer be acquired. Between the mid-1960s and mid-1970s, there was something of an explosion of knowledge in the relevance of many loudspeaker performance characteristics *vis-à-vis* human perception of sound. Much of this additional knowledge pointed the way to new concepts of system design; at the same time, practical, affordable, high-power amplifiers became available, enabling low-efficiency loudspeaker designs to be constructed. They had until then been severely restricted in their applications, power amplifiers being generally limited to around 50 watts – and those were considered large!

The low efficiency 'bookshelf' loudspeakers which began to appear once amplifier power became cheaper per watt allowed hitherto unheard of frequency ranges to be realised from very small boxes. The studios had for many years used a poor quality loudspeaker in the mixing console for a 'what will it sound like on the radio' reference, but as home systems improved, the tendency grew towards using small, reasonable quality domestic loudspeakers for the 'domestic' reference. It soon also became apparent that monitoring on small loudspeakers at close range gave other benefits, such as less disturbance by poor room acoustics, and a somewhat easier task in the judgement of vocal balances and the appropriate levels of reverb. When home recording became a more practical reality, largely due to the above-mentioned decline in the real cost of the electronics and tape machines, these second-string monitor systems, used as domestic references in professional studios, became the front-line monitors in the new domestic recording boom.

The majority of reasons for this were obvious: the sizes were more suitable to use in the bedroom or garage, the performance was more similar to what the home recordists were used to, and, of course, the prices were affordable. On the other hand, the familiarity with domestic systems, which led to their use as monitors in the domestic recording world, was at the expense of a lack of understanding about the true reasons why, as explained in Chapter 21, they had first been used as secondary studio monitors.

22.9.2 Monitoring trends

Studio monitors are usually designed to be able to provide 'real life' sound pressure levels when required, and to some extent the presumption is made that they will be used in rooms with at least some degree of acoustic control. They are not necessarily always the most pleasant loudspeakers to listen to, but that should not be surprising when one considers that 'smoothing over the cracks' is the last thing that one would want in a quality control operation. The quality control used to be in the hands of trained professionals who knew the strengths and weaknesses of all of their equipment, and the large and small monitors were used when appropriate, according to the specific details for which the engineering staff were listening. The home recording development took much of the recording away from trained professionals, and into the hands of what were, in truth, amateur engineers. The subsequent success of much of the music so produced soon took the 'amateurs' into the more professional studios, but so many of their untrained practices unfortunately followed them. Many of these people were unable to 'interpret' the large studio monitors, so held tightly to the 'security blanket' of their small domestic loudspeakers.

Of course mixing is easier on many small domestic hi-fi loudspeakers; the low frequency performance is usually too poor to be truly representative of what is being listened to, and the transparency and general neutrality is often such that they fail to show so many of the undesirable phase and harmonic distortion products which may mar the recordings. So, this 'bury your head in the sand' attitude makes mixing easier for the untrained recordists, simply because 'what you cannot hear, you do not need to worry about'. Unfortunately, this attitude causes endless grief to those who decide to purchase first quality hi-fi systems, and who in turn may listen a little more critically. All of this has led to the current situation where despite the ever improving technology, the tonal balance discrepancies, and the harshness and aggression of so many modern recordings, are, if anything, worse than ever.

22.9.3 Influence on signal paths

Bad monitoring is causing problems not only in the domestic end product; it is also allowing so many console manufacturers to get away with totally sub-standard circuitry. Again and again, an open, transparent sound, when heard direct to monitors, becomes a closed, tight, dull representation of itself when heard via the console monitor returns. This point was discussed at length in Chapter 14, but I must emphasise again that the degree of difference noticed is directly proportional to the quality of the monitoring and the neutrality of the room. If this is the effect just by passing through the monitor returns, then what degradation goes on in the channels of many consoles? On a pair of NS10s or similar, the differences are frequently all but undetectable.

22.9.4 The rise of the NS10

Precisely how, I am not sure that anybody can truly explain, but the situation has arisen whereby, to many people, Yamaha NS10s have become *the* reference. Many people, quite sensibly, use the NS10s as a point of reference, whilst not in any way using them as gospel. Indeed, I am all in favour of their use as a reference, which is much better than the situation which existed before, where no such universally 'known' reference was in use. Unfortunately, however, there are far too many people in the industry who do use the NS10 and others of its genre as gospel, even if they refuse to admit it in experienced company.

As far as I can surmise, the NS10 succeeded because it had the punch which excited musicians during the recording process, it was relatively robust in daily recording use, and the musical balance of mixes done on them 'travelled' reasonably well to play back on other units of reasonably similar shape and size, which constituted a large part of the domestic market. Oh, and of course, they were relatively cheap! On the down side, distortions are relatively high. The phase response is not too good, the low frequencies are virtually non-existent, and the fine detail resolution leaves much to be desired. In short, to try to seek recorded neutrality and transparency from such units is a waste of time, and any subtleties which require high quality monitoring characteristics will not be realised. Any problems created by the use of NS10s are by no means the fault of Yamaha, any more than Rolls Royce could be held responsible for the lack of traction when one of their cars was being used to plough a field. The NS10 was not designed to do what it is being asked to do, and even the 'studio' version was produced by Yamaha largely because of customer demand, and not from their own design concepts. A point worth making in their defence is that I for one do not know how to make anything significantly more representative for the same price.

22.9.5 The cost of reality

My best attempt yet at producing a 'quality' close-field monitor system capable of resolving fine detail was the Reflexion Arts 250; but a pair of them with crossovers and amplifiers would not leave much change out of £3000 ($4500), not far short of six times the price of a pair of NS10s and a reasonable amplifier. Numerous other companies are now producing good quality, and even very high quality, close-field monitors, yet all remain in roughly the same price bracket. Even the ATC SCM10s, which were recently introduced at fractionally under £1000, need roughly the same amount spending again on a good quality 300 W/channel amplifier if their full potential is to be realised. Loudspeaker developments have leapt forward in much greater steps than has microphone technology over the same time period, but unlike the forward leaps in the electronics world, the advancements have not brought significant cost reductions in real terms.

Good loudspeakers are still relatively expensive propositions because, as yet, in this field, nobody seems to have found a way to make 'good', cheaply.

It is just one of those facts of life at the moment that good monitors are expensive, but as they are so crucial to the assessment of the whole recording process, why should they so often be relegated to some sort of second division status when considering equipment budgets? I am sure that in any reasonable recording set-up, I could dispense with one or two effects, spend the money on better monitors, and subsequently achieve much better overall recording facilities. Psychologically, though, because the loudspeakers are not perceived to be in the actual recording chain, to so many inexperienced people they seem to be less important. Realistically, though, they *are* in the recording chain, as they have a direct influence on the person or persons controlling that chain. Try fitting a racing car with an inaccurate set of instruments, then telling the driver that they do not actually affect the performance of the car! If a 'professional' cabinet maker arrived at your house with a bent, smudged and worn out tape measure, there would be only two conclusions to draw: either you had a Zen woodworker, or the cabinets were going to be somewhat rough.

The fact that good monitors are expensive is no reason why they should be ignored. There is simply no substitute if predictable results are to be delivered to the paying public. It is amazing that the public have accepted for so long the appalling disparity in the frequency balances of so many CDs. The anomalies of vinyl disc pick-ups and the vagaries of cassette reproduction took the brunt of this for some time, but with the widespread use of CD, no such scapegoat exists. The reality is that the sub 60 Hz frequencies are largely out of control, with a large section of the recording industry simply not monitoring them at all. Low frequency tone control is all too often being adjusted for its effect at, say, 150 Hz, with its corresponding effect at 40 Hz being unheard and ignored. The result of this is that people buying expensive hi-fi systems, which used to be necessary to appreciate what the production staff were hearing in the studio, are now merely showing up what the production staff *should* have been hearing in the studio – and should have been correcting before it reached the shops!

22.9.6 The hard and the soft of it

In the mid and high frequency ranges, so many recordings now exhibit an unacceptable degree of harshness. Many conflicting opinions have been aired on this point. Some people have been blaming the 'softer' sound of dome mid-range units for allowing harder sounds to be bearable in the studio. Others have laid the blame firmly at the door of horn monitors, citing their usually narrow lower mid-range directivity as the source of the problem, by producing a reverberant field which is light on mid-frequencies. Personally, I do not feel that it is a matter of horns, domes, cones, or whatever else – it is a matter of good or bad!

So many signal processing devices use or cause phase distortions, and even harmonic distortions. When 'monitors' are used whose phase accuracy or harmonic distortion content are inadequate for the purposes of accurate, detailed monitoring, the processing cannot be monitored without the additional 'processing' by the loudspeakers. Consequently, the excessive use of delays and reverbs can create such phase chaos that good positional images are lost, and harshness and aggression can creep into the mix as a result of clashing effects. Poor monitors with bad phase resolution will often not allow this to become apparent at the time of mixing. It is like trying to judge when a picture is in focus by looking at it through an out of focus lens; sometimes the responses sum, sometimes they subtract, but nobody can be quite sure when.

One obvious indication that something was 'wrong' with NS10s was the widespread use of toilet paper over the tweeters, the major controversy being that of one sheet or two! Considering it carefully, it must surely be apparent to anybody with even a slight inkling towards electro-acoustics that a sheet of toilet paper cannot mirror the perfect inverse of the problem. Whenever a mid-range response such as this exists, it is simply not possible to judge when 'enough' of any effect becomes 'too much', especially when many such effects cause interactions when they are summed at the mixing stage. This is the time when the problems creep in insidiously, so only unforgiving high quality monitoring will be able to separate the desirable effects from the unwanted distortions that lead to the all-too-common harshness.

I do not want to be seeming to be too harsh on the NS10. In many ways it is a victim of its own success, for as I have stated before, the fault lies not in the loudspeaker itself, nor with its designers, but in its misuse as a cheap alternative for something which it cannot properly substitute. Many other loudspeakers of similar price and size are equally unsuitable for front line monitoring, although a number of specialist companies are now manufacturing compact monitors of excellent performance. It is totally unfair to the record buying public to continue to put out recordings of arbitrary quality, when a little more attention to detail could 'clean up the act' considerably. In fact, the public are being seriously short-changed while people continue to use monitors costing not much more than 10 or 20 CDs. It should also be plain to all concerned that, given the huge worldwide market, if small, high quality monitors *could* be produced cheaply, then somebody would certainly be doing it by now. In the meantime, if so many people claim to be recording professionals, then they should get used to accepting professional realities, such as good monitors being essential tools, even if they do not come cheaply!

22.9.7 Inter-system compatibility

It may sound to some like this is some sort of sales pitch on behalf of my own particular side of the industry – acoustics and monitoring – but this is not the case. What drives me to write all of this is frustration which I know is shared by most of my fellow designers, and indeed most of the magazine editors who I know. The current attitudes of many people towards monitoring are blocking progress and doing a general disservice to all concerned, from musician to record buyer.

Large and small monitors should be used in conjunction with each other. It is virtually impossible, by using one system only, to gauge accurately what the results will be on the other. This is where the close-field, small monitor originated – as the reference for the small radio loudspeakers. It is a function of the psycho-acoustics that when more bass is present, it may give a sensation of a lack of top. Furthermore, when one monitor system is heard, even only slightly louder than an identical one, the louder one will subjectively appear to have more lows and highs than the quieter one. The overall tonal balance at one volume level will not be the same at any other level. A lack of distortion will often be perceived as a lack of loudness, as can a lack of phase anomalies. All of these things are direct functions of human auditory perception and there is nothing that any system designer can do about that. Allowances for this can only be achieved by the skill and experience of the individuals involved in the recordings, and none of these problems can be resolved by use of any one small pair of inexpensive loudspeakers.

22.9.8 Summary

How and why has the inexpensive small monitor philosophy become 'the word of God' to so many users? Possibly because as far as averages go, they may represent a good average in terms of both the 'monitor' sound and the 'hi-fi' sound, and also by sitting at a good average position in terms of size. This probably means that they do have a very useful role to play in being the loudspeaker to be referred to for that average; in other words, in the original concept of using a domestic reference. Somehow or other, though, that reference role, probably due to many of the reasons discussed above, has gradually in all too many cases assumed the dominant role.

Unfortunately, in such situations, too much bad sound escapes detection unless a far more telling pair of monitors is used in tandem with them. I have been engineering and/or producing now for over 25 years, and I would be the first person to accept that working practices in an artform such as recording cannot be dictated. On the other hand, how can an artistic photographer work with an unintentionally foggy lens? How can a painter work with dirty paints and poor light? How can a sculptor work with a blunt chisel? No artist(e) could produce fine detail working with such limited tools, and no recording personnel can truly know what they are producing with inadequate monitors – no matter how easy they may be to use!

Even given the great differences in human opinions as to what may sound right, it is inconceivable that those differences could be solely responsible for so many of the widespread differences in tonal balance from one CD to another. Much of the blame for this must be in poor monitoring conditions. If monitor transparency limitations are accepted as normal, then it completely removes the demand for console manufacturers to get *their* circuitry into shape. I strongly recommend that recording staff should try the following test for themselves: using a high quality pair of monitors, large or small, play a CD direct to monitors, then via the console stereo returns. If any difference in sound quality or spacious transparency is heard, then try again with some commonly used small references. If no difference can be heard on the small ones, think hard!

In terms of the problems of the harshness of so many recordings, there is now reason to believe that hard monitoring can result in hard sounds. Initially, on thinking about the problem, it would be reasonable to presume that hard monitors would tend to produce softer mixes to compensate, just as a top light monitor could be expected to produce a top heavy mix. These can be the effects during short term use, such as when a person spends a few hours working on an unfamiliar system. Long term however, we seem to acclimatise to these things, so we can get used to a hard sound, accepting it as normal. Further harshness which may be added unintentionally, due to the over-use of signal processing, may not always be heard as such if it does not grossly exceed the harshness of the monitor itself. Toilet paper which may be used to control the harshness of a loudspeaker may also be effective in hiding further harshness from the engineers.

The conflicting long and short term effects of monitor performance on perception suggest that only the use of 'neutral' monitors will ensure that these tendencies can be removed from the equation. The only seemingly insoluble problem is that of tailoring a large system to the sound pressure level at which it will be used. These operating levels differ from engineer to engineer, and producer to producer, and there is no mean setting which will be optimum for all levels, while still being capable of producing tonal balance compatibility with small loudspeakers at domestic levels. It is because that last reference is so important that the close-field monitoring has gained its place in the process; but that place must be one of subservience, not dominance, if we are to be fair to all the record buying public – and, after all, they are ultimately the people who pay the wages of the whole recording industry!

Bibliography

As this book deals with complex interrelationships, and is not as sectional-ised as a conventional text book, a general bibliography is included here to give readers a guide to further investigation into the concepts. Most of the bibliography is concerned with 'key' books or papers, most of which contain in themselves an abundance of further references. The prefix number in square brackets adjacent to each book or reference is for use as a guide to the chapters of this book to which each relates most strongly: [G] indicates general reference. The author wishes to acknowledge the enormous contribution to this work from these papers, not all of which, it must be said, are in total agreement with each other.

Books

1 [4,15,16] Beranek, L., *Acoustics*, McGraw-Hill, London (1974)
2 [19,20,21] Briggs, G. A., *Cabinet Handbook*, Wharfedale Wireless Works, Idle, Yorkshire (1962)
3 [G] Briggs, G. A., *Loudspeakers*, 5th edn, Wharfedale Wireless Works, Idle, Bradford (1972)
4 [G] Briggs, G. A., *Sound Reproduction*, 3rd edn, Wharfedale Wireless Works, Idle, Bradford (1953)
5 [G] Borwick, J., *Loudspeaker Handbook*, Butterworth (1985)
6 [G] Colloms, M., *High Performance Loudspeakers*, 3rd edn, Pentech Press, London (1985)
7 [18] Haykin, S., *Adaptive Filter Theory*, Prentice Hall (1986)
8 [G] Rayleigh, *Theory of Sound*, Dover Press, London (1945)
9 [7,18] Temes, G. C. and Mitre, S. K., *Modern Filter Theory and Designs*, Wiley, New York (1973)
10 [G] Tremaine, H. M., *Audio Cyclopedia*, 2nd edn, Howard Sams, New York (1974)
11 [18] Widrow, B. and Stearns, S., *Adaptive Signal Processing*, Prentice Hall (1985)

Papers

12 [7,20] Acoustical, 'The Quad Comparative Amplifier Tests'. Acoustical Manufacturing Co Ltd, Huntingdon, England (1978)
13 [4,5,15] Allison, R. F., 'The Influence of Room Boundaries on Loudspeaker Power Output'. *Journal of the Audio Engineering Society*, Vol 22, 314–320 (June 1974)

14 [4,15,18] Allison, R. F. and Berkovitz, R., 'The Sound Field in Home Listening Room'. *Journal of the Audio Engineering Society*, Vol 20 (July/Aug 1972)

15 [7,20,21] Ashley, Robert J. and Kaminsky, Allan L., 'Active and Passive Filters as Loudspeaker Crossover Network'. *Journal of the Audio Engineering Society*, Vol 19, 494–502 (June 1971)

16 [4,7,13,20,21] Ashley, R. J., 'Group and Phase Delay Requirements for Loudspeaker Systems'. Proc. IEEE Int. Conf. on Acoustics, Speech and Signal Processing, Denver, Colorado, Vol 3, 1030–1033 (April 1980)

17 [7,9] Ashley, Robert J., 'On the Transient Response of Ideal Crossover Networks'. *Journal of the Audio Engineering Society*. Presented October 1961, 12th Annual Convention of AES

18 [4,7,9,12,13,19,20,21] Augspurger, George, 'Monitor Equalisation'. *Studio Sound* (February 1977)

19 [4,7,9,12,13] Bauer, B.B., 'Audibility of Phase Distortion'. *Wireless World*, Vol 80, 27–28 (March 1974)

20 [7] Baekgaard, Erik, 'A Novel Approach to Linear Phase Loudspeakers Using Passive Crossover Networks'. *Journal of the Audio Engineering Society*, Vol 25, Number 5 (May 1977)

21 [13,15,18] Bridges, S., 'Effect of Direct Sound on Perceived Frequency Responses of a Sound System'. 66th Convention of the Audio Eng. Soc. preprint No. 1644 (May 1980)

22 [7,20,21] Brociner, V., 'Problems of Matching Speakers to Solid-State Amplifiers'. *Electronics World*, 77 (January 1967)

23 [18] Clarkson, P. M., Mourjopoulos, J. and Hammond, J. K., 'Spectral, Phase and Transient Equalisation for Audio Systems'. *Journal of the Audio Engineering Society*, Vol 33, 127–132 (1985)

24 [7,20] Collins, Andrew R., 'Testing Amplifiers with a Bridge'. *Audio*, Vol 25 (March 1972)

25 [4,6,13] Craig, J.H. and Jeffress, L. A., 'The Effect of Phase on the Quality of a Two Component Tone'. *Journal of the Acoustical Society Am.*, Vol 34, 1752–1760 (1962)

26 [10,11] Dinsdale, J., 'Horn Loudspeaker Design'. *Wireless World*, 80, Nos 1459, 1461, 1462 (1974)

27 [18] Elliott, S. J., Stothers, I. M. and Nelson, P. A., 'A Multiple Error Least Mean Squares Algorithm and its Application to the Active Control of Sound and Vibration'. *Proc. IEEE, Transactions on Acoustics, Speech and Signal Processing*, ASSP 35, 1423–1434 (1987)

28 [7,18] Elliott, S. J. and Nelson, P. A., 'Multiple Point Least Squares Equalisation in a Room'. ISVR Technical Report No 165 (1988)

29 [4,15] Ellis, R. J. G. and White, J., 'Problems in Designing Small Control Rooms for Monitoring Television Stereo Sound'. *Proceedings of the Institute of Acoustics*, Vol 11, Part 7, 145–169 (1989)

30 [15] Fahy, F. J. and Schofield, C., 'A Note on the Interaction Between a Helmholtz Resonator and an Acoustic Mode of an Enclosure'. *Journal of Sound and Vibration*, 72(3), 365–378 (1980)

31 [18] Farnsworth, K. D., Nelson, P. A. and Elliot, S. J., 'Equalisation of Room Acoustic Responses Over Spatially Distributed Regions'. *Proc. Institute of Acoustics, Reproduced Sound, Windermere* (1985)

32 [4,7,9,12,13] Haeslett, A. M., 'Phase Distortion in Audio Magnetic Recording'. *Journal of the Audio Engineering Society*, Vol 24 (Dec 1976)

33 [18] Hamada, Hareo, Nelson P. A. and Elliott, S. J., 'Multiple Point Least Squares Equalisation for Sound Reproduction Systems'. Workshop on Application of Signal Processing to Audio and Acoustics (1989)

34 [7] Harms, Wilfred, 'The Butterworth Connection'. *Hi-Fi News and Record Review* (Nov 1982)

35 [4,7,9,12,13,15] Harwood, H. D., 'Audibility of Phase Effects in Loudspeakers'. *Journal of the Audio Engineering Society*, Vol 18, No 1 (1974)

36 [4,7,13,20,21] Heyser, Richard C., 'Determination of Loud-speaker Signal Arrival Times'. Parts 1, 2 and 3. *Journal of the Audio Engineering Society*, October/November/December 1971

37 [7,13,18] Heyser, Richard C., 'Loudspeaker Phase Characteristics and Time Delay Distortion'. Parts 1 and 2. *Journal of the Audio Engineering Society* (January/April 1969)

38 [10,11] Holland, K. R., Fahy, F. J., Morfey, C. L. and Newell P. R., 'The Prediction and Measurement of the Throat Impedance of Horns'. *Proceedings of the Institute of Acoustics*, Vol 11, Part 7, 247–254 (1989)

39 [4,15] Ishii, S. and Mizatini, T., 'A New Type of Listening Room and its Characteristics – A Proposal for a Standard Listening Room'. 72nd Convention of the Audio Eng. Soc. (1982)

40 [7,20] Keftopics, 'Crossover Filters – An Integral Part of Overall System Engineering'. *Keftopics*, Vol 4, No 2, Kef Electronics, Tovil, Maidstone, Kent, UK (1980)

41 [10,11] Kinoshita, S., Yoshimi, T., Hamada, H. and Locanthi, B.N., 'Design of 48 mm Beryllium Diaphragm Compression Driver'. Pioneer Electronic Corporation, Tokorozawa, Japan

42 [7] Linkwitz, Siegfried H., 'Active Crossover Newtorks for Non-Coincident Drivers'. *Journal of the Audio Engineering Society* (January/February 1976)

43 [7] Lipshitz, Stanley P. and Vanderkooy, J., 'A Family of Linear-Phase Crossover Networks of High Slope Derived by Time Delay'. *Journal of the Audio Engineering Society*, Vol 31, Nos 1 and 2 (January/February 1983)

44 [4,7,9,13,15,18,20,21] Lipshitz, Stanley, P., Pocock, M. and Vanderkooy, J., 'On the Audibility of Mid-Range Phase Distortion in Audio Systems'. *Journal of the Audio Engineering Society*, Vol 30, No 9 (September 1982)

45 [7] Long, E. M., 'A Time-Align Technique for Loudspeaker System Design'. *Journal of the Audio Engineering Society* (July/August 1976)

46 [4,7,9,12,13] Madsen, E. R. and Hansen, V. E., 'Threshold of Phase Detection by Hearing'. 44th Convention of Audio Eng. Soc. Rotterdam (February 1973)

47 [2,12,13] Masters, Ian G., 'The Audibility of Distortion'. *Stereo Review* (January 1989)

48 [7] Mitchell, R. M., 'Transient Performances of Loudspeaker Dividing Networks'. *Audio*, 48, 24 (January 1964)

49 [4,7,9,12,13] Moller, H., 'Loudspeaker Phase Measurements, Transient Response and Audible Quality'. Bruel and Kjaer, Application Notes. Presented at the 48th Audio Engineering Society Convention, California (1975). Expanded Version

50 [18] Mourjopoulos, J., 'On the Variation and Invertibility of Room Impulse Response Functions'. *Journal of Sound and Vibration*, Vol 102, 217–228 (1985)

51 [18] Neely, S. T. and Allen, J. B., 'Invertibility of a Room Impulse Response'. *Journal of the Acoustical Society of America*, Vol 66, 165–169 (1979)

52 [7,18] Nelson, P. A. and Elliott, S. J., 'Least Squares Approximation to Exact Multiple Point Sound Reproduction'. ISVR Memorandum No 683 (1988)

53 [9,12,13] Newell, P. R. and Holland, K. R., 'Impulse Testing of Monitor Systems'. *Proceedings of the Institute of Acoustics*, Vol 11, Part 7, 269–275 (1989)

54 [4] Ohm, G. S., 'Uber die Definition des Tones, nebst daren geknupfter Theorie der Sirene und ahnlicher tonbildener Vorrichtungen'. *Ann. Phys. Chem.*, Vol 59, 513–565 (1843)

55 [4] Ohm, G. S., 'Noch ein Paar Worte uber die Definition des Tonnes'. *Ann. Phys. Chem.*, Vol 62, 1–18 (1844)

56 [4,7,9,10,12,13] Oppenheim, A. V. and Lim, J. S., 'The Importance of Phase in Signals'. *Proc. IEEE*, Vol 69, 529–541 (May 1981)

57 [7] Orr, T., 'Active Filter Design'. *ETI* (July 1980)

58 [4,6,7,9,12,13] Plomb, R. and Steeneken, H. J. M., 'Effects of Phase on the

Timbre of Complex Tones'. *Journal of the Acoustical Society Am.*, Vol 46, 409–421 (1969)

59 [4,7,9,15,18,20,21] Preis, D., 'Phase Distortion and Phase Equalisation in Audio Signal Processing – A Tutorial Review'. *Journal of the Audio Engineering Society*, Vol 30, No 11 (November 1982)

60 [7,20,21] Von Recklinghausen, D. R., 'Mismatch Between Power Amplifiers and Loudspeaker Loads'. *Journal of the Audio Engineering Society*, Vol 6, No 4 (1958)

61 [18] Sakamoto, N., Gatok, T., Kogure, T. and Shimbo, M., 'Controlling Sound Image Localisation in Sterophonic Sound Reproduction'. *Journal of the Audio Engineering Society*, Vol 29, 794–799 (1981) and Vol 30, 719–721 (1982)

62 [7,9,12,13] Schaumberger, A., 'Impulse Measurement Techniques for Quality Determination in Hi-Fi Equipment with Special Emphasis on Loudspeakers'. *Journal of the Audio Engineering Society*, Vol 19, 101–107 (February 1971)

63 [4,6,7,9,12,13] Schroeder, Manfred A., 'Models of Hearing'. *Proceedings of the IEEE*, Vol 63, No 9 (September 1975)

64 [7] Small, Richard H., 'Constant-Voltage Crossover Network Design'. *Journal of the Audio Engineering Society*. Originally published in *Proceedings IREE, Australia*, 31 (1970)

65 [10,20,21] Small, Richard H., 'Direct Radiator Loudspeaker System Analysis'. *Journal of the Audio Engineering Society*, 20, No 5 (1972)

66 [7,20,21] Small, Richard H., 'Phase and Delay Distortion in Multiple-Driver Loudspeaker Systems'. *Journal of the Audio Engineering Society*

67 [7] Smith, A.P., 'Electronic Crossover Networks and their Contribution to Improved Loud-speaker Transient Response'. *Journal of the Audio Engineering Society*, Vol 19, 674–679 (Sept 1971)

68 [7,10,11,13] Smith, D., Keele Jr, D. B. and Eargle, J., 'Improvements in Monitor Loudspeaker Systems'. Presented at 69th Convention, May 1981, Los Angeles

69 [4,7,9,12,13] Suzuki, H., Momla, S. and Shindo, T., 'On the Perception of Phase Distortion'. *Journal of the Audio Engineering Society*, Vol 28, 570–574 (Sept 1980)

70 [7] Theile, A. N., 'Optimum Passive Loudspeaker Dividing Networks'. *Proc. IREE, Australia*, Vol 36, 220–224 (July 1975)

71 [7] Vanderkooy, J. and Lipshitz, S. P., 'Is Phase Linearization of Loudspeaker Crossover Network Possible by Time Offset and Equalisation?' *Journal of the Audio Engineering Society*, Vol 32, No 12 (December 1984)

72 [7] Wall, P. K., 'Active and Passive Loudspeaker Crossover Networks Without Transient Distortion'. 50th Convention of Audio Eng. Soc., London (1975)

Index

Schultz, 278
Slant plates, 157
Small, Richard, 132
Soares, Luis, 283, 288, 293, 294, 295
Son Audax HR100, 163–224
Sound field distortion, 69, 70
SoundField microphones, 9, 12, 58
Southampton University, 7, 328
Spencer-Allen, Keith, 74
Sticklen, Derek, 349
Stylisation of sound, 11
Subjective pitch change of instruments, 60
Swift, David, 101

TAD TD2001, 323, 340
Tannoy Dual Concentric, 16, 18, 154, 163–224, 370, 379
Taylor, Professor, 228
Three channel stereo, 32
Toole, Floyd, 2, 9, 26, 256
Trondheim University, 154

UREI 815, 225
UREI loudspeakers, 84, 85, 87, 96, 206
Uzzle, Ted, 278, 284

Valve amplifiers, 4
Viemeister, Neal F., 9
Virgin Records, 280
Volt-amps, 81, 83

Wattless power, 81, 83
Weather frequencies, 3
Wharfedale, 105
Wolf, Dr, 142
Wood, speed of sound in, 159
Wrightson, 299

Yamaha NS10, 352, 353, 381, 383
Yamaha NS40, 353
Yamaha NS1000, 64

Zobel networks, 105